Practical Physical Chemistry

Practical Physical Chemistry

Third Edition

A. M. James, MA DSc
and
F. E. Prichard, PhD

Longman

**Longman
1724-1974**

LONGMAN GROUP LIMITED
Burnt Mill,
Harlow,
Essex CM20 2JE

Distributed in the United States of America by Longman Inc., New York

Associated companies, branches and representatives throughout the world

First and second editions by J. and A. Churchill Ltd, 1961, 1967
Third edition 1974

ISBN 0 582 44259 1

Library of Congress Catalog Card Number: 73—85687

*Printed in Great Britain by
J. W. Arrowsmith Ltd, Bristol*

Preface

This revision of the Laboratory Manual of Physical Chemistry differs from the previous edition in two main aspects, firstly, the use of SI units throughout, and secondly, more guidance in the handling of experimental data and discussion of results.

The object, as before, is to provide the average student with good, basic experiments in Physical Chemistry which illustrates fundamental principles or the applications of these principles. Additionally such exercises fulfil other purposes which are not as apparent in the other branches of chemistry, such as the realization of the importance of making precise measurements, the choice of the best available technique, the care necessary in the design and completion of the experiment, the collection, tabulation and handling of data and finally, the presentation of a report. Suggestions of the more important topics for discussion are included; it is hoped that this will provide the stimulus for further experimentation.

The book is self-contained and does not rely on detailed reference to the literature for experimental techniques. However, students are encouraged to refer to standard textbooks and data source books for detailed theoretical treatments and accepted values of physical quantities respectively. The literature value (together with its source) of the quantity measured must always be included in the report of any experiment.

The experiments have been selected so that in general they can be completed in a one, three- or five-hour period, some longer ones may be either carried over into a second period or conducted as a class experiment in which the students share their data for calculation and discussion before final report writing. There has been considerable re-grouping of experiments; this has made it possible to give a more adequate coverage of appropriate theory and has lessened the need for extensive cross-referencing within the book.

Additional experiments involving the use of spectrometers (including magnetic resonance spectrometers) have been included; details of design and operation of such instruments have been omitted since they are always adequately described in the instruction manual which accompanies the instrument. Details of methods of measuring the fundamental quantities of mass, length, volume and time and the calibration of the appropriate measuring devices have been ommitted. Since these are dealt with extensively in specialized texts and BSS publications it was felt that further repetition was unnecessary. Despite this omission, however, it is imperative that all measuring instruments be calibrated at regular intervals and that the instruments be always used under the conditions used for calibration and as specified by the manufacturer. It is only in this way that the best results can be obtained from an instrument. Further, in a physical chemical laboratory all glass-ware must be scrupulously clean and all the chemicals used must be of the highest purity available (eg. ethanol contaminated with benzene is of no use as a spectro-scopic solvent in the ultraviolet region). Many of the experiments will not give meaningful results if the chemicals are impure. Radiochemical experiments have not been included as there is already excellent coverage of these in other more specialized books.

SI units have been used throughout this book. In the experimental procedures the most convenient practical units are used, eg. concentrations are expressed in $mol\,dm^{-3}$ (equal to the old 'molarity') in preference to the rigorous $mol\,m^{-3}$; volumes in cm^3 and not m^3; pressures in mmHg (the measured quantity) and not in $N\,m^{-2}$. The term 'absorbance' is used throughout the book in place of the more cumbersome term 'decadic absorbance' formerly called 'optical density'. Students however, are encouraged and advised to present their final results rigorously in the accepted SI units.

We have borne in mind the difficulties encountered by teaching staff in many colleges and by past generations of students and we are grateful for their comments and criticisms. We are indebted to Mr T. F. McCombie of BDH Chemicals Ltd, for permission to reproduce the tables of Spectroscopic Solvents and Materials, pp. 320, 322 and 323 from the 1973 edition of their catalogue. The task of writing or rewriting a book consumes a large amount of free time; we wish to thank our respective spouses for their patience, help and encouragement during the produc-tion of this new edition. Finally we wish to thank Miss Bobbi Gouge of Longman, for all her help and painstaking efforts on our behalf.

AMJ   FEP

Warning

Students are reminded that all chemicals are potentially dangerous and that they must always be handled with care. Additional precautions, eg. wearing protective clothing, working in a fume chamber, are essential for those special chemicals which have been shown to be particularly toxic or hazardous in use.

Cleanliness is of paramount importance in any laboratory, the more so in a physical chemical laboratory. Any contamination of chemicals, which would invalidate experi-mental results, must be eliminated; spillage of liquids, mercury and other chemicals, which could cause permanent damage to instruments, interfere with precise measure-ments or be a potential health hazard, must be cleared up immediately. Students are strongly advised to use pipetting devices, to wear protective clothing and to observe safety instructions pertaining to each laboratory.

In this manual we have not singled out chemicals which are known to be toxic or hazardous, because the lists of these are continually being extended. The particular danger of a chemical is usually indicated on the manufacturer's label and any handling instructions given there must be rigorously followed.

Contents

List of experiments

The experiments listed provide a basis for courses in Practical Physical Chemistry for students at all levels. It is not expected that any one student will carry out all the experiments, but a sufficient number is given to allow choice at the discretion of the supervisor.

Experiment

List of principal symbols

a	activity; a_A, activity of A
c	concentration (mol dm^{-3}), number of components
f	number of degrees of freedom
g	gramme
g	gravitational constant
h	hour
i	current
k	rate constant
k_c	cryoscopic constant
k_e	ebullioscopic constant
l	l_+, l_- ion conductance
m	molality (g solute per kg solvent)
m	mass
n	refractive index; number of electrons involved in electrode process; degrees of freedom (statistics)
p	pressure; vapour pressure; number of phases
$p^{\ominus}$	vapour pressure of pure solvent
s	second
t	time; temperature $^{\circ}$C
t_+, t_-	transport number of anion, cation
u_+, u_-	ionic mobility
v	volume; electrophoretic or electro-osmotic velocity
$\bar{v}$	electrophoretic or electro-osmotic mobility
w	weight
x	mole fraction
$\bar{x}$	mean value of x (statistics)
z_+, z_-	charge on an ion
A	area: Debye–Hückel constant; absorbance (decadic)
Å	Ångström unit (10^{-10} m)
B	barometric pressure, Debye–Hückel constant
C	capacitance of condenser
D	distribution coefficient

E	energy; electromotive force; energy of activation; extinction coefficient
$E_{0.5}$	half-wave potential
E_X	electrode potential of X
$E_X^{\ominus}$	standard electrode potential of X
F	surface pressure
G	Gibbs' free energy ($H-TS$); conductance
H	enthalpy
I	current; intensity of radiation; ionic strength
J	cell constant of conductance cell
K	equilibrium constant
K_a	acid dissociation constant
K_b	basic dissociation constant
K_h	hydrolysis constant
K_s	thermodynamic solubility product
K_{therm}	thermodynamic equilibrium constant
K_w	ionization product of water
L_e, L_f	heat of vaporization, fusion
M_r	relative molecular mass
R	gas constant; resistance
R_F	ratio of distance moved by component to distance moved by front in paper partition chromatography
S	entropy; standard deviation (statistics)
S^2	variance (statistics)
T	temperature K.; transmittance
V	electrical potential
V	volt
α	fraction of electrolyte dissociated or hydrolysed
α_m	specific optical rotatory power
α_n	molar optical rotatory power
γ	surface tension; activity coefficient (γ_+, γ_- or $\gamma_{\pm}$)
ϵ	molar extinction coefficient; electron; relative permittivity
ζ	electrokinetic potential
η	coefficient of viscosity; over potential
θ	angle of contact
κ	conductivity (electrolytic)
λ	wavelength
μ	micro; $\mu m = 10^{-6}$ m; $\mu g = 10^{-6}$ g
ν	kinematic viscosity; frequency
$\bar{\nu}$	wavenumber
π	ratio of circumference to diameter of circle, 3.14159
ρ	density; resistivity
τ	time of half reaction
Γ	surface excess concentration (mole m^{-2})
Δ	increase in eg., ΔT, ΔS, etc.

Λ molar conductivity

Σ sum

$\approx$ approximately equal to

$\log x = \log_{10} x$

$\ln x = \log_e x = 2.303 \log x$

bp. = boiling point

fp. = freezing point

vp. = vapour pressure

1
Errors and the mathematical treatment of results

Experimental physical chemistry, that is the investigation of the physical properties of chemical substances and their relation to chemical reactions, is mainly concerned with quantitative measurements. From such measurements it is generally possible to advance a hypothesis to explain the phenomenon or reaction under study. Where quantitative observations are not always possible, eg. in parts of colloid science, detailed qualitative observations still prove of value.

In this book, experiments are described to illustrate the physical methods and techniques involved in:

(a) assessing the purity and often the nature of a substance;
(b) the determination of the physical constants of pure substances;
(c) the study of the changes of physical properties which occur in either physical changes or chemical reactions, with a view to the understanding of the mechanism;
(d) the determination of concentrations.

1 The accuracy of measurements

Any measurement, whether it be the determination of weight, volume, time, density, etc., is only of limited accuracy. The result will differ slightly from the truth and will, in all probability, differ from replicate measurements made under apparently identical conditions. An experienced worker can usually say qualitatively whether such differences lie within the limits of experimental error, but it is necessary to be able to attach to a result some index of its reliability. This is obtained from statistical methods.

It should be obvious that all experiments do not require the same degree of accuracy, thus the accuracy required in the determination of the order of a reaction need not be so high as that required in a gravimetric determination. The *accuracy* is best defined as the closeness of agreement between an experimental value and the true value, whilst the *precision* is the closeness of agreement between replicate experimental values. The closer these values the more precise is the method; the values may not however agree with the true value. Thus we can have the situation of a highly precise but inaccurate method.

Various sources of error occur in experimental procedures, the following should be recognized and precautions taken, where possible, to eliminate or reduce them.

(a) Mistake on the part of the experimentalist, such as the misreading of a scale, failure to conform exactly to the described procedure, and incorrect calculation of results. This type of error, while very serious, is easily overcome by care and patience.

(b) Instrument errors are common to all instruments as each has only a limited accuracy. It is the responsibility of the observer to find the limits of accuracy of the instrument. Tables are generally provided by the manufacturer quoting the reliability of the instrument under specified conditions of test. Checks on the calibration of any instrument should be made at regular intervals using standards calibrated by a reputable authority; the National Physical Laboratory will calibrate most physical equipment. Single checks on the scales of burettes, thermometers,

Table 1.1   The diameter of a tube measured, by two observers, with a travelling microscope

	Observer A		Observer B
n	Diam./cm	Mean diam. of first n readings	Diam./cm
1	1.512	1.512	1.512
2	1.510	1.511	1.515
3	1.515	1.512_3	1.513
4	1.517	1.513_5	1.516
5	1.514	1.513_6	1.514
6	1.516	1.514	1.511
7	1.514	1.514	1.513
8	1.512	1.513_7	1.514
9	1.512	1.513_6	1.512
10	1.514	1.513_6	1.510
11	1.515	1.513_7	1.514
12	1.514	1.513_7	1.512
13	1.513	1.513_7	1.511
14	1.514	1.513_7	1.512
15	1.513	1.513_7	1.513
16	1.513	1.513_6	1.514
17	1.514	1.513_6	1.512
18	1.514	1.513_6	1.513
19	1.514	1.513_7	1.513
20	1.513	1.513_7	1.512
21	1.515	1.513_7	1.514
22	1.513	1.513_7	1.512
23	1.515	1.513_7	1.512
24	1.511	1.513_6	1.513
25	1.513	1.513_6	1.513

spectrophotometers, etc., are useless, as this assumes that the scale is equally and correctly calibrated.

It is often possible, when the instrument cannot be used under the correct conditions giving greatest accuracy, to make so called 'instrument corrections'. Such a correction must be made if it will have a significant effect on the final reading.

Other errors which occur in the use of instruments are well known and can be eliminated by proper use. Thus in any instrument in which a micrometer screw (fine adjustment) is incorporated, eg. travelling microscope, Abbé refractometer, polarimeter, there will be a certain amount of backlash. This can give rise, particularly after prolonged use, to large errors due to wear on the moving parts. It is largely eliminated if the final adjustment is always made in the same direction. Zero errors arise through a displacement of the scale. In polarimeters, the eccentricity of the scale can be eliminated by taking readings on both sides of the scale, again always approaching the null point from the same direction.

(c) A random error is one that is individually unpredictable, but after a long series of determinations averages zero. Thus suppose that the length of an object is measured with a scale graduated in tenths of a scale division. The error on any determination may be in the range +0.05 to −0.05, but the average of a large number of such measurements will be virtually correct. The random error does, however, decrease with an increase in the number of observations as the operator becomes more familiar with the technique. A series of readings by two observers of the diameter of a tube repeatedly measured across the same diameter with a travelling microscope (Table 1.1) shows the overall arithmetic mean of twenty-five readings to be between 1.5135 and 1.5140 cm for A and between 1.5125 and 1.5130 cm for B. This further shows the change in the mean value of n readings as n increases, until after some twenty-five readings the value can be quoted accurately to 0.001, there still being some doubt as to the fourth decimal place.

2   The application of statistical methods

It is now possible to consider briefly the application of statistical methods of analysis to this set of readings to establish the variability of each observer and test if the experimental difference is significant. First a histogram (Fig. 1.1) can be plotted showing the distribution of the readings. The maximum of the histogram, 'the most probable value', does not agree exactly with the arithmetic mean. If a very large number of readings, and not a sample of twenty-five, had been taken the histogram would have been symmetrical and would have approximated closely to a smooth distribution curve. If the readings were completely random about the mean, the frequency distribution would follow a normal or Gaussian curve (Fig. 1.2), the equation to which is

$$y = \frac{1}{\sigma\sqrt{2\pi}} \exp\left[-(x-\bar{x})^2/2\sigma^2\right] \tag{1.1}$$

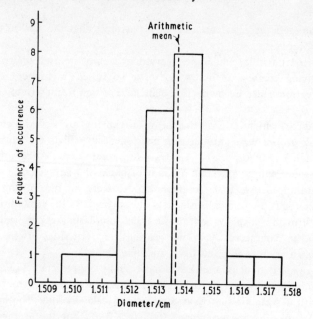

Fig. 1.1   Histogram showing the distribution of readings for Observer A

where σ is the standard deviation for an infinitely large sample (defined later).

y has the significance of a probability coefficient, any element under the curve between x and $(x + dx)$ gives the probability that a single sample drawn from the batch will have a value of x in this range. Equation (1.1) is in the form where the total area under the curve is unity. In the sampling the most probable value to occur is that at the maximum of the curve, this corresponds, for normal distributions, to the arithmetic mean.

The arithmetic mean is defined as $\bar{x} = \Sigma\, x/n$, where n is the number of observations denoted by x and Σ is the summation symbol, thus

$$\Sigma\, x = x_1 + x_2 + x_3 \ldots$$

The variability of a set of observations can be assessed in several ways.

(a) The *range* defined as the difference between the largest and smallest reading of the sample.

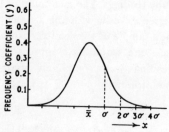

Fig. 1.2   Normal distribution or error curve

(b) The *deviance, d*, defined as the difference (either positive or negative) between the observation and the mean, thus:

$$d = (x - \bar{x}) \tag{1.2}$$

While the sum of the deviances must, by definition, be zero, the sum of the squares of the deviances is always positive. The

$$mean\ deviation = \frac{\Sigma d}{n} = \frac{1}{n} \Sigma |x - \bar{x}| \tag{1.3}$$

where no regard is paid to the sign of the deviance.

(c) The *variance*, usually denoted by σ^2 for a very large sample, is a measure of the dispersion of observations around the mean value. It is defined as the mean value of the square of the deviances, thus

$$\sigma^2 = \frac{1}{n} \Sigma (d^2) = \frac{1}{n} \Sigma (x - \bar{x})^2 \tag{1.4}$$

In actual practice the use of this equation tends to underestimate the variance, because in calculating $\Sigma(x-\bar{x})^2$ we should be calculating the sum of the squares of the deviations from the true mean of the population from which the sample is drawn. As the value of the sample mean will not in general be identical with the true mean, equation (1.4) is modified by replacing the number of observations (n) in the denominator by the *number of degrees of freedom* ($n-1$). Thus the variance for a sample (S^2) is given by

$$S^2 = \frac{\Sigma (x - \bar{x})^2}{n - 1} \tag{1.5}$$

It is a laborious procedure to calculate the mean value of all the observations and hence the deviance for each reading, before squaring and summing. Equation (1.5) can be rearranged to the more convenient form:

$$S^2 = \frac{\left[\Sigma (x^2) - \dfrac{(\Sigma x)^2}{n} \right]}{n - 1} \tag{1.6}$$

The substitution of numerical values of x into this equation is now an easy matter on a calculating machine.

(d) The *standard deviation, S*, is the square root of the variance:

$$S = \left(\frac{\Sigma (x - \bar{x})^2}{n - 1} \right)^{1/2} = \left(\frac{\left[\Sigma (x^2) - \dfrac{(\Sigma x)^2}{n} \right]}{n - 1} \right)^{1/2} \tag{1.7}$$

σ in the normal frequency equation or S for a sample is a constant for a given set of observations and is a measure of the spread of the curve. This is often referred to as the standard error.

The range is little used except as a means of control of quality of production (quality control chart). The mean deviation, while simply calculated, is now considered obsolete since its properties are such that it is difficult to handle mathematically. The best criterion of variability is the variance; this yields more information about the variability from the data than any other quantity. Its properties are well established and various tests of significance based upon it are available. One important property of the variance is that it is additive, this means that if a process is subject to a number of variances S_1^2, S_2^2 ... due to independent causes, the total variance is the sum of the separate variances. This is useful for it makes possible the breakdown of the total variance into its separate parts, thereby revealing the significance of each. From this it may be possible to modify the experimental procedure, thereby reducing the more important source(s) of error.

Table 1.2 gives the analysis of the results in Table 1.1.

Table 1.2   Analysis of results from Table 1.1

	Definition	Observer A	Observer B
Range	Largest−smallest observation	0.007	0.006
Number of readings	n	25	25
Degrees of freedom	$n-1$	24	24
Sum of readings	Σx	37.840	37.820
Arithmetic mean, $\bar{x}$	$\Sigma x/n$	1.5136	1.5128
Sum of square of readings	$\Sigma (x^2)$	57.274680	57.214138
Square of sum of readings	$(\Sigma x)^2$ $(\Sigma x)^2/n$ $[\Sigma (x^2) - (\Sigma x)^2/n]$	1431.865600 57.274624 0.000056	1430.352400 57.214096 0.000042
Variance, S^2	$\dfrac{\left[\Sigma (x^2) - \dfrac{(\Sigma x)^2}{n}\right]}{n-1}$	0.00000233	0.00000175
Standard deviation, S	$\left(\dfrac{\left[\Sigma (x^2) - \dfrac{(\Sigma x)^2}{n}\right]}{n-1}\right)^{1/2}$	0.00153	0.00133
Mean deviation, d	$\dfrac{1}{n}\Sigma |x - \bar{x}|$	0.00117	0.00102
Coefficient of variation	$\dfrac{\text{Standard deviation}}{\text{Mean}} = \dfrac{S}{\bar{x}}$	0.1%	0.09%
Accuracy of estimate for 95% probability limit	$\text{Mean} \pm \dfrac{2.06S}{\sqrt{n}}$	1.5136 ± 0.0006	1.5128 ± 0.0005

3   Accuracy of results

The standard error of the mean, defined as $\bar{x} \pm \sigma_{\bar{x}}$, is only valid for an infinite number of readings. For a finite number of readings on a small sample the Student Test must be applied. The value of the mean is then quoted as:

$$m = \bar{x} \pm tS_{\bar{x}} \tag{1.8}$$

where $S_{\bar{x}}$ the standard deviation of the mean is defined as $S/\sqrt{n}$. The value of t, obtained from Tables (Table 1.3), depends on the number of degrees of freedom (ie. a measure of the size of the sample) and the level of probability. The probability levels generally listed are 0.5, 0.1, 0.05, 0.02 and 0.01, it is customary to use the 0.05 level to determine the limits; this means that 95% of the readings will be within the limits. For high levels of significance, used mainly in biological tests on drugs, the 0.01 level is used; this means that 99% of the readings will be within the limits.

Table 1.3   Values of t for three levels of probability

Degrees of freedom	Level			Degrees of freedom	Level		
	0.5	0.05	0.01		0.5	0.05	0.01
1	1.0	12.71	63.66	25	0.684	2.06	2.79
5	0.727	2.57	4.03	30	0.683	2.04	2.75
10	0.700	2.23	3.17	50	0.679	2.01	2.68
15	0.691	2.13	2.95	100	0.677	1.98	2.63
20	0.687	2.09	2.84	∞	0.674	1.96	2.58

The results quoted in Table 1.4, calculated with equation (1.8), show the variation of the limits with the number of readings and the level of probability for Observer A (Table 1.1). While the mean values for the different number of observations are the same for the first five, ten and twenty-five readings, the limits at any level decrease as the number of readings increases, and for a given number of readings, increases as the level of probability increases from $P = 0.5$ to $P = 0.01$.

Table 1.4   Variation of limits with number of readings and level of probability

Number of readings	5	10	25
$\bar{x}$	1.5136	1.5136	1.5136
Standard deviation, S	0.00269	0.00211	0.00153
Limits: 0.5	± 0.0009	± 0.0005	± 0.0002
0.05	± 0.0034	± 0.0015	± 0.0006
0.01	± 0.0055	± 0.0022	± 0.0009

In reporting a mean value it is only necessary to quote the mean, $\bar{x}$, the standard deviation, S, and the number of observations from which these were calculated. From these data, the limits at any desired level can then be calculated by any interested worker. Thus Observer A could simply report his results as:

$$\bar{x} = 1.5136 \text{ cm}, \qquad S = 0.00269, \qquad n = 5$$
$$\bar{x} = 1.5136 \text{ cm}, \qquad S = 0.00211, \qquad n = 10$$
$$\bar{x} = 1.5136 \text{ cm}, \qquad S = 0.00153, \qquad n = 25$$

4 Comparison of the significance between two sets of observations using the student *t* test

To test the reproducibility of a given method it is necessary to record several sets of observations and compare the results by statistical methods to find if there is a significant difference between them. The method to be described is primarily for a comparison of two sets of readings which may have been made by the same observer at different times, by two different observers, or even by one observer using two different experimental techniques or conditions. The use of this method to compare more than two sets of results, while possible, is extremely laborious and other methods, not discussed here, are recommended.

If $x_1 \ldots$ and $x_2 \ldots$ are the individual observations of two experiments of means $\bar{x}_1$ and $\bar{x}_2$ respectively then it can be shown that:

$$t = \frac{\bar{x}_1 - \bar{x}_2}{\left(\dfrac{\Sigma(x_1 - \bar{x}_1)^2 + \Sigma(x_2 - \bar{x}_2)^2}{n_1 + n_2 - 2} \right)^{1/2}} \left(\frac{n_1 n_2}{n_1 + n_2} \right)^{1/2} \qquad (1.9)$$

where n_1 and n_2 are the number of observations in the two sets of results respectively. For simplicity in calculation, it should be remembered that

$$\Sigma (x_1 - \bar{x}_1)^2 = [(\Sigma (x_1^2) - (\Sigma x_1)^2/n_1],$$

etc. Thus the value of t is proportional to the difference of the two means, inversely proportional to the error and directly proportional to the size of the experiment (last term). The fraction $(n_1 n_2/n_1 + n_2)$ is a maximum when $n_1 = n_2$, thus for preference the two samples should be of the same size. The experimental values are substituted into equation (1.9) and the value of t calculated. The position of this value of t is looked up in the appropriate table for the number of degrees of freedom $(n_1 + n_2 - 2)$ and hence the significance of the two sets of readings is determined. If the calculated value is less than the value at 0.05, the difference is most probably due to chance and is therefore not significant; if it is greater than the 0.05 but less than the 0.01 value then there is a significant difference between the sets of readings, and if it exceeds the 0.01 value the difference is highly significant. A comparison of the data of Observers A and B (Table 1.1) illustrates this.

$$t = \frac{1.5136 - 1.5128}{\left(\dfrac{0.000056 + 0.000042}{48} \right)^{1/2}} \left(\frac{25 \times 25}{50} \right)^{1/2}$$

$$= 1.981$$

For 48 degrees of freedom t at $0.5 = 0.679$, at $0.1 = 1.68$, at $0.05 = 2.01$, therefore this difference is not significant.

It is not suggested that the experiments described in this book are all suitable for such statistical analysis. Nevertheless, to stress its importance and application and to obtain practice of the method, it is suggested that the results of certain selected experiments should be treated in this way.

5 Combination of errors

A physical measurement is meaningless without an estimate of its accuracy and precision. It is not always possible (or necessary) to take sufficient readings to permit a statistical analysis of the results. An estimate of random errors, which affect the precision of measurements, can often provide the required information.

When a physical quantity is not measured directly but is obtained by calculations in which a number of measured quantities are combined, then the precision of the final result cannot be greater than that of the least precise measurement in the investigation. It is important to determine how the errors in individual measurements affect the final result.

The errors quoted after a final result are not definite limits but are statistical indices. Thus statistical theory is used to combine the uncertainties of the various measurements. The following rules can be applied to determine the index of precision or the probable error.

General case

Suppose the measured quantities X, Y, Z ... are used to evaluate the quantity Q, ie. $Q = f(X, Y, Z \ldots)$. If the uncertainties in the measurements of X, Y, Z ... are $\pm\Delta X$, $\pm\Delta Y$, $\pm\Delta Z$, ... then the uncertainty in Q, $\pm\Delta Q$, is given by

$$(\Delta Q)^2 = \left(\frac{\partial Q}{\partial X}\right)^2 (\Delta X)^2 + \left(\frac{\partial Q}{\partial Y}\right)^2 (\Delta Y)^2 + \left(\frac{\partial Q}{\partial Z}\right)^2 (\Delta Z)^2 \ldots \qquad (1.10)$$

For specific cases equation (1.10) may be simplified.

Sums or differences

If $\qquad Q = X \pm Y \pm Z$

Then $\quad Q \pm \Delta Q = (X \pm \Delta X) \pm (Y \pm \Delta Y) \pm (Z \pm \Delta Z) \ldots$

and $\qquad \Delta Q = (\Delta X^2 + \Delta Y^2 + \Delta Z^2)^{1/2}$ $\qquad (1.11)$

The error on a sum or difference is greater than that of one of the quantities but less than the sum of the errors.

Products or quotients

If $\quad Q \pm \Delta Q = (X \pm \Delta X)(Y \pm \Delta Y)(Z \pm \Delta Z) \ldots$

or $\qquad Q \pm \Delta Q = \dfrac{(X \pm \Delta X)}{(Y \pm \Delta Y)} (Z \pm \Delta Z) \ldots$

then $\qquad \dfrac{\Delta Q}{Q} = \left[\left(\dfrac{\Delta X}{X} \right)^2 + \left(\dfrac{\Delta Y}{Y} \right)^2 + \left(\dfrac{\Delta Z}{Z} \right)^2 \ldots \right]^{1/2}$ (1.12)

The *fractional* error on a product (or quotient) is equal to the square root of the sum of the squares of fractional errors on the terms multiplied (or divided).

In most cases it is found that there are one or two principal sources of error and that the remaining measurements incur negligible errors.

Consider the following examples.

1 The use of a travelling microscope, which can be read with an uncertainty of ± 0.003 cm, to measure the height of a column of liquid. The height is the difference between two readings of equal accuracy, ie.;

$$\Delta H = (H_2 \pm 0.003) - (H_1 \pm 0.003)$$

the uncertainty on the length is given by equation (1.11) as

$$\Delta H = [(0.003)^2 + (0.003)^2]^{1/2} \approx \pm 4 \times 10^{-3} \text{ cm.}$$

2 The probable error in the determination of the density of carbon tetrachloride ($\rho \approx 1.6 \times 10^3$ kg m^{-3}) in a 25 cm^3 pyknometer of weight 10 g. If it is assumed that the only significant sources of error are the weighings and the uncertainty of these is $\pm 10^{-4}$ g,

$$\rho = \frac{(\text{wt. of pyknometer} + CCl_4) - (\text{wt. of pyknometer})}{(\text{wt. of pyknometer} + H_2O) - (\text{wt. of pyknometer})}$$

Uncertainty in numerator (ie. weight of CCl_4) by equation (1.11)

$$= \left[\left(\frac{1.6 \times 25 + 10}{10^4} \right)^2 + \left(\frac{10}{10^4} \right)^2 \right]^{\frac{1}{2}} \approx \pm 5 \times 10^{-3} \text{ g}$$

Uncertainty in denominator, weight of water

$$= \left[\left(\frac{25 + 10}{10^4} \right)^2 + \left(\frac{10}{10^4} \right)^2 \right]^{\frac{1}{2}} \approx \pm 4 \times 10^{-3} \text{g}$$

Then by equation (1.12)

$$\frac{\Delta \rho}{\rho} = \left[\left(\frac{5 \times 10^{-3}}{50} \right)^2 + \left(\frac{4 \times 10^{-3}}{25} \right)^2 \right]^{\frac{1}{2}}$$
$$\approx \pm 1.8 \times 10^{-4}$$

The error on the density is thus less than 2 in 10^4 despite the fact that four weighings each with an uncertainty of 1 in 10^4 are involved.

3 Consider the error involved in the determination of a relative molecular mass using the ebullioscopic method in which the uncertainty in a single reading of the Beckmann thermometer is ± 0.002 K. The weight of solvent w_A is 40×10^{-3} kg

($\Delta w_A = 10^{-2} \times 10^{-3}$ kg) and the weight of solute w_B is 1×10^{-3} kg ($\Delta w_B = 10^{-4} \times 10^{-3}$ kg).

The uncertainty in a value of T of 0.400 K by equation (1.11) is therefore

$$[(0.002)^2 + (0.002)^2]^{1/2} \approx \pm 0.003 \text{ K}$$

Since
$$M_r(B) = \frac{k_e w_B}{\Delta T_e w_A} \tag{3.18a}$$

Using equation (1.12)

$$\frac{\Delta M_r(B)}{M_r(B)} = \left[\left(\frac{\Delta w_B}{w_B} \right)^2 + \left(\frac{\Delta w_A}{w_A} \right)^2 + \left(\frac{\Delta(\Delta T_e)}{\Delta T_e} \right)^2 \right]^{\frac{1}{2}}$$

$$= \left[\left(\frac{10^{-4} \times 10^{-3}}{1 \times 10^{-3}} \right)^2 + \left(\frac{10^{-2} \times 10^{-3}}{40 \times 10^{-3}} \right)^2 + \left(\frac{0.003}{0.400} \right)^2 \right]^{\frac{1}{2}}$$

$$= [10^{-8} + 6.25 \times 10^{-8} + 56.25 \times 10^{-6}]^{1/2}$$

$$\approx \pm 7.5 \times 10^{-3}$$

It can be seen that the errors in the weighing are insignificant and can be ignored. In a calculation of this type where several readings of ΔT_e for different weights of solute have been made it is usual to calculate the average fractional error.

6   Calculation of the best line through a series of experimental points

In an experiment it soon becomes obvious that there is a regular trend in the values of y with different values of the independent variable x. Any such relationship may be written $y = f(x)$; the treatment of the results will give the nature of the function. The finally obtained equation summarizes the experimental results more conveniently than tables, and further is more correct since in its derivation all the experimental points are used. It is therefore desirable to be able to draw the best line, whether straight or otherwise, through the points. This line is not always easy to decide on a graph, but is readily obtained by calculation which eliminates personal error. The 'method of least squares' is generally used, in which the sum of the squares of the deviations of the experimental points from the chosen line is as small as possible. Whenever possible it is desirable that the plotted graph should be linear, this is of help in interpolation and extrapolation. The graph of log (vapour pressure) against T^{-1} which is linear is far easier for interpretation than the corresponding $p-T$ graph.

The equation to any line may be expressed in the form:

$$y = a + bx + cx^2 + dx^3 \ldots \tag{1.13}$$

where a, b, c, d are the constants to be determined. If N readings have been made then:

$$\Sigma y = Na + b \Sigma x + c \Sigma x^2 + d \Sigma x^3 \ldots$$

Further
$$\Sigma\, xy = a\,\Sigma\, x + b\,\Sigma\, x^2 + c\,\Sigma\, x^3 + d\,\Sigma\, x^4 \ldots$$
$$\Sigma\, x^2 y = a\,\Sigma\, x^2 + b\,\Sigma\, x^3 + c\,\Sigma\, x^4 + d\,\Sigma\, x^5 \ldots$$
$$\Sigma\, x^3 y = a\,\Sigma\, x^3 + b\,\Sigma\, x^4 + c\,\Sigma\, x^5 + d\,\Sigma\, x^6 \ldots$$

thus by determining Σy, Σxy, $\Sigma x^2 y \ldots \Sigma x^6$ for the experimental points and substituting in these simultaneous equations the constants can be calculated.

For a straight line the procedure is somewhat simplified.

$$y = a + bx \tag{1.14}$$

whence
$$\Sigma\, y = Na + b\,\Sigma\, x$$

and
$$\Sigma\, xy = a\,\Sigma\, x + b\,\Sigma\, x^2$$

It can then be shown that:

$$a = \frac{\Sigma\, x\,\Sigma\, xy - \Sigma\, x^2\,\Sigma\, y}{(\Sigma\, x)^2 - N\,\Sigma\, x^2} \tag{1.15a}$$

and
$$b = \frac{\Sigma\, x\,\Sigma\, y - N\,\Sigma\, xy}{(\Sigma\, x)^2 - N\,\Sigma\, x^2} \tag{1.15b}$$

two formulae which are simply solved using a calculating machine.

7   Significant figures and calculations

Great care should be exercised in deciding the number of figures to be used to express a result. It is a common, but deplorable, experience to find the concentration of a solution quoted to four decimal places (eg. 0.9775 mol dm^{-3}) when the titration has been carried out using uncalibrated glassware! Any measurement should be noted to as many figures as are significant, this generally means that one uncertain figure is retained. In a weighing, for example, the weights are determined to the nearest 0.1 mg, thus 1.7654 g means that the weight is less than 1.7655 g but more than 1.7653 g. A weight of 1.760 g signifies that the weight has been recorded to the nearest mg and that it is nearer 1.760 g than to either 1.759 or 1.761 g.

The following rules give some general guidance.

(a) Only retain as many significant figures as will give one uncertain figure. Thus the recorded volume of a solution lying between 25.7 and 25.9 should be 25.8 cm^3 and not 25.80 cm^3. The latter figure means that the volume lies between 25.79 cm^3 and 25.81 cm^3.

(b) When numbers have to be rounded off, add one to the last figure retained if the figure dropped is five or more. In rounding off numbers, if the dropped figure can result in a loss of accuracy, then it is common practice to record it as a subscript; thus the implication in writing 65.17_4 is that the accuracy is certainly better than ±0.01 but not so good as ±0.001.

(c) In addition or subtraction use only the number of significant figures as there are in the least accurately known quantity. The final answer cannot be more precise than the quantity with the largest uncertainty. Thus the subtraction $175.43 - 1.0167$ should be computed as $175.43 - 1.02$.

(*d*) It is the relative and not the absolute error that matters in multiplication or division. Thus retain in each number one more significant figure than that in the number with the greatest uncertainty. Thus the multiplication $176.21 \times 0.0176 \times 2.6543$ should be computed as $176.21 \times 0.0176 \times 2.654$ and the result expressed to three significant figures.

(*e*) In making calculations with logarithms, the four-figure tables are sufficiently precise if interpolation is used; this interpolation is avoided with five-figure tables. A 10 inch slide rule which is useful for checking calculations is accurate to about 0.25%.

(*f*) Estimate the errors involved and calculate uncertainty as indicated in section 5, quote the uncertainty with the final answer.

(*g*) In certain cases it is more convenient to work in and calculate using cgs units. However, the final answer should in all cases be converted into the accepted SI unit.

8   Presentation of report

Experimental work should be recorded at once in fullest possible detail, in a bound laboratory notebook. Some of the material recorded will never be used because only final, successful measurements will be adopted. The final result of the work will usually be presented as a report. Depending on the time available and the purpose of the exercise two types of report are acceptable. The first type, the *laboratory record* contains all the essential details without going into the theory, experimental details or comparison with other methods, etc.; these are left to the second type, the *formal report*.

The numerical results where possible should be presented in tabular form, supplemented, where necessary, by a summary of other data necessary to the calculations. This is especially important where long calculations are involved. The columns of the tables should be adequately labelled, with sign, units and factors.

For example, ebullioscopic data for acetone as solvent.

$\Delta T_e/\mathrm{K}$	$w_B/10^{-3}\,\mathrm{kg}$	$w_A/10^{-3}\,\mathrm{kg}$	$\dfrac{w_B}{w_A}$
0.135 ± 0.005	0.3460 ± 0.0001	37.20 ± 0.01	0.0093

This means the temperature is in Kelvin, the weight w_B is 0.3460×10^{-3} kg with an uncertainty of ± 0.0001.

There is no need to show arithmetical computations which may be carried out by slide rule, logarithm tables or calculating machine, depending on the accuracy.

Graphical presentation of results is desirable. Normally two variables are involved: one, the independent variable is under the control of the experimentalist; the other, the dependent variable is not. Use the convention that a graph is plotted with the appropriate function of the independent variable as the abscissa (x-axis) and the appropriate function of the dependent variable as the ordinate (y-axis). It is important that axes are labelled correctly and if several curves are drawn on one sheet these should be labelled or a key provided. Cross references to the Tables must be made. The size of the symbols (circles, squares, crosses, etc.) used to mark the points should, where possible, be approximately equal to the limits of error on the results.

If the calculation involves measuring the slope of a line, as in reaction kinetics, or the intercept, as in the determination of a pK value, the equation to the best line through the points should be calculated (equations (1.15a) and (1.15b)). Any points showing a large deviation from the rough line should be ignored.

At the end of the report there should be a summary of the data obtained, where possible; this should include the accepted literature value obtained from reference books or original papers. The questions or topics listed at the end of each experiment are a guide to the important points to be included in the discussion.

Section 9 contains a formal report and is an attempt to improve the general presentation of reports. It is not claimed to be a perfect model but from experience it is markedly better than the vast majority of reports presented. The results are from a student's notebook.

9   Model formal report   10 January 1973

A study of the variation of the vapour pressure of carbon tetrachloride with temperature

Theory: The saturation vapour pressure, or vapour pressure, of a liquid is the pressure at which liquid and vapour can co-exist in equilibrium at a given temperature. From the phase rule for a one-component system in equilibrium in 2 phases, the number of degrees of freedom is 1. The vp. is independent of the relative amounts of liquid and vapour, but depends on the temperature. For such an equilibrium, the free energy of the substance is the same in both phases (liquid I, and vapour II), ie. $G_I = G_{II}$ or for an infinitesimal change, in which equilibrium is maintained:

$$dG_I = dG_{II} \tag{1.16}$$

since
$$dG = V\,dP - S\,dT \tag{1.17}$$

it follows that at constant T and P that:

$$dG_I = V_I\,dP - S_I\,dT \tag{1.17a}$$

and
$$dG_{II} = V_{II}\,dP - S_{II}\,dT \tag{1.17b}$$

Combining equations (1.16), (1.17a) and (1.17b) gives:

$$(V_{II} - V_I)\,dP = (S_{II} - S_I)\,dT \tag{1.18}$$

The entropy change for vaporization is given by:

$$\Delta S = \Delta H/T = L_e/T \tag{1.19}$$

where $L_e/J\ mol^{-1}$ is the molar heat of vaporization. Combining equations (1.18) and (1.19) and rearranging gives:

$$\frac{dP}{dT} = \frac{L_e}{T(V_{II} - V_{I})} \tag{1.20}$$

where V_{II} and V_I are the molar volumes of the substance in the vapour and liquid states respectively. Since $V_{II} \gg V_I$ the volume of the liquid can be neglected in comparison to the volume of the vapour. Assuming further that the vapour obeys the ideal gas equation, equation (1.20) becomes:

$$\frac{d \ln P}{dT} = \frac{L_e}{RT^2} \tag{1.21}$$

This is the Clausius–Clapeyron equation, which may be integrated over a small temperature range, ie. assuming that L_e does not vary with temperature, to give:

$$\ln P = \frac{-L_e}{RT} + \text{Const.} \tag{1.22a}$$

or
$$\log P = \frac{-L_e}{2.303\ RT} + \text{Const.} \tag{1.22b}$$

Equation (1.22b) predicts that the graph of $\log P$ against T^{-1} is linear and of slope $-L_e/2.303R$. In this experiment the vp. of carbon tetrachloride is measured over a range of temperatures and the value of L_e obtained from the slope of the log $P - T^{-1}$ plot.

In the isoteniscope method (a static method) for the determination of the vapour pressure of a liquid, the pressure is measured when a small amount of liquid is vaporized into an evacuated container. The vp. of the liquid is balanced by an adjustable air pressure so that the levels of the liquid in the two arms of the U-tube are equal.

Apparatus and materials used: Isoteniscope (to avoid repetition this is not included; see Fig. 2.10), thermostat variable between 0 and 80 °C, pure carbon tetrachloride, pump and manometer.

Procedure: The isoteniscope was first thoroughly cleaned and dried. Carbon tetrachloride was then introduced until bulb A was half-filled and the U-tube filled to a height of about 5 cm. The isoteniscope was then connected to the mercury manometer, stabilizing vessel and the pumping system by joint D. Then the isoteniscope was placed in an ice-bath at 0 °C, taking care that it was completely immersed. With tap F closed and tap E open, the system was evacuated on a water pump; the liquid in A boiled, and when it was thought that all the air had been displaced, tap E was closed. Tap F was then carefully opened to let some air in through the capillary until the liquid levels in B were the same. The pressure difference on the manometer and the atmospheric pressure were recorded. The pumping operation was repeated, keeping the temperature constant, until a constant manometer reading was obtained. This was to ensure the complete removal of air from the system and to ensure that there were no leaks in the apparatus. Once this had been achieved, the temperature was increased by 5 °C, without further pumping. When the liquid in the isoteniscope had attained the new temperature, air was carefully admitted by opening tap F until the levels of the liquid in B was equal. The manometer reading was recorded. This operation was repeated at about 5 °C intervals until the bath temperature was 76.5 °C. The barometric pressure was read during the course of the experiment, but was found to be constant.

Results and calculations
Barometric pressure = 765.5 mmHg

Vp. of CCl_4/mmHg = Barometric press./mmHg $-$ Manometer reading/mmHg

$T/°C$	T/K	$10^3 K/T$	Manometer rdg./mmHg	P/mmHg	Log(P/mmHg)	P/N m^{-2}
0.0±0.1	273.2	3.660±0.001	727.0±0.5	38.5±0.5	1.585±0.006	5.133±0.07
5.0	278.2	3.59$_4$	721.0	44.5	1.648	5.933
15.0	288.2	3.470	687.0	78.5	1.895	10.466
20.0	293.2	3.41$_1$	664.0	101.5	2.007	13.532
25.0	298.2	3.35$_3$	652.0	113.5	2.055	15.132
30.0	303.2	3.29$_8$	620.5	145.0	2.161	19.332
35.0	308.2	3.24$_5$	582.0	183.5	2.264	24.465
40.0	313.2	3.19$_3$	553.0	212.5	2.327	28.330
45.0	318.2	3.14$_3$	529.0	236.5	2.374	31.531
50.0	323.2	3.09$_4$	448.0	317.5	2.502	42.330
55.0	328.2	3.04$_7$	390.0	375.0	2.575	49.996
60.0	333.2	3.00$_1$	326.0	439.5	2.643	58.595
65.0	338.2	2.95$_7$	244.0	521.5	2.717	69.527
70.0	343.2	2.91$_4$	154.0	611.5	2.786	81.526
76.5	349.7	2.860	0.0	765.5	2.884	102.058

Calculated slope of line (Fig. 1.4) = -1.641×10^3 K

Hence $L_e = -$slope $\times 2.303 \times 8.314$ J mol^{-1}

                         $= 1.641 \times 10^3 \times 2.303 \times 8.314 = 31.42$ kJ mol^{-1}

Uncertainty in the slope = $\pm 0.003 \times 10^3 = \pm 3$, thus the uncertainty in L_e is ± 0.06 kJ mol^{-1}. The literature value of L_e at 311 K = 31.99 kJ mol^{-1}.

Discussion
The measured mean molar heat of vaporization of carbon tetrachloride is 31.42 kJ mol^{-1} with an uncertainty of ± 0.06 kJ mol^{-1}; this is in good agreement with the accepted literature value. L_e is independent of the pressure units used for measurement since it is dependent only on the slope of the log $P - T^{-1}$ line and not the position of the line, thus:

$$\log (P/\text{mmHg}) = -L_e/2.303 \, RT + \text{Const.}$$

or $$\log (P/\text{N m}^{-2}) = -L_e/2.303 \, RT + \text{Const.} - \log 133.322$$

Conversion of the pressure units from mmHg to N m^{-2} merely adjusts the position of the line.

Figure 1.3 is a phase diagram for a one-component system; in the area above the line only the liquid phase exists, ie. $p = 1$. Applying the phase rule, $p + f = c + 2$, gives a value of $f = 2$; thus to define a system in this area completely it is necessary to state both the temperature and the pressure. A similar consideration applies to the vapour phase which exists below the line. The line represents the $P-T$ conditions under which liquid and vapour co-exist in equilibrium, thus $c = 1$, $p = 2$ and so $f = 1$. Thus a system in which liquid and vapour are in equilibrium may be completely defined by specifying one variable, either the temperature or the pressure. This $P-T$ curve only extends to the critical temperature of the liquid.

Any air left in the carbon tetrachloride will be displaced during the experiment and so will contribute to the total pressure, ie. the measured pressure will not be due to carbon tetrachloride alone.

The graph of $\log P$ against T^{-1} is linear over the temperature range studied; this suggests that, within the limits of experimental error, L_e is independent of the temperature and the approximations made in the simplification of equation (1.20) are valid.

Trouton's rule states that the molar heat of vaporization divided by the normal bp. (L_e/T), ie. the molar entropy of vaporization, is approximately 90 J K^{-1} mol^{-1} for most normal liquids (ie. those which are not associated or dissociated). For carbon tetrachloride the normal bp. obtained from Fig. 1.3 is 349.6 K and thus

$$\Delta S = \frac{31.42 \times 10^3}{349.6} = 89.87 \text{ J K}^{-1} \text{mol}^{-1}.$$

Carbon tetrachloride thus conforms to Trouton's rule and can therefore be classified as a normal liquid.

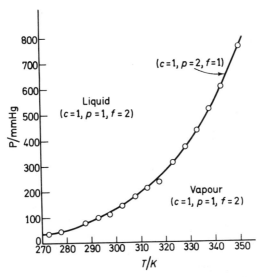

Fig. 1.3   Variation of the vapour pressure of carbon tetrachloride with temperature

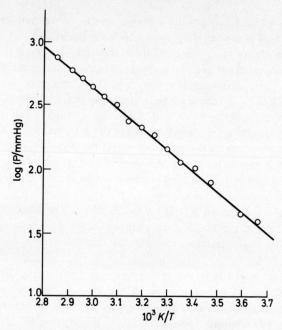

Fig. 1.4　Log (vp/mmHg) as a function of T for carbon tetrachloride

2

General properties of liquids

1 Density

The density (ρ) of a liquid is the mass in kg of 1 m^3.

The density of a liquid may be determined either by weighing a known volume of the liquid in a density bottle or pyknometer or by a buoyancy method (Archimedes' principle). If sufficient liquid is available the density can be approximately determined by a hydrometer or more accurately by a Westphal balance.

Experiment 2.1   Determine the density of a liquid using a Westphal balance

Theory: The Westphal balance provides a rapid but accurate method for determining the density of a liquid relative to that of water at the same temperature. The method involves determining the upthrust acting on a small sinker immersed in a liquid. Ten graduations divide the balance beam equally and, on one or more of these marks can be hung one or more of the four riders provided. The weights of these riders are in the ratio $1:0.1:0.01:0.001$. The balance is first adjusted and calibrated with distilled water at the measured temperature.

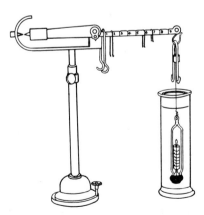

Fig. 2.1   The Westphal balance

Requirements: Westphal balance (Fig. 2.1) complete with riders and sinker, thermostat at 25.00 °C pure benzene or carbon tetrachloride.

Procedure: Set up the balance and suspend the glass sinker (which generally incorporates a thermometer) from the end of the balance beam, so that the sinker is completely immersed in the water. For preference the vessel containing the water should be thermostatted. Now place the unit rider on position ten (or on the hook at the end if only nine positions are marked) and adjust the small counterpoise screw on the end of the beam until the pointer at the opposite end of the beam is in line with the fixed pointer. Record the temperature and the depth of immersion of the sinker below the water surface. The adjustment is now complete.

Dry the sinker and suspend in the liquid under observation, at the same temperature, taking care that the sinker is immersed to the same depth as it was in water. For liquids of density less than 1, adjust the position of the unit rider on the beam and add other riders, in order of decreasing weight, at appropriate marks on the beam, until balance is again achieved. For liquids of density greater than 1, one unit rider must be on position 10 (or on hook) and the remainder on the beam.

The density relative to water at the same temperature is the sum of the products obtained by multiplying the weight of each rider by its fractional distance along the beam (ie. the sum of the clockwise moments).

Treatment of experimental data and discussion
1.   Explain the principles on which the method is based.
2.   Why must the sinker be completely immersed and not touch the sides of the vessel?

Experiment 2.2   Determine the density of a liquid by pyknometry

Theory: The most convenient method involves the measurement of the volume of a suitable vessel, called a pyknometer, by weighing it first empty and then filled with

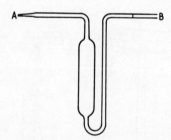

Fig. 2.2   Pyknometer

water. The pyknometer is then weighed when filled with the liquid. Before each weighing the volume of liquid is adjusted so that it fills the vessel from the mark on one arm to the drawn-out tip of the other. The density varies considerably with temperature and hence every filling must be carried out in a thermostat at the required temperature.

Requirements: Pyknometer, thermostat at $25.00 \pm 0.05\ °C$, boiled water, pure benzene or carbon tetrachloride.

Procedure: Wash the pyknometer (Fig. 2.2) successively with a mixture of nitric and chromic acids, distilled water, ethanol and a little ether. Dry by drawing a current of filtered air (cotton-wool plug) through the vessel. Attach a thin piece of nickel or platinum wire to hang it on the balance.

As it is essential that the outer glass surface should always be in the same state of dryness, dip the pyknometer (not the capillary ends) into the thermostat water, wipe carefully with a dry cloth. Allow the vessel to hang in the balance case for 15 min and weigh. Repeat this operation until the weight is constant to 0.25 mg.

Fill the pyknometer with air-free distilled water, by dipping the end A into the water and sucking through a rubber tube attached to B. Before adjusting the volume of the water, suspend the pyknometer in the thermostat (controlled to $\pm 0.05\ °C$) for 15–20 min. Remove the excess water by touching the end A with a piece of filter paper and suck out the water until the meniscus is at the mark on B. Should there be insufficient water, place a rod carrying a drop of water against A when water will be sucked in by capillarity. The filled pyknometer acts as a very sensitive thermometer and so it is easy to see if it has attained the thermostat temperature by observing if the meniscus remains on the mark after adjustment.

When the adjustment is completed, wipe the pyknometer and weigh after standing for 15 min in the balance case. Repeat the filling operations until the variations in weight do not exceed 0.25 mg. Exercise extreme care during the drying operation otherwise water will be expelled from the pyknometer by the heat of the hand. This is particularly serious for density determinations below room temperature.

Empty the pyknometer, dry, fill with the liquid under investigation and weigh as before. Repeat the process until a concordant result is obtained. Use pure benzene or carbon tetrachloride for the determination.

Treatment of experimental data and discussion

1.   Let m_1, m_2 and m_3 be the weights of pyknometer empty, full of water, and of liquid respectively at $t\ °C$, and ρ_t the density of water at $t\ °C$. Neglecting the weight of air displaced, the volume of the pyknometer at $t\ °C = (m_2 - m_1)/\rho_t$.

Thus the density of the liquid at $t\ °C = (m_3 - m_1)/(m_2 - m_1) \times \rho_t$.

The weight of air displaced by the liquid and water is corrected for as follows:

Volume of liquid at $t\ °C = (m_2 - m_1)/\rho_t$.

Thus weight of displaced air $= m_4 = (m_2 - m_1)\rho_{air}/\rho_t$ where ρ_{air} is the density of air under the conditions of weighing.

Hence the true volume of pyknometer $= (m_2 + m_4 - m_1)/\rho_t$ and the corrected density of the liquid at $t\ °C$

$$= \frac{m_3 + m_4 - m_1}{m_2 + m_4 - m_1} \times \rho_t. \tag{2.1}$$

2.   What factors limit the accuracy of this method of measuring density?
3.   Why is it necessary to correct for the air displaced?
4.   Why is the distilled water first boiled and allowed to cool?
5.   Give an account of the additional precautions necessary if the density of the liquid is to be measured at a temperature considerably below room temperature.

Experiment 2.3   Determine the molar volume of ethanol and its partial molar volume at 25 °C in dilute aqueous solution

Theory: The quantitative study of solutions has been greatly advanced by the concept of partial molar quantities first introduced by G. N. Lewis. Extensive properties of solutions (eg. volume, energy, free energy, etc.) are not truly additive when pure components are mixed; thus when ethanol and water are mixed a decrease in volume occurs. Expressed mathematically, the partial molar volume of component A,

$$\overline{V}_A = \left(\frac{\partial V}{\partial n_A}\right)_{T.P.n_B}$$

and of component B,

$$\overline{V}_B = \left(\frac{\partial V}{\partial n_B}\right)_{T.P.n_A}$$

where n_A and n_B are the amounts (mole) of components A and B.

The total volume V of the solution is equal to the volumes of all the components in the solution; the volume of each component is equal to the product of the number of moles of that component and its partial molar volume, thus

$$V = n_A\overline{V}_A + n_B\overline{V}_B \tag{2.2}$$

For ideal solutions the volume of a component in solution is identical with its volume in the pure state. If ethanol and water formed an ideal solution, then the molar volume of a solution in which the mole fraction of each component is 0.5 would be given by:

$$V = 0.5(18 \times 10^{-6}) + 0.5(58.6 \times 10^{-6})$$
$$= 38.3 \times 10^{-6} \, m^3 = 38.3 \, cm^3$$

where $\overline{V}_m(H_2O) = 18 \times 10^{-6} \, m^3 \, mol^{-1}$ and $\overline{V}_m(C_2H_5OH) = 58.6 \times 10^{-6} \, m^3 \, mol^{-1}$.

When such a solution is prepared the volume is 36.9 cm^3. The cause of the change is difficult to establish, but it is obvious that the two components have changed in character on mixing. Compound formation may have occurred or there may be a change in the equilibrium between different types of associated molecules

Requirements: Pyknometer, thermostat at 25.00 ± 0.05 °C, absolute ethanol.

Procedure: Prepare about 25 cm^3 samples of solutions of ethanol and water containing approximately 0, 20, 40, 60, 80 and 100% (wt/wt) of ethanol, by accurately weighing the two components into a stoppered bottle. Determine the density of each solution at 25.00 °C as described in Expt. 2.2. Two pyknometers may be used with advantage.

Treatment of experimental data and discussion

1. Calculate the molar volume of ethanol defined as:

$$\frac{\text{Relative molecular mass}}{\text{Density}} = \frac{46.07 \times 10^{-3}}{\text{Density}} \ \text{m}^3 \, \text{mol}^{-1}$$

2. Calculate the specific volume (ie. reciprocal density) of each solution and plot against the percentage by weight of the ethanol. Draw a smooth curve through the points, and draw tangents to the curve at different concentrations. The intercepts of these lines on the 0% and 100% ordinates gives the partial specific volumes of water and ethanol respectively at the various concentrations. Hence, calculate the partial molar volume, ie. partial specific volume x relative molecular mass.

3. Draw two curves showing the variation of (a) the partial molar volume of ethanol, and (b) the partial molar volume of water with mole fraction.

4. Explain, in words, the exact meaning of a partial molar quantity, such as partial molar volume.

5. Explain why the partial molar volume varies with concentration. Is this variation typical of all liquid mixtures?

2 Viscosity

The viscosity or internal friction of a liquid is the property whereby it resists the relative motion of its parts. When a force, which causes a disturbance in a liquid, ceases to act then the disturbance will die away owing to the viscosity of the liquid. Newton deduced that the viscosity produces retarding forces proportional to the velocity gradient (du/dx) and to the area (A) of contact between moving sheets of liquid. Thus the retarding force

$$F \propto A \, \frac{du}{dx} = \eta A \, \frac{du}{dx} \tag{2.3}$$

where η is the coefficient of viscosity or dynamic viscosity. This law is valid for all homogeneous liquids but not for colloidal solutions and suspensions — non-Newtonian liquids. η is defined as the force per unit area (N m^{-2}) required to maintain a unit difference of velocity (1 m s^{-1}) between two parallel layers 1 m apart. It has the dimensions $m \, l^{-1} \, t^{-1}$ and is expressed in units of N s m^{-2}.

The viscosity of a liquid always decreases with increasing temperature and many empirical equations representing the dependence have been proposed. Those of the exponential type,

$$\eta = A e^{B/RT} \tag{2.4}$$

where A and B are constants for the given liquid, are preferred since they most probably have a theoretical basis.

The viscosity of a liquid depends on such factors as molecular size (particularly length) and intermolecular forces. Non-polar liquids generally have low viscosities, whilst liquids such as glycerol, where directed bonding can occur between molecules, have very high viscosities.

The experimental methods for the determination of liquid viscosities are similar to those for gases: (i) flow through capillary tubes, (ii) torque on rotating cylinders — Couette viscometers, (iii) fall of solid spheres through liquids, and (iv) flow of liquid through an aperture in a plate.

When a liquid flows through a long capillary of a material which it wets, emerging with a small velocity, the volume *(v)* which passes a section of the tube in a time *t*/s is given by Poiseuille's equation:

$$v = \frac{\pi p t r^4}{8 l \eta} \tag{2.5}$$

where p is the pressure difference between the ends of the tube of length l and radius r. The determination of the absolute viscosity of a liquid thus involves the accurate measurement of p, t, r and l. It is usually sufficient to compare the viscosity of the liquid with that of water or other standard liquid, by measuring the times taken for equal volumes of the two liquids to flow through the same capillary under pressures due to their own weights. The densities of both liquids must be known. The absolute viscosity of the liquid can then be obtained knowing the viscosity of the standard liquid.

The British Standard (BSS No. 188 (1957)) form of Ostwald viscometer (Fig. 2.3*a*), consists of a capillary (about 10 cm long) with a bulb above. There are two etched marks at B and C. A definite volume of liquid, to ensure constant hydrostatic head, sufficient to fill the tube from a point above C to F is introduced into the wide arm at D by a pipette. By applying pressure, the liquid is forced up so that the upper surface is above C, and then allowed to flow back through the capillary under a pressure due to its own weight. The time is recorded for the meniscus to pass from C to B. The difference in level is always the same and hence the hydrostatic pressure of each liquid is proportional to its density. If η_1 and η_2 are the viscosities of the two liquids of densities ρ_1 and ρ_2, and t_1 and t_2 are the times of flow it follows from equation (2.5) that

$$\frac{\eta_1}{\eta_2} = \frac{t_1 \rho_1}{t_2 \rho_2} \tag{2.6}$$

Thus, if the absolute viscosity of one liquid is known that of the given liquid can be determined.

The suspended level viscometer (Fig. 2.3*b*), depending on the same principle as the Ostwald viscometer, is of great use in studying the variation of viscosity with the concentration of a solution, as the solution can be diluted *in situ*. 20 cm^3 of the solution is transferred to the bulb A, and, with the side arm, D, closed this is pumped through the capillary tube F until the meniscus is above C. When D is

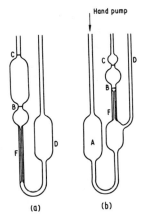

Fig. 2.3 Typical capillary tube viscometers. *(a)* U-tube viscometer, *(b)* Suspended level viscometer

opened the liquid below F falls back and the head of liquid CF is held suspended. The time is recorded for the meniscus to pass from C to B. Since the head of liquid CF is always the same it is independent of the volume of liquid in the viscometer. A known volume of solvent is now pipetted into A, the solution well mixed, and the time of flow from C to B again determined. This may be repeated many times.

The velocity of flow of a liquid through a capillary is proportional to the fourth power of the radius of the tube. To cover the wide range of viscosities it is essential to have a series of viscometers with tubes of different radii. The primary standard for the calibration of viscometers is the kinematic viscosity ($v = \eta/\rho$) of distilled water at 20 °C, = $1.0038 \times 10^{-6}\,\mathrm{m^2\,s^{-1}}$, ie. $\eta = 1.002 \times 10^{-3}\,\mathrm{N\,s\,m^{-2}}$.

Experiment 2.4   Determine the viscosity of a pure liquid (nitrobenzene) and its variation with temperature

Requirements: Viscometer, pyknometer, stop-watch, glass-fronted thermostats at different temperatures regulated to ±0.05 °C, pure nitrobenzene.

Procedure: The accuracy of a viscosity determination depends largely on the cleanliness and care in drying the capillary tube. Particles of dust and grease cause erratic results. Choose the correct viscometer for the liquid so that the time of outflow is not less than 100 seconds.

Thoroughly clean and dry the viscometer (as in Expt. 2.2) and clamp vertically (using a plumb line) in the thermostat in such a position that it can be viewed easily. Thermostat a sample of the calibrating liquid (water) and the test liquid in separate stoppered tubes. Pipette 5 cm^3 (or volume specified on viscometer) of the calibrating liquid into the arm D of the viscometer. When temperature equilibrium has been established (about 20 min) force the liquid (by means of a tube plugged with cotton wool attached to D) into the chamber above the capillary so that the meniscus is above mark C. Release the pressure, thus allowing the liquid to flow through the capillary and record the time, to the nearest 0.1 s, taken for the meniscus to pass between the marks C and B. Repeat the experiment several times, until there is agreement to within 0.2 s on 100 s. Empty the viscometer, wash and dry as previously described, fill with the unknown liquid and again determine the flow time.

Determine the density of the liquid at the same temperature (Expt. 2.1).

Repeat the determination of both time of outflow and density at different temperatures in the range 25–80 °C.

Treatment of experimental data and discussion

1. Determine the viscosity of the liquid at the different temperatures using equation (2.6) and plot a graph of log η against T^{-1}. Test the validity of equation (2.4) and determine the values of the constants A and B. What are the units of A and B? Is there any theoretical basis for an equation of the type shown in (2.4)?
2. Why is it essential that the viscometer be mounted vertically in the thermostat?
3. Can differences of surface tension of the different liquids introduce errors into the value of η measured by this method?

Experiment 2.5   Determine the relative viscosities of mixtures of ethanol and water

This experiment is best carried out in conjunction with Expt. 2.3 for the determination of densities.

Procedure: Prepare mixtures of ethanol and water containing approximately 0, 20, 40, 60, 80 and 100% (wt/wt) of ethanol.

Determine the density and the viscosity of each of the liquid mixtures at 25 °C as previously described (Expt. 2.3 and 2.4).

Treatment of experimental data and discussion

1. Plot the viscosity of each liquid mixture relative to that of one component (solvent) against the mole fraction of that component.
2. Discuss the shape of the curve in terms of the changes which occur when two liquids are mixed.

3   Surface tension

In the interior of a liquid a given molecule, completely surrounded by other molecules, is attracted equally in all directions. On the other hand, a molecule at the surface will be attracted inwards since the number of molecules per unit volume is greater in the liquid than in the vapour phase. As a direct result of this inward pull, the surface of a liquid has the tendency to contract to the smallest possible area. Thus free drops of liquid and gas bubbles assume a spherical shape for then the area/volume ratio is a minimum. Work is necessary to bring molecules from the bulk of the liquid into the surface; the *surface free energy* (J m^{-2}) is defined as the work required to increase the area by 1 m^2. Since there is this tendency for a surface to contract, it behaves as if it were in a state of tension. It is thus possible to give a numerical value to the surface tension which is the same at every point and in all directions along the surface of the liquid. The *surface tension* (γ), numerically identical with the surface free energy, is the force (N) acting at right angles to any line of 1 m length on the surface (units: N m^{-1}).

In a similar way, the *interfacial free energy* is the work required to enlarge the surface of separation of immiscible or partially miscible liquids by 1 m^2. Generally the interfacial tension between two liquids is less than the larger surface tension.

The surface tensions of liquids (apart from liquid crystals and some metals) decrease with increasing temperature. The van der Waals equation

$$\gamma = \gamma_0(1 - t/t_c)^n \tag{2.7}$$

where γ_0 is a constant, t_c the critical temperature and n a constant (≈ 1.23), describes this change for non-associated liquids. Katayama proposed the equation:

$$\gamma \left\{ \frac{M}{\rho_l - \rho_g} \right\}^{\frac{2}{3}} = kt_c(1 - t/t_c) \tag{2.8}$$

where M is the relative molecular mass, ρ_l and ρ_g the densities of the liquid and vapour respectively at temperature t (°C), and k a constant in which the molecular volume is taken into account. This equation gives a better representation of the experimental facts at temperatures near the critical temperature than does equation (2.7).

Static and dynamic methods may be used to determine the surface tension of a liquid. Static methods include the measurement of capillary pull, capillary rise and the shape or size of hanging drops; the dynamic methods include bubble pressure and vibrating jet. The dynamic methods have the advantage that the liquid surface is constantly renewed.

It is rarely necessary to use a laborious absolute method for the determination of the surface tension of a liquid; if the apparatus is calibrated with a liquid of known surface tension, most of the experimental difficulties can be overcome.

Experiment 2.6   Determine the surface tension of a pure liquid (water, benzene or carbon tetrachloride) at 25 °C by the capillary rise method

Theory: When a clean grease-free capillary tube is dipped into a liquid, it rises to a height (h) above the horizontal free surface of the liquid (Fig. 2.4a).

The column of liquid can be regarded as held up by the force of surface tension; which, at equilibrium balances the weight of the column of liquid.

Thus
$$2\pi r \gamma \cos \theta = \pi r^2 g \rho h$$

or
$$\gamma = \frac{g \rho h r}{2 \cos \theta}$$

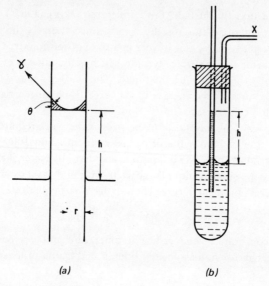

(a)   (b)

Fig. 2.4   (a) Surface tension forces in a capillary tube. (b) Apparatus for measuring surface tension by the capillary tube method

If the liquid wets the glass $\theta = 0$ and $\cos \theta = 1$ whence

$$\gamma = \frac{g\rho hr}{2}$$

(2.9)

The weight of liquid in the meniscus (shaded) has not been taken into account. If the meniscus is so small that it may be considered to be a hemisphere, its volume is given by: $\pi r^3 - \frac{2}{3}\pi r^3 = \frac{1}{3}\pi r^3$, and equation (2.9) becomes:

$$\gamma = \tfrac{1}{2}g\rho r(h + \tfrac{1}{3}r)$$

(2.10)

For even greater accuracy, when the meniscus is assumed to be an ellipsoid of revolution, the equation becomes:

$$\gamma = \tfrac{1}{2}g\rho hr \left\{ 1 + \tfrac{1}{3}\left(\frac{r}{h}\right) - 0.1288 \left(\frac{r}{h}\right)^2 + 0.1312 \left(\frac{r}{h}\right)^3 \right\}$$

(2.11)

Requirements: Capillary tubes of about 25 cm length and 0.2 mm diameter, travelling microscope, boiling tube (5 cm diameter), thermostat at 25 °C, pure water, benzene or carbon tetrachloride.

Procedure: Clean and dry the capillary tube (as in Expt. 2.2). Pour a quantity of the liquid under investigation into the large boiling tube and assemble the apparatus (Fig. 2.4b). Clamp the apparatus in a glass-sided thermostat at 25 °C, and allow temperature equilibrium to be established (20 min). Blow gently through a piece of rubber tubing attached to X, thus raising the liquid up the capillary, release the pressure and allow the liquid to fall back to its equilibrium level. Determine the height, h between the meniscus and the free surface of the liquid using a travelling microscope. Depress the meniscus by suction, release and again determine the height, h. If the capillary is clean, the readings after depressing and raising the meniscus will be the same. Repeat the determination of h five or six times and if there is wide disagreement, remove and clean the capillary tube.

Thoroughly clean the apparatus and determine h for different liquids, making sure that the meniscus is always at approximately the same place in the tube. This is accomplished by either raising or lowering the capillary in the rubber bung so that the meniscus is at a marked place.

The radius of the tube may be determined by *(a)* calibration using a liquid (eg. water) of known surface tension using equation (2.9); *(b)* direct measurement with the travelling microscope, focusing first on the far inside surface and then on the near inside surface of the dry tube; or *(c)* direct measurement with a thread of mercury. In the last method partly fill the dry capillary tube with clean, dry mercury and determine the length of the thread in different parts of the tube. Pour the mercury out into a weighed receiver and weigh. The radius of the tube can now be calculated and the bore checked for uniformity.

Treatment of experimental data and discussion

1. Using equation (2.9), calculate the surface tension of the liquid at 25 °C.
2. Estimate the error involved in using equation (2.9) instead of equation (2.10) or equation (2.11).
3. Why should the meniscus be at approximately the same place in the tube for the different liquids?
4. Discuss the relative merits of the static and dynamic methods (Expt. 2.8) for the determination of the surface tension.

Experiment 2.7   **Determine the surface tension of a liquid by the double capillary method at different temperatures**

Theory: The method previously described (Expt. 2.6) for accurate work is slow and laborious and suffers from many disadvantages. The double capillary method overcomes most of these objections. In this method the difference, Δh, in the heights to which a liquid rises in two capillary tubes, of radii r_1 and r_2, is measured. For each tube when equilibrium is established, it follows from equation (2.9) that:

$$\gamma = \frac{g\rho h_1 r_1}{2} \quad \text{and} \quad \gamma = \frac{g\rho h_2 r_2}{2}$$

whence

$$\gamma = \Delta h \rho \left[\frac{g}{2} \left(\frac{r_1 r_2}{r_2 - r_1} \right) \right] \tag{2.12}$$

The term in square brackets is a constant for the apparatus and is determined by calibration with a liquid of known surface tension and density. For greater accuracy, terms similar to those in equations (2.10) and (2.11) can be included.

Requirements: Two capillary tubes, internal diameters 0.3—0.4 mm and 1.5—2.0 mm, boiling tube (or U-tube (Fig. 2.5*b*) with similar size capillary tubes), thermostat which can be regulated over a range of temperatures, travelling microscope, pure methyl acetate (ethyl acetate, chloroform, or chlorobenzene) and benzene.

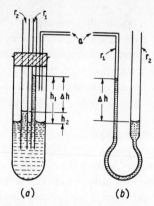

Fig. 2.5   *(a)* and *(b)* Different types of apparatus for measurement of surface tension by the differential method

Procedure: Thoroughly wash and dry the capillary tubes (as in Expt. 2.2). Assemble the tubes in the rubber bung so that they can be easily moved up or down; mount them as close as possible so that they can be viewed simultaneously.

Pour pure benzene into the boiling tube and clamp the apparatus in a constant temperature bath at 20.0 ± 0.1 °C. When temperature equilibrium has been reached, apply slight pressure by blowing through tube a, release the pressure, and, when the equilibrium position has been reached in both tubes, determine the difference in height (Δh) of the meniscus in the two tubes. Repeat the measurement after depressing the levels by suction at a. This ensures that the liquid wets the glass above the meniscus and gives a zero angle of contact. These readings of Δh should agree if the glassware is clean and free from grease. It is not necessary to determine the absolute values of h_1 and h_2. To avoid subsequent errors due to possible non-uniformity of bore, the meniscus must always be brought to approximately the same position in the tubes; mark this position on both the tubes. The calibration is now complete.

Thoroughly wash and dry the apparatus and refill with the other pure liquid. Adjust the positions of the tubes so that the meniscus is at the same position as for benzene. Determine Δh for the liquid, as previously described at different temperatures in the range 20–50 °C. The density of the liquid required at each temperature, may be determined as in Expt. 2.1 or obtained from tables.

Treatment of experimental data and discussion

1.   From the readings of Δh for benzene at 20 °C and the values of the surface tension of benzene (28.88×10^{-3} N m^{-1}) and its density (0.8794×10^3 kg m^{-3}), calculate the apparatus constant of equation (2.12). Hence calculate the surface tension of the liquid at the different temperatures.

2.   Plot a graph showing the variation of the surface tension of the liquid with temperature. Discuss possible reasons for this variation.

3.   List the disadvantages of the single capillary tube method (Expt. 2.6) and explain how these are overcome in the differential method.

4. State the limitations of the differential method, and explain how the accuracy of the method could be improved.
5. Why is it inadvisable to calibrate the apparatus with water, when it is to be used to measure the surface tension of an organic liquid?

Experiment 2.8   Determine the surface tension of chlorobenzene at different temperatures by the bubble pressure method and hence determine the critical temperature

Theory: If a bubble is blown at the bottom of a tube dipping vertically into a liquid, the pressure in the bubble increases at first as the bubble grows and the curvature diminishes. The smallest radius of curvature and the maximum pressure occur when the bubble is hemispherical, further growth results in a diminution of pressure so that air rushes in and the bubble is detached from the tube. At the instant before this, the pressure in the bubble is given by $P = g\rho h + 2\gamma/r$, where h is the depth of the tube below the surface, r its radius, ρ and γ the density and surface tension of the liquid respectively. This equation formed the basis of Jaeger's method for the measurement of surface tension. The method suffers from two practical disadvantages: (1) the exact height of the bottom of the tube below the liquid level is difficult to determine, and (2) a rather large volume of liquid is required. These difficulties are overcome in the method due to Sugden, in which the difference in pressure, Δp, required to blow bubbles at the ends of tubes of different radii is measured. Empirically it has been established that for an apparatus of suitable dimensions the surface tension is given to an accuracy of 0.1% by:

$$\gamma = A\,\Delta p \left(1 + 0.69 r_L \frac{g\rho}{\Delta p}\right) \tag{2.13}$$

where r_L is the radius of the larger tube, ρ is the density of the liquid under investigation, $\Delta p = p_1 - p_2$, p_1 and p_2 the pressures required to form bubbles in the smaller and larger tubes respectively and A the apparatus constant determined by calibration. The radius of the smaller tube does not enter this equation explicitly, but is included in the constant A.

The method is independent of the contact angle, and, as a fresh liquid surface is formed with each bubble, there is little risk of local contamination effects. The density of the liquid under test does not have to be accurately known; an error of 1% in the density introduces an error of less than 0.1% in the final value of the surface tension.

Requirements: Apparatus (Fig. 2.6) with capillary tubes of diameters 0.05–0.10 mm and 1.0–2.0 mm. Manometer filled with dibutyl phthalate, large beaker of water used as thermostat bath, pure benzene and chlorobenzene.

Procedure: The apparatus consists of a pressure gauge A connected to the capillary tubes dipping into liquid in vessel B and also to a source of pressure. The larger tube may be taken from a selected length of quill tubing; the smaller is made by drawing out a capillary. The ends of the tubes must be cut and ground until they are

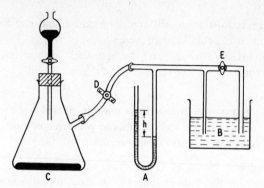

Fig. 2.6   Maximum bubble pressure apparatus for the determination of surface tension

smooth, sharp edged and perpendicular to the sides. Mount the tubes as shown so that they are immersed to the same depth and remain in the same position relative to one another. Pressure is applied by slowly dripping mercury from the tap funnel into the flask C. If water is used in place of mercury, it is necessary to include a calcium chloride drying tube before the screw clip D.

Using a travelling microscope, determine the internal radius of the larger tube (r_L). Thoroughly clean the tubes and jets with cleaning mixture, water, alcohol and finally the liquid to be used. Pour some benzene into B and place in the bath at 25 °C for 15–20 min. Generate pressure in flask C with clip D closed. With tap E closed (ie. smaller tube in use) open D slowly until bubbles just form and break away from the end of the tube. Record the manometer pressure, ie. the difference in the height of the liquid in the two arms, with a scale or travelling microscope. Repeat this several times and obtain an average maximum bubble pressure reading. Open the tap E and repeat the above procedure to determine the pressure required for bubbles to form and just break from the larger tube. Carry out these determinations for the two tubes at three temperatures in the range 0–60 °C, controlling the bath temperature to ±0.1 °C. This completes the calibration.

Wash out the capillary tubes and repeat the above determinations of pressure every 10 °C in the range 0–80 °C for chlorobenzene. The density of benzene and chlorobenzene at the different temperatures may be either measured (Expt. 2.1) or obtained from tables.

Treatment of experimental data and discussion

1.   Convert the manometer readings (measured in mmHg) into pressure expressed in N m^{-2}, ie. $\Delta p = g\rho_{man} \Delta h$, where Δh/m is the difference between the manometer readings when using the small and the large tube and ρ_{man} is the density of the manometric liquid (for dibutyl phthalate $\rho = 1.046 \times 10^3$ kg m^{-3}). Using the data for benzene, calculate the value of the constant A in equation (2.13) over a range of temperatures.

2.   With this calibration calculate the surface tension of chlorobenzene at the different temperatures. For non-associated liquids the variation of surface

Table 2.1 Density and surface tension of benzene and the density of chlorobenzene at different temperatures

Temperature °C	10^{-3} x Density $(\rho^t/\text{kg m}^{-3})$		10^3 x Surface tension of benzene/N m^{-1}
	Benzene	*Chlorobenzene*	
0	0.9001	1.1278	31.58
10	0.8894	1.1171	30.22
20	0.8788	1.1063	28.88
30	0.8684	1.0956	27.56
40	0.8575	1.0847	26.26
50	0.8468	1.0738	24.98
60	0.8361	1.0628	23.72
70	—	1.0518	22.48
80	—	1.0406	21.26

tension with temperature may be expressed empirically as in equation (2.7).
Taking logs of this equation and differentiating gives:

$$t = \frac{n\gamma}{d\gamma/dt} + t_c \qquad (2.14)$$

Plot a graph of temperature against $\dfrac{\gamma}{d\gamma/dt}$ and determine the constant n and
the critical temperature t_c. (It is sufficient to take differences of γ over steps
of 10 °C and divide the difference by the temperature increment; this gives
the value of $d\gamma/dt$ at the mean temperature.)

3. What is the significance of the critical temperature of a substance? What other
methods are available for determining t_c?

4. Why is tap E arranged so that, when closed, air only passes to the smaller tube
and when open air passes to both tubes?

5. List the advantages and disadvantages of this method compared with the
capillary tube method.

Experiment 2.9 Determine the surface tension of benzene and water at 25 °C by the drop-weight method

Theory: This method depends on the equilibrium between the mass of a drop of
liquid (m) just about to fall from a vertical tube of external radius, r, and the surface or interfacial tension (γ) acting round the periphery of the drop. The weight of
a drop that falls should be given by:

$$mg = 2\pi r\gamma \qquad (2.15)$$

In actual fact, only a portion of the drop falls; Harkins and Brown proposed that
the drop of weight ($2\pi r\gamma$) given by equation (2.15) be called the 'ideal drop'. The
fraction of the ideal drop which falls was determined by these workers in an

extensive series of experiments. The form of the maximum stable hanging drop is a function of $rv^{-1/3}$, where r is the outer radius of the tube and v the volume of the drop (determined from the mass of the drop and the density of the liquid). This determines the fraction $f(rv^{-1/3})$ of the drop which falls. The actual weight of the drop which falls is thus:

$$mg = 2\pi r\gamma f(rv^{-1/3}) \tag{2.16}$$

Hence

$$\gamma = \frac{mg}{2\pi r f(rv^{-1/3})} = \frac{Fmg}{r} \tag{2.17}$$

Thus, to determine the surface tension of a liquid it is only necessary to determine the mass of one drop of the liquid, calculate its volume v from a rough density determination, multiply this by $1/r^3$, look up the value of F (Table 2.2) and finally multiply by mg/r. The density of the liquid is involved only in determining the value of the correction which is not very sensitive to density differences.

Using extreme precautions, the precision obtained is 0.02−0.03%, but with a simplified apparatus at a greatly reduced cost, a precision of 0.3% can be obtained.

Requirements: Drop-weight surface tension apparatus with ground tip (Fig. 2.7), three weighing bottles, travelling microscope, thermostat at 25 °C, pure benzene.

Procedure: The simple apparatus (Fig. 2.7) illustrates the principle of the method and gives results of reasonable accuracy.

Table 2.2   Correction factors (F) for the drop-weight method

vr^{-3}	F	vr^{-3}	F
∞	0.159	3.433	0.2587
5000	0.172	2.995	0.2607
250	0.198	2.0929	0.2645
58.1	0.215	1.5545	0.2657
24.6	0.2256	1.048	0.2617
17.7	0.2305	0.816	0.2550
13.28	0.2352	0.692	0.2499
10.29	0.2398	0.570	0.243
8.19	0.2440	0.512	0.2441
6.662	0.2479	0.455	0.2491
5.522	0.2514	0.403	0.2559
4.653	0.2542		

Thoroughly clean (as in Expt. 2.2), dry and assemble the glassware. Place a dry weighed weighing bottle under the tip and replace the cork. Fill the bulb and capillary tube with water and carefully adjust the water level in the bulb so that a drop forms and breaks away after 4−5 min (1 cm head of liquid is usually sufficient for this). Allow the first drop to form slowly and so saturate the space within the container. Now increase the pressure so that subsequent drops form and break about every 30−40 s. Ensure that the drop falls away by gravity only and not as a result of vibration. Collect 25−30 drops and weigh the collected water. It is

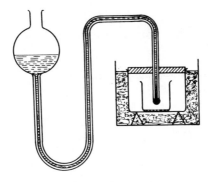

Fig. 2.7   Drop-weight apparatus for surface tension measurement

advisable, particularly with more volatile liquids, to cool the collecting bottle in ice water before weighing.

Repeat the determination with benzene.

Measure the outside radius of the jet with a travelling microscope.

Treatment of experimental data and discussion

1. Calculate the volume of the drop v which forms on the jet of radius r and hence the value of vr^{-3}. Obtain the appropriate value of F (Table 2.2) and using equation (2.17) (m in kg and r in m) determine the surface tension of water and of benzene.
2. Explain why it is necessary for the drop to form slowly.
3. Why must the container always contain a small amount of the liquid under investigation in the capillary tube?
4. Summarize the experimental errors and uncertainties in the theory.

Experiment 2.10   Determine the interfacial tension between benzene and water at 25 °C by the drop-weight method

Theory: The drop-weight method (Expt. 2.9) is suitable for the measurement of interfacial tensions. The drops which form are now considerably larger than those which form in air. Their size can thus be accurately determined by a direct measurement of their volume using a micrometer syringe with a piston of known cross-sectional area. Such a syringe is capable of measuring to 10^{-4} cm^3.

Requirements: 'Agla' micrometer syringe (Burroughs Wellcome Co. Ltd), hypodermic needle, pure benzene and water, thermostat at 25 °C.

Procedure: Select a syringe needle with a large jet; outside diameter 0.7–1.0 mm, internal diameter 0.4–0.5 mm. The jet must be ground cylindrical with sharp edges. Measure the external diameter with a travelling microscope.

Completely fill the clean syringe and needle with water and clamp so that the jet is immersed in benzene maintained at 25 °C. Record the initial reading on the micrometer. Slowly and without causing any vibration, force a drop of water out by screwing down the micrometer; the delivery of the last 5–10% of the drop should take at least 1 min. Record the final reading on the micrometer and hence determine the volume of the drop. Repeat for 10–15 drops and determine the mean volume of a drop.

Treatment of experimental data and discussion

1. Determine the interfacial tension (γ_{12}) between benzene and water as described in Expt. 2.9. The mass of the drop (m) in equation (2.17) is now replaced by $v(\rho_1 - \rho_2)$ where v is the recorded volume and ρ_1 and ρ_2 the densities of water and benzene respectively.

Thus
$$\gamma_{12} = \frac{Fvg(\rho_1 - \rho_2)}{r} \tag{2.18}$$

2. Test the validity of Antonoff's rule 'That the interfacial tension between two liquids is approximately equal to the difference between the surface tensions of the two liquids', ie.

$$\gamma_{12} = \gamma_1 - \gamma_2 \tag{2.19}$$

Experiment 2.11 Determine the surface tension of mixtures of ethanol and water by the ring or torsion balance method

Theory: This method is probably more widely used than any other for the determination of surface tension. It is rapid, simple and when carried out under correct conditions has an accuracy as high as ±0.25%.

The method depends essentially on the determination of the force required to detach a circular ring of platinum wire from the surface of a liquid with an angle of contact of 0°.

The shape of the liquid in contact with the water is determined by the surface tension and density of the liquid and the radius of the ring (R') and the wire (r). A simplifying, though false, assumption is that the ring holds up a hollow cylinder of liquid with vertical walls of radii R' and $R' + 2r$ (Fig. 2.8a). Thus the total pull, P, which is equal to the weight of liquid (mg) suspended is:

$$P = mg = 2\pi R'\gamma + 2\pi\gamma(R' + 2r)$$
$$= 4\pi\gamma(R' + r) = 4\pi\gamma R \tag{2.20}$$

where R is the mean radius of the ring.

Table 2.3 Correction factors (F) for the ring method

R^3V^{-1}	$Rr^{-1} = 32$	$Rr^{-1} = 42$	$Rr^{-1} = 50$
0.3	1.018	1.042	1.054
0.5	0.946	0.973	0.9876
1.0	0.880	0.910	0.929
2.0	0.820	0.860	0.8798
3.0	0.783	0.828	0.8521

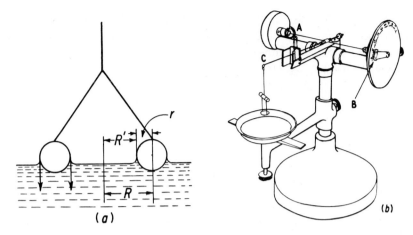

Fig. 2.8   Determination of surface tension by du Nouy's method. *(a)* Principle of the method. *(b)* Torsion balance

Without this simplification, the measured pull must be multiplied by a correction factor (F), determined experimentally by Harkins and Jordan. Table 2.3 gives the values of F as a function of Rr^{-1} and $R^3 V^{-1}$, where V is the volume of liquid raised calculated from $P/\rho g$. The corrected equation is thus:

$$PF = 4\pi R\gamma$$

or
$$\gamma = \frac{PF}{4\pi R} \qquad (2.21)$$

Requirements: Torsion balance, platinum ring of known R and r, absolute ethanol.

Procedure: As it is generally not possible to provide thermostatic control, the measurements should be carried out at room temperature (record this). Thoroughly clean the platinum wire with hot acid cleaning mixture, rinse with distilled water and dry by momentarily heating in a luminous flame. Without touching the ring suspend it from the hook on the beam C. Turn the knob until the pointer B is at zero and adjust the torsion of the wire by A so that the beam C lies in the horizontal position and just clear of the arm. Raise the platform holding the watch-glass and liquid until the liquid just touches the ring. Now simultaneously turn the knob which twists the torsion wire and lower the platform by means of the screw (underneath) until the ring is torn from the surface. By this method the beam C remains horizontal all the time, thus preventing high readings. Record the position of the pointer on the scale at which the ring is pulled free. Repeat several times for each liquid mixture until concordant results are obtained.

Calibrate the instrument over the range of scale readings required as follows: dry the ring and suspend it on the hook. Place a small weighed square of paper on the ring, adjust the pointer to the zero position and also the torsion of the wire so that the beam is horizontal. Add a fractional gram weight to the paper and again twist the wire until the beam is horizontal, record the pointer reading on the scale.

Repeat for other fractional weights and confirm that the weight (including the paper) divided by the scale reading is constant. If the torsion of the wire is not proportional to the angle through which the pointer turns then it is necessary to plot a calibration curve. Thus the value of the pull in Newtons can be calculated per scale division (total weight $\times$ g)/scale reading.

Using this procedure, determine the surface tension of pure water, absolute ethanol and 1.0, 0.1, 0.01 and 0.001 mol dm^{-3} aqueous ethanol mixtures.

Treatment of experimental data and discussion

1. For each liquid calculate the force P, from the pointer reading and the calibration of the wire. Knowing the dimensions of the ring, Rr^{-1} and the calculated value of $R^3 V^{-1}$ ($V = P/\rho g$) look up the value of F (Table 2.3). Calculate the surface tension of each liquid mixture (equation (2.21)).

2. Plot a graph showing the variation of the surface tension with the concentration of ethanol. Explain this variation in terms of the surface excess concentration of one component.

3. Comment on the use of this method as an absolute method of determining γ.

4   Vapour pressure

The saturation vapour pressure, or the vapour pressure, is the pressure at which liquid and vapour can coexist in equilibrium at a given temperature. The vapour pressure of a pure liquid in contact with vapour depends on the temperature and is independent of the relative amounts of liquid and vapour present.

The Clapeyron–Clausius equation, relating the vapour pressure (p) with the molar heat of vaporization (L_e) and the temperature (T), may be written:

$$\frac{d \ln p}{dT} = \frac{L_e}{RT^2} \tag{2.22}$$

assuming that the volume of liquid is negligible compared with that of the vapour, and that the vapour obeys the ideal gas equation.

Integration of equation (2.22) gives

$$\log p = \frac{-L_e}{2.303\,RT} + \text{Const.} \tag{2.23}$$

which is valid over a limited range of temperature only, since L_e is assumed to be independent of temperature.

There are three methods available for the determination of vapour pressures of liquids.

1. *Static methods.* The pressure exerted when a small amount of liquid is vaporized into an evacuated container is measured at a definite temperature.

2. *Dynamic methods.* The boiling point of a liquid under a fixed external pressure is determined.

3. *Transpiration method.* A known volume of gas is passed through a liquid

at constant temperature and the amount of liquid carried off is measured. This method, of limited application, will not be illustrated by an experiment.

Experiment 2.12   Study the variation of the vapour pressure of a pure liquid with temperature

Requirements: Apparatus (Fig. 2.9) mercury manometer, pure water, carbon tetrachloride, chloroform, butan-2-ol.

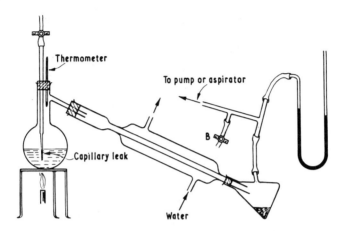

Fig. 2.9   Simple apparatus for measurement of the vapour pressure of volatile liquids

Procedure: The simplified apparatus (Fig. 2.9), suitable for the measurement of the vapour pressure of volatile liquids over a range of temperatures, consists of an ordinary distillation unit, evacuated with a vacuum pump and trap.

Pour about 200 cm^3 of pure liquid into the flask (500 cm^3) and evacuate the system as far as possible with the bleeder valve B closed. When the liquid starts to boil, stop evacuating and open B a little, thereby raising the pressure. Close B and heat the flask with a very small flame until the liquid just boils and distils at the rate of about 1 drop a second. Bumping is prevented by a fine capillary dipping in the liquid. As soon as the temperature and pressure are stabilized at this distillation rate, read and record both. Now admit some air through B thereby raising the pressure, close B and again apply heat until the steady distillation rate is again achieved. Record the temperature and pressure. Repeat over the complete range of pressures. Record the barometric pressure; the vapour pressure of the liquid is the difference between the barometric pressure and the pressure difference on the manometer.

Treatment of experimental data and discussion
1.   Tabulate the vapour pressure at the various temperatures (*a*) in mmHg, and (*b*) in N m^{-2}. Plot graphs of p against T and log p against T^{-1}. Calculate the mean heat of vaporization over the temperature range studied (equation (2.23)).
2.   Explain why the value of L_e is independent of the pressure units used.

3.   How far may the $p - T$ curve be extended?
4.   The $p - T$ curve is a simple phase diagram. State the number of components present, and use the phase rule to determine the number of degrees of freedom for a system lying in an area and for a system defined by a point on the $p - T$ curve.

Experiment 2.13   Study the variation of the vapour pressure of a pure liquid with temperature, using an isoteniscope

Theory: The principle of this method, using an isoteniscope, is to balance the vapour pressure of the liquid by an adjustable pressure of air so that the levels of the liquid in the two arms of the U-tube are equal.

Requirements: Isoteniscope (Fig. 2.10) mercury manometer, thermostat, pure carbon tetrachloride, chloroform, butan-2-ol.

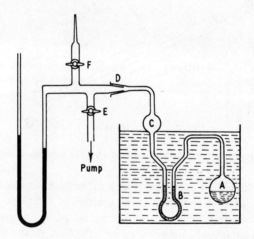

Fig. 2.10   Isoteniscope

Procedure: Half fill bulb A (2 cm diameter) and the U-tube B (3–5 cm high) with the liquid and connect to the pumping system and manometer by joint D. The vessel C merely acts as a buffer to stabilize the pressure. Place the isoteniscope in an ice bath at about 0 °C and with F closed and E open, evacuate the system so the liquid in A boils and displaces all the air. The liquid levels in the two arms of B will now be different. Turn off tap E and admit small amounts of air through the capillary by carefully opening tap F, until the liquid levels in B are the same. Keep the bath temperature constant until temperature equilibrium has been attained and ensure that no change in the levels in B occurs. Read the pressure difference on the manometer. Repeat the pumping out operation to confirm that all the air has been removed from the liquid in A; finally, admit air again and record the pressure difference. Without further pumping out, raise the temperature of the bath, allow the liquid to attain the new temperature, cautiously admit air until the liquid levels in the U-tube are the same and record the pressure difference. Repeat every 5–10 °C up to the boiling point. Record the barometric pressure. The vapour pressure of the liquid is the difference between the barometric pressure and the pressure difference on the manometer.

Treatment of experimental data and discussion

1. Tabulate the vapour pressure at the various temperatures (*a*) in mmHg, and (*b*) in $N\,m^{-2}$. Plot graphs of *p* against *T* and log *p* against T^{-1}. Calculate the mean heat of vaporization over the temperature range studied (equation (2.23)).
2. Explain why the value of L_e is independent of the pressure units used.
3. The *p* − *T* curve is a simple phase diagram. Use the phase rule to discuss the diagram.
4. Why is it necessary for all the air to be boiled out of the liquid?
5. In the derivation of equation (2.23), it is assumed that L_e is independent of temperature. Comment on this assumption and its effect on the accuracy of your relation for *p* as a function of *T*.
6. Determine the bp. of the liquid. State Trouton's rule and use your data to test its validity for the liquid studied.

Experiment 2.14   Study the variation of the vapour pressure of a liquid with temperature by the method of Ramsay and Young

Theory: The method requires very little liquid and the boiling point of the liquid is determined at the pressure registered on the manometer, unaffected by super-heating or any slight impurities.

Requirements: Apparatus (Fig. 2.11), mercury manometer. Pure ethyl bromide, carbon tetrachloride or chloroform.

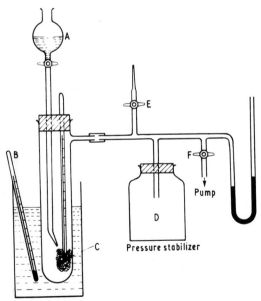

Fig. 2.11   Apparatus of Ramsay and Young for vapour pressure measurements

Procedure: Cover the thermometer bulb (C) with a thin layer of muslin or cotton wool; this acts as a wick, kept moist with the liquid, which is slowly dripped through tap funnel A. Completely evacuate the system and turn off tap F. Make sure that the manometer reading now remains constant; if it is not, look for leaks in the system. Open the tap on the separating funnel and allow the liquid to drop on to the cloth surrounding the thermometer bulb, at a rate of 4—5 drops a minute. Now adjust the temperature of the bath B so that it is always maintained at a temperature of 20 °C higher than that on C. Adjust the flow of liquid so that each drop evaporates before the next one falls.

The liquid on the cloth is heated by condensation of the vapour and by radiation from the outer tube and cooled by evaporation until a steady temperature is registered. This is the boiling point of the liquid under the reduced pressure shown on the manometer.

When the thermometer and manometer are steady, record these values. Open E and admit some air and increase the pressure by about 100 mmHg. Repeat the determination of the steady temperature at the new pressure. Carry out at least four more determinations at pressures up to one atmosphere. Record the barometric pressure.

Treatment of experimental data and discussion

1.   Tabulate the vapour pressure at the various temperatures (*a*) in mmHg, and (*b*) in N m^{-2}. Plot graphs of p against T and log p against T^{-1}. Calculate the mean heat of vaporization over the temperature range studied (equation (2.23)).

2.   Does the graph of log p against T^{-1} show any evidence for the variation of L_e with T? Review the approximations made in the derivation of equation (2.23).

3.   Use your experimental data to test the validity of Trouton's rule for the liquid.

4.   Is this method applicable for the measurement of vapour pressures exceeding 760 mmHg?

3

Determination of relative molecular mass

An accurate value of the relative molecular mass of a compound is of great importance in determining its chemical constitution. Methods are available for the determination either in the gas or vapour phase (density, vapour density) or in solution (elevation of boiling point and depression of freezing point of solvent).

1 Relative molecular masses in the gaseous state

The normal density of a gas is the weight in kg of 1 m^3 of gas measured at 0 °C and at a pressure of 760 mmHg at sea level in latitude 45° (STP). The relative, or vapour density is the ratio of the normal density of the gas to that of hydrogen. Assuming ideal gas behaviour, it can be shown that the relative molecular mass is approximately twice the relative density.

When accurate results are required, the normal density should be determined and corrected for deviations from ideal behaviour; the resulting value is the limiting density (ρ_0). It follows that the relative molecular mass is $2.241\ 4 \times 10^{-2}\rho_0$ (molar volume = $2.241\ 4 \times 10^{-2}\ m^3$).

For approximate work the gas laws may be assumed:

$$pv = nRT = \frac{w}{M_r}RT$$

or

$$M_r = \frac{wRT}{pv} \tag{3.1}$$

where v/m^3 is the volume of w kg of gas of relative molecular mass M_r at pressure $p/N\ m^{-2}$ and temperature T. Thus M_r may be evaluated by determining the volume of a given mass of gas, or the mass of a known volume at known p and T. This method will not be described owing to the difficulties involved in handling large volumes of gas in teaching laboratories.

Experiment 3.1   Determine the relative molecular mass of chloroform by Victor Meyer's method

Theory: In this method the volume of vapour produced from a weighed amount of liquid is determined by displacement of an equal volume of air.

Requirements: Victor Meyer's apparatus (Fig. 3.1) (the outer heating vessel, B, can with advantage be of metal), pure chloroform.

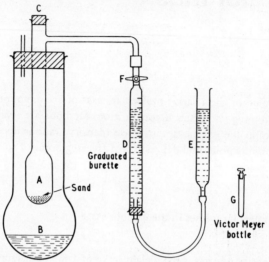

Fig. 3.1   Victor Meyer's apparatus for determination of vapour densities

Procedure: The displaced air is best collected in a Hempel burette as shown, but it may be collected in an inverted graduated tube. Weigh a Victor Meyer bottle, G, with ground-glass stopper, introduce about 0.1 g chloroform and reweigh. Remove stopper C, close tap F and bring the water in B to the boil (this liquid must have a boiling point 50 °C higher than that of the test liquid). When at temperature equilibrium, open F and adjust the water level in the burette, D, to the fifty mark by raising or lowering E, then clamp E. Insert stopper C; the level of water in D is slightly depressed but will not continue to fall if the temperature in B is steady. Equalize the levels in D and E and read the burette — initial reading. Remove C, again adjust the water to the fifty mark and drop in the weighed bottle (loosen stopper before introducing) and immediately replace C. Collect the displaced air in the burette at approximately atmospheric pressure by lowering E. When no further depression of the level occurs, close F and detach the burette at the rubber joint. Remove the burette from the vicinity of the heating jacket, equalize the water levels and measure the volume of air displaced; record the temperature and barometric pressure.

Treatment of experimental data and discussion

1.   From the volume of moist air collected (V/cm^3) measured at t °C and barometric pressure, B mmHg, calculate the volume of dry air (V_0/m^3) at STP.

$$V_0 = \frac{V}{10^6} \times \frac{273}{273 + t} \times \frac{B - p}{760}$$

where p/mmHg is the vp. of water at t °C. From the mass of liquid (m/kg) calculate the normal density ρ_0/kg m^{-3} and hence M_r ($CHCl_3$).

2. Discuss the sources of error involved in this experiment.
3. What factors determine the size of sample to be used?
4. Why is it necessary to equalize the water levels in the two arms of the Hempel burette before reading the volume of air displaced?
5. How would the result be affected if the substance dissociated?

Experiment 3.2   Determine the relative molecular mass of chloroform by Dumas' method

Theory: In this method, a spherical glass flask, of measured volume, is filled with vapour under atmospheric pressure and at a known temperature, and the weight of the vapour determined. The method is very suitable for the study of dissociating vapours, since it gives accurate results and the temperature of determination can be varied.

Requirements: Glass balloon (Fig. 3.2), water bath, pure chloroform.

Fig. 3.2   Dumas glass balloon for vapour density measurements

Procedure: Boil about 500 cm^3 of distilled water in a flask for 10–15 min, to expel the dissolved air, and allow it to cool. Thoroughly clean and dry the bulb, wipe with a clean chamois leather and hang on the balance for 5–10 min until it attains constant weight. To minimize errors due to absorption of moisture, and, changes in the buoyancy of air due to barometric fluctuations, use a counterpoise bulb of approximately the same dimensions on the opposite balance arm. If this is not possible, make a buoyancy correction (p. 303).

Introduce 5–10 cm^3 of chloroform by inverting the projecting arm in a beaker of chloroform and alternately heating and cooling the bulb. Immerse the bulb, mounted in a heavy metal frame-work, as completely as possible in a water-bath at 50 °C. Heat the bath to boiling point and keep at this temperature for at least 10 min until most of the chloroform has vaporized. Read the barometer and from tables determine the temperature of the bath.

While the bath is still boiling vigorously, without moving the bulb, seal off the narrow projection with a gas jet and retain the drawn-off portion. When sealed, remove the bulb from the bath, allow it to cool to room temperature, dry and weigh (including the drawn-off portion).

Insert the arm under the surface of the cooled air-free water and break the seal with crucible tongs. Water immediately enters and fills the bulb; a small drop of chloroform (<1 cm^3) should remain. Determine the weight of the full bulb and all glass pieces to the nearest 0.1 g. Record the temperature of the water.

Treatment of experimental data and discussion

1. From your weighings calculate the weight of chloroform and the volume of the bulb. Hence evaluate the vapour density and relative molecular mass of chloroform.

2. Why is it necessary for some chloroform to remain in the bulb after filling with water?

3. What possible error is avoided by using deaerated water to fill the bulb?

4. List the advantages and disadvantages of this method for the determination of relative molecular masses.

Experiment 3.3   Determine the relative molecular mass, the viscosity and the molecular diameter of hydrogen and nitrogen by effusion

Theory: The passage of a gas through a small orifice (pin-hole) in a thin wall is known as effusion. If the diameter of the hole is small compared to the mean free path (ie. the distance a molecule moves between successive collisions), then when a molecule reaches the opening it will pass through and is unlikely to collide with another molecule. Under these conditions the number of molecules passing through the orifice is almost equal to the number that would normally strike an area of wall equal to that of the hole. The rate of effusion, μ/kg m^{-2} s^{-1} is given by:

$$\mu = \rho(RT/2\pi M_r)^{1/2} = (p\rho/2\pi)^{1/2} \tag{3.2}$$

the density ρ is:

$$\rho = M_r p/RT \tag{3.3}$$

where M_r is the relative molecular mass of the gas and p the pressure. The rate of effusion defined as μ/ρ (units: m s^{-1}) is:

$$\mu/\rho = (p/2\pi\rho)^{1/2} \tag{3.4}$$

Thus at constant pressure and temperature the rate of effusion, like the rate of diffusion, is inversely proportional to the square root of the density and hence (from equation (3.3)), inversely proportional to the square root of M_r. Thus the ratio of the times (t_A and t_B), required for equal volumes of two gases to escape through the same pin-hole under identical conditions of temperature and pressure is:

$$\frac{t_A}{t_B} = \left(\frac{M_r(A)}{M_r(B)}\right)^{1/2} \tag{3.5}$$

or
$$(M_r(A))^{1/2}/t_A = (M_r(B))^{1/2}/t_B = k$$

where the constant, k, depending on the design of the apparatus, the pressure and temperature, is determined with a pure gas of known M_r.

If the gas escapes through a long narrow capillary tube, instead of the pin-hole, then the viscosity and hence the mean free path and molecular diameter can be calculated. The time required t' for a given volume of gas to escape is now proportional to the coefficient of viscosity, η, ie.

$$\eta_A/\eta_B = t'_A/t'_B \tag{3.6}$$

The viscosity coefficient is related to the mean free path, L, by the equation:

$$\eta = \tfrac{1}{3}\rho \bar{c} L \tag{3.7}$$

where the mean velocity of the particles, $\bar{c}$, is

$$\bar{c} = \left(\frac{8RT}{\pi M_r}\right)^{\frac{1}{2}} \tag{3.8}$$

Thus from equations (3.3), (3.7) and (3.8):

$$\frac{L_A}{L_B} = \frac{\eta_A}{\eta_B}\left(\frac{M_r(B)}{M_r(A)}\right)^{\frac{1}{2}} \tag{3.9}$$

The molecular diameter, σ, can be calculated from the mean free path:

$$\sigma^2 = \frac{1}{\sqrt{2}\pi n L} \tag{3.10}$$

where n is the number of molecules per cm^3. Thus for equal volumes of two gases, from equations (3.9) and (3.10):

$$\frac{\sigma_B}{\sigma_A} = \left(\frac{L_A}{L_B}\right)^{\frac{1}{2}} = \left(\frac{\eta_A}{\eta_B}\right)^{\frac{1}{2}}\left(\frac{M_r(B)}{M_r(A)}\right)^{\frac{1}{4}} \tag{3.11}$$

Requirements: Apparatus (Fig. 3.3); this is best mounted rigidly on a board. The thin, foil A, containing the pin-hole is cemented on to one arm of two-way tap, E; the capillary tube, J, fused on the other arm is 50–60 cm long with a bore of about 0.5 mm. The size of the pin-hole and the diameter and length of J should be adjusted to give a time of efflux of hydrogen through A or J of 2–3 min. Bulb B is of about 500 cm^3 capacity and is connected to the twin bulbs, D, by strong plastic tubing. Cylinders of pure oxygen, nitrogen and hydrogen, and a stop-watch are also required.

Procedure: Connect the tube H to the needle valve of an oxygen cylinder. Open tap E (to allow gas to pass through A), and close screwclip I and tap C. Ensure that the needle valve is closed and open the main valve on the cylinder; now cautiously open the needle valve so that gas blows slowly out through the safety valve G. Pass the gas through G for a few minutes to flush out any foreign gas. Open C and carefully close G by placing a finger on the outlet tube. Some of the gas escapes through A (thereby flushing that part out), the remainder displaces mercury in B. Before the mercury is completely displaced, close E and, when the mercury level is 1–2 cm

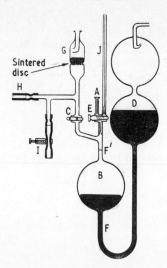

Fig. 3.3   Apparatus for gas effusion and viscosity studies

below the mark F, close C. Close the needle valve and remove the finger from G. Open E, to the tube leading to A and, when the mercury level passes the mark F, start the stop-watch and record the time for the level to reach the mark F′. Repeat at least twice more with oxygen to obtain concordant results. Using the same procedure, determine the time taken for the gas to escape through capillary J.

Repeat the entire process with hydrogen and with nitrogen. Record the temperature.

Treatment of experimental data and discussion

1.   Use the following data for oxygen to calibrate the apparatus:
       $M_r(O_2) = 32.00$; viscosity at 23 °C = 0.2039 kg m^{-1} s^{-1}; mean free path = 10.5 × 10^{-6} m; molecular diameter = 2.98 × 10^{-10} m. From the experimental data for the efflux of each gas through the diaphragm and capillary tube calculate the relative molecular mass (equation (3.5)), the viscosity (equation (3.6)), the mean free path (equation (3.9)) and the molecular diameter (equation (3.11)) of H$_2$ and N$_2$.

2.   Discuss the errors inherent in this method for the determination of η, M_r, L and σ.

3.   Calculate the ideal size of the pin-hole to give the most accurate results.

4.   This is a relative method for the determination of a gas viscosity. Briefly describe an absolute method.

2   Relative molecular mass in solution

The osmotic pressure (π) of a solution is a measure of the difference of the free energy of the solvent in the pure state and in solution. For a dilute solution of a non-volatile solute B in a volatile solvent, A, it can be shown that:

$$\pi = \frac{1000 \, \rho_A R T}{M_r(A)} \ln \frac{p^\ominus}{p} \tag{3.12}$$

where $p^\ominus$ and p are the vapour pressures above the solvent and solution respectively, and ρ_A the density of the solvent. Similar equations relate the elevation of the bp. and the depression of fp. of a solvent with the vapour pressure.

Raoult's law states that 'the relative lowering of the vp. of a solvent is equal to the mole fraction of the solute', thus:

$$\frac{p^\ominus - p}{p^\ominus} = x_B = \frac{n_B}{n_A + n_B} \tag{3.13}$$

where x_B is the mole fraction of solute and n_A and n_B the number of moles of solvent and solute respectively, In very dilute solutions $n_A \gg n_B$ and equation (3.13) reduces to:

$$\frac{p^\ominus - p}{p^\ominus} = \frac{n_B}{n_A} = \frac{w_B}{M_r(B)} \times \frac{M_r(A)}{w_A} \tag{3.14}$$

where w_A and w_B are the weights of solvent and solute respectively.

Theory of cryoscopic and ebullioscopic methods

Dilute solutions approximately obey the van't Hoff osmotic pressure equation:

$$\pi = cRT = \frac{w_B R T}{M_r(B) V} \tag{3.15}$$

where $c/\text{mol m}^{-3}$ is the concentration of solute, w_B/kg is the weight of solute contained in volume $V \, \text{m}^3$ of solvent.

The use of osmotic pressure measurements for molecular weight determinations is now limited to substances of high molecular weight, cg. proteins, starches. It will not therefore be further discussed.

The other colligative properties of solutions, which depend on the number and not the nature of the molecules present, such as the elevation of the bp. and the depression of the fp. find extensive use in the determination of relative molecular masses. The direct consequence of the lowering of the vapour pressure of a volatile solvent by a non-volatile solute is that the bp. (fp.) of a solution must be higher (lower) than that of the pure solvent (Fig. 3.4). The elevation of the bp. (ΔT_e) and the depression of the fp. (ΔT_c) are related to the lowering of the vapour pressure by equations (3.15a) and (3.15b):

$$\Delta T_e = \frac{R T_{0B}^2}{L_e} \ln \frac{p^\ominus}{p} ; \quad \Delta T_c = \frac{R T_{0F}^2}{L_f} \ln \frac{p^\ominus}{p} \tag{3.15}$$

$$(a) \qquad\qquad\qquad (b)$$

where T_{0B} and T_{0F} are the boiling and freezing points and L_e and L_f the molar heats of vaporization and fusion of the solvent respectively.

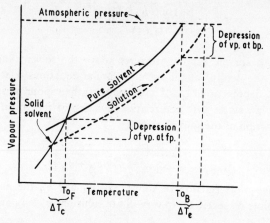

Fig. 3.4   The vapour pressure—temperature curves for solvent and solution

For dilute solutions, assuming the validity of Raoult's law (equation (3.13)); equations (3.15a) and (3.15b) become:

$$\Delta T_e = \frac{cRT_{0B}^2}{l_e}; \qquad \Delta T_c = \frac{cRT_{0F}^2}{l_f} \qquad\qquad (3.16)$$

$$(a) \qquad\qquad\qquad (b)$$

where c/(mol solute per kg of solvent) is the concentration and l_e/J K^{-1} kg^{-1} and l_f/J K^{-1} kg^{-1} the heats of vaporization and fusion of the solvent respectively.

The ebullioscopic constant, k_e, and the cryoscopic constant, k_c, are defined as the elevation of the bp. and depression of fp. respectively when the concentration of solute, c, is 1 mole per kg of solvent. It thus follows from equations (3.16a) and (3.16b) that:

$$k_e = \frac{8.314 T_{0B}^2}{l_e}; \qquad k_c = \frac{8.314 T_{0F}^2}{l_f} \qquad\qquad (3.17)$$

$$(a) \qquad\qquad\qquad (b)$$

Thus where a choice of solvent is possible, the best to use is that with the lowest heat of vaporization or fusion; in this way the largest change in temperature for a given concentration of solute is obtained (Table 3.1).

If w_A and w_B are the weights (in kg) of solvent and solute respectively then equations (3.16) and (3.17) can be combined and written:

$$M_r(B) = \frac{k_e w_B}{\Delta T_e w_A}; \qquad M_r(B) = \frac{k_c w_B}{\Delta T_c w_A} \qquad\qquad (3.18)$$

$$(a) \qquad\qquad\qquad (b)$$

Thus from a knowledge of ΔT_e or ΔT_c for a given solution it is possible to calculate the relative molecular mass of the solute.

When the solute ionizes, each ion acts to increase ΔT_e or ΔT_c and the experimentally observed value of the relative molecular mass will be reduced. Assuming that, for weak electrolytes, the magnitude of the effect is a measure of the degree of dissociation, α, then:

$$\alpha = \frac{M_0 - M_r(B)}{M_r(n - 1)}\qquad(3.19)$$

where M_0 is the theoretical (formula) relative molecular mass, and n is the number of ions into which one molecule dissociates.

Similarly, if the solute associates in the solvent, the values of ΔT_e and ΔT_c will be reduced and the apparent relative molecular mass increased. If n moles of solute associate, the degree of association, α, is given by:

$$\alpha = \frac{M_r(B) - M_0}{M_r(B)[1 - 1/n]}\qquad(3.20)$$

The theory underlying the equations quoted is only valid for dilute solutions; under these conditions, the temperature change, which is too small to be measured with an ordinary thermometer, is measured either with a Beckmann thermometer or with a thermistor. A Beckmann thermometer with a large bulb, and a range of some 6° graduated in 0.01°, registers temperature differences and not absolute values of temperature. The thermometer can be adjusted, by transferring mercury from a reservoir at the top of the capillary tube or vice-versa, to read at any desired temperature. A thermistor is a resistive circuit component with a large negative temperature coefficient of resistance. Thermistors are made of various mixtures of oxides of Mn, Ni, Co, Cu, U, Fe, Zn, Ti and Mg; the temperature coefficient is determined by the relative proportions of the different oxides in the mixture. Over a temperature range of 70–700 K the resistivity may change from 10^5 to $10^{-2}\,\Omega$ m. The large temperature coefficient of resistance is ideal for accurate temperature measurement, a sensitivity of 5×10^{-4} K is readily attained. Since thermistors are small and rapidly attain the equilibrium temperature of the system, they are ideal for measuring the elevation of bp. or depression of fp. of a solvent.

Table 3.1   Ebullioscopic and cryoscopic constants for some common solvents

Solvent	bp./°C	k_e /K kg mol^{-1}	l_e /kJ kg^{-1}	Barometric correction* (per mmHg)	fp./°C	k_c /K kg mol^{-1}	l_f /kJ kg^{-1}
Acetic acid	118.3	3.07	446.9	0.0008	16.6	3.9	187.0
Acetone	56.1	1.71	510.8	0.0004	—	—	—
Benzene	80.2	2.5	394.8	0.0007	5.4	5.12	126.8
Chloroform	61.5	3.86	246.0	0.0009	—	—	—
Ethanol	78.3	1.15	844.7	0.0003	—	—	—
Nitrobenzene	—	—	—	—	5.3	7.2	93.72
Water	100	0.513	2 257.4	0.0001	0	1.858	333.5

* The correction in this column gives the number to be subtracted for each mmHg of difference between the barometric pressure and 760 mmHg.

Cryoscopic methods

Experiment 3.4   Determine the relative molecular mass of camphor in benzene by the cryoscopic method

Requirements: Beckmann thermometer, apparatus (Fig. 3.5), pure benzene, camphor (or naphthalene).

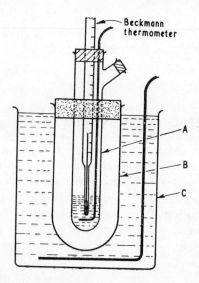

Fig. 3.5   Beckmann's fp. depression apparatus

Procedure: Dry a small quantity of resublimed camphor in a desiccator and roll it into pellets with a spatula or pellet press. Do not compress excessively or it will not readily dissolve. Purify the benzene by fractional crystallization. Adjust the Beckmann thermometer so that when immersed in a bath at 5 °C the mercury is towards the top of the scale.

Fill the cooling vessel C with iced water containing pieces of ice; the temperature of this bath should be about 2 °C. Clean and dry the freezing tube A, fit with two corks, and, by means of a thin wire, suspend it from the arm of a balance. Record the weight to the nearest 0.01 g and introduce sufficient benzene to cover the thermometer bulb, when this is 1 cm from the bottom of A. Stopper the tube and reweigh. Now place the Beckmann thermometer and stirrer in A and insert A in the dry air jacket B. Finally place the assembled apparatus in the iced water in C.

The benzene must separate in fine individual crystals. A transparent coating of solid benzene must not form on the wall of the tube or the bulb of the thermometer.

Stir the benzene gently (up and down movement about once a second) as the temperature falls. To ensure the formation of fine crystals the temperature must not be allowed to fall more than 0.5 °C below the true fp. When solidification occurs, the temperature rises to the true fp. Using a hand lens, make several estimations of the temperature reached, stirring the solvent all the while. Always tap a Beckmann thermometer just before a reading, to overcome sticking of the mercury in the very fine capillary. Remove the tube A and allow the benzene to melt; replace in the air jacket and again determine the freezing point. The readings should agree to within $0.005°$.

If the benzene fails to crystallize, tube A may be removed and immersed directly in C (care must be taken that no solid forms on the walls of the vessel) and then replaced in B after thorough drying.

Now remove A and allow the solid benzene to melt. Through the side arm introduce the weighed amount of camphor to give an approximately 1% solution. Stir well to dissolve the camphor and make several determinations of the freezing point as previously described. Add further camphor to give 2, 3 and 4% solutions and repeat the freezing-point determination.

Treatment of experimental data and discussion

1. Plot a graph of ΔT_c against the weight concentration; this should be linear and pass through the origin. Determine the slope of this line, and, using equation (3.18b) and the appropriate value of k_c (Table 3.1) calculate the mean relative molecular mass of camphor.

2. Discuss the errors involved in the experiment and suggest how they may be reduced. Explain why it is essential that a film of solid benzene must not form on the bulb of the thermometer.

3. Give a thermodynamic derivation of equation (3.18), carefully explaining the thermodynamic reasons for stipulating that the solution must be dilute.

4. For accurate results the value of k_c should be determined for the solvent using a pure compound of known relative molecular mass. Discuss this statement.

Experiment 3.6 Determine the cryoscopic constant for water using substances of known relative molecular mass

Requirements: Beckmann thermometer, apparatus (Fig. 3.5), pure sucrose, mannitol, and urea.

Procedure: Using the method described in Expt. 3.5, determine the freezing point of pure water and water containing small varying amounts of sucrose, mannitol and urea.

Treatment of experimental data and discussion

1. From the slope of the ΔT_c − concentration curve for the various solutes and their known relative molecular masses, calculate k_c for water.

2. What are the factors which determine the value of k_c?

3. Discuss possible sources of error or reasons why the experimental values do not agree with the theoretical value of k_c obtained from equation (3.17b).

4. What value of k_c would have been obtained if sodium chloride had been used as solute?

Experiment 3.7   Determine the degree of association of benzoic acid in benzene

Theory: The relative molecular mass of a large number of acids and hydroxy-compounds dissolved in benzene, determined by the cryoscopic method, is higher than the theoretical value. This is due to association, ie. two or more molecules combining to give a larger molecule.

Requirements: Beckmann thermometer, apparatus (Fig. 3.5), pure benzene and benzoic acid.

Procedure: Using the method of Expt. 3.5, determine the depression of the fp. of benzene with varying amounts of benzoic acid.

Treatment of experimental data and discussion

1.   Plot a graph of ΔT_c against the weight concentration, and from the slope of this line and the appropriate value of k_c (Table 3.1) determine the mean relative molecular mass of benzoic acid (equation (3.17b)).
2.   Assuming that two molecules of benzoic acid associate calculate the degree of association of benzoic acid in benzene (equation (3.20)).
3.   Is there any experimental evidence to suggest that α varies with the concentration?
4.   Discuss the reasons for the dimerization of benzoic acid in benzene.

Experiment 3.8   Determine the apparent relative molecular masses of strong and weak acids by the equilibrium method

Theory: This method depends upon the determination of the temperature at which the equilibrium between a solution and excess solid solvent is established. It is suited to aqueous solutions and is capable of high accuracy.

Requirements: Dewar vacuum flask (250 cm^3) fitted with a rubber bung carrying a Beckmann thermometer and two short tubes. Through one of these tubes passes a ring stirrer; the other must be large enough to admit a 10 cm^3 pipette. Approximately 0.25 mol dm^{-3} solutions of hydrochloric, acetic and monochloracetic acids, 1 dm^3 of standard 0.1 mol dm^{-3} sodium hydroxide solution.

Procedure: Introduce about 60 cm^3 of chilled distilled water into the flask and then practically fill the flask with broken ice prepared from distilled water. Replace the rubber bung and thermometer and plug the pipette tube with cotton wool. Stir vigorously, and record the equilibrium temperature as accurately as possible on the pre-set Beckmann thermometer.

Drain out the water and replace with about 60 cm^3 of a chilled solution of 0.25 mol dm^{-3} hydrochloric acid. Stir the mixture vigorously until a constant temperature is attained. Record this temperature and simultaneously remove a 10 cm^3 sample of the solution with a pipette. Drain the pipette into a weighed bottle and reweigh. Titrate the acid with the standard sodium hydroxide solution. Repeat this titration on a second sample.

Dilute the solution in the flask with chilled distilled water to give approximately half the concentration of acid. Again determine the equilibrium temperature and the concentration by titration.

Wash out the flask and repeat using acetic and monochloracetic acids.

Treatment of experimental data and discussion

1. From the weight of each sample and the titration with standard alkali determine the weight of acid in a given weight of water, and hence the relative molecular mass of each acid at the two different concentrations (equation (3.18b)).

2. Explain the results for the different acids in terms of the degree of dissociation (equation (3.19)).

3. Discuss the advantages of this method over the conventional method (Expt. 3.4).

Experiment 3.9   Determine the relative molecular mass of acetanilide by Rast's method

Theory: In this method, camphor is used as the solvent. The molar freezing point constant is about 40 K; thus a 10% solution in camphor may have a fp. many degrees lower than that of pure camphor. In consequence, an ordinary thermometer is adequate. It is always necessary to determine the fp. and k_c for a given sample of camphor, as commercial samples vary from batch to batch.

The method is only applicable to the determination of relative molecular masses of substances which are soluble in camphor and which do not decompose below 180 °C or react chemically or form eutectic mixtures or solid solutions with camphor. The method can be adapted to a micro-determination using as little as 0.2 mg of solute.

Requirements: Melting-point apparatus, 0–200 °C thermometer in 0.1 °C, resublimed camphor, pure naphthalene and acetanilide.

Procedure: Carefully powder some camphor on a piece of unglazed porcelain and transfer a portion to a melting-point capillary tube. (Camphor may be more easily powdered after moistening with ether.) Attach the tube to a thermometer and set up in the melting-point apparatus. Heat the oil bath with continual stirring until the camphor melts (about 180 °C), allow to cool until the camphor resolidifies and then slowly raise the temperature (about 1 °C per minute) and record the temperature at which the last traces of solid disappear. Repeat at least three times to obtain concordant results.

Accurately weigh a test tube and add about 200 mg of pure naphthalene and reweigh. Add approximately 2 g of powdered camphor and weigh again. Cork the tube and quickly melt the contents by placing in an oil bath at 180 °C swirl the tube to mix the contents and obtain a clear solution. (Avoid heating the solution longer than 1 min to minimize sublimation of camphor.) When cool remove the contents of the tube and grind to give a homogeneous powder. Determine the melting point of this solid as previously described. Repeat with further samples until concordant results are obtained.

Now prepare a solution of about 200 mg of acetanilide in 2 g of camphor, as described for naphthalene, and determine its melting point.

Treatment of experimental data and discussion

1,   From the values of ΔT_c with naphthalene determine the cryoscopic constant for camphor (equation (3.17*b*)), and hence the relative molecular mass of acetanilide.

2.   Explain why camphor has such a high value for k_c.

3.   Discuss the limitations of this method and the errors inherent in the procedure.

4.   Has the method any advantages?

Ebullioscopic methods

The true bp. of a liquid is the temperature at which the liquid and vapour are in equilibrium at a pressure of 760 mmHg. In the determination of the bp. of a solution the thermometer must be immersed in the solution; it is therefore of great importance to eliminate superheating. This has been achieved (compare Beckmann's original method), firstly, by passing solvent vapour into the solution thereby raising the solution to its bp. (Landsberger's method) and, secondly, by boiling the solution and making it pump itself over the bulb of a thermometer or thermistor held in the vapour (Cottrell's method).

Experiment 3.10   Determine the relative molecular mass of benzoic acid in a polar and a non-polar solvent using the Cottrell apparatus

Theory: Even in the absence of superheating the boiling points of most solvents are about 0.1 °C higher at a depth of 4 cm than at the surface, owing to hydrostatic pressure. Thus no matter how thoroughly equilibrium is established between the liquid and bubbles of vapour, the thermometer registers only a mean boiling

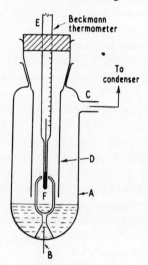

Fig. 3.6   Cottrell apparatus

temperature. The Cottrell apparatus marks an important advance in the determination of such boiling points. In this, the boiling liquid pumps itself over the bulb of the thermometer, so that a thin layer of solution, which readily comes to equilibrium with the vapour at atmospheric pressure, covers the bulb of the thermometer. This then registers the true bp. eliminating errors due to superheating and hydrostatic pressure.

Requirements: Cottrell apparatus (Fig. 3.6), Beckmann thermometer, pure benzene, acetone and benzoic acid.

Procedure: The thick platinum wire B sealed into A facilitates uniform boiling. The side tube C leads to a condenser and the sheath D prevents cold condensate dripping on to thermometer E. When the liquid is boiled, bubbles and liquid trapped by the funnel are pumped through F (2 or 3 arms) on to the thermometer.

Set the Beckmann thermometer to the required temperature range. Thoroughly clean and dry A, stopper the top and side arm and weigh. Introduce some pure acetone and weigh to the nearest 0.1 g. Assemble the apparatus and wrap in asbestos paper; if gas heating is employed, set up further asbestos sheets to shield the apparatus from draughts. With a *small* flame heat the liquid so that it pumps over the thermometer bulb. Make a preliminary investigation of the effect of varying the rate of heating, by recording the rates at which drops fall from the thermometer without altering its reading. (Suppose that the temperature remains steady over the range of 8—16 drops per minute; then select the rate of heating that gives 12 drops per minute.) Once selected, this rate must be adhered to for the rest of the experiment. Tap the thermometer lightly and record the bp. of acetone.

Allow the apparatus to cool and introduce a weighed pellet of benzoic acid (to give a final concentration of about 1%) through the side arm. Determine the bp. of the solution as previously described. Further pellets of benzoic acid may now be added and the new bp. determined.

Repeat the determination using benzene as the solvent.

Record the barometric pressure.

Treatment of experimental data and discussion

1. In determining the exact concentration, some difficulty is encountered in that an unknown quantity of solvent is in the condenser. This can be allowed for by assuming that 0.2 g of solvent remains in the condenser. A better method is to determine an empirical value of k_e for the solvent in the apparatus, using a solute of known M_r. Plot graphs of ΔT_e against the concentration for the two solvents and hence calculate the relative molecular mass of benzoic acid in each of the two solvents (equation (3.18a)).

2. What general conclusions can be drawn from the experimental values of M_r in the two solvents? Assuming that benzoic acid dimerises in benzene, calculate the degree of association (equation (3.20)).

3. What is understood by superheating? Why must it be avoided in this experiment?

4. Is it necessary to record the barometric pressure?

Experiment 3.11   Determine the relative molecular mass of benzoic acid in a polar and a non-polar solvent using an ebulliometer

Theory: The semi-microebulliometer (available commercially) makes use of a Cottrell pump to pump a mixture of liquid and vapour over a thermistor. The thermistor, on account of its size, readily comes to temperature equilibrium with the thin layer of solution and vapour covering it. Instead of measuring ΔT_e the decrease in resistance of the thermistor is measured.

Requirements: Ebulliometer (Fig. 3.7), heating unit, resistance bridge network, galvanometer, pure benzene, acetone, naphthalene and benzoic acid.

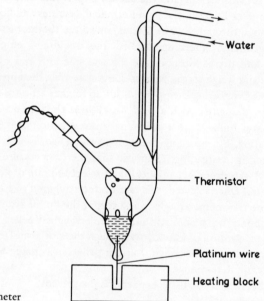

Fig. 3.7   Ebulliometer

Procedure: Clip the ebulliometer body into the frame so that the platinum wire is in the heating block, connect the cold finger to a cold water supply and drain, and the thermistor to the bridge. Follow the detailed instructions in the manual for balancing the bridge. Pipette 10 cm³ of the solvent into the vessel and replace the cold finger; the tip must rest against the side of the ebulliometer. Switch on the heater and adjust the control to give a steady rate of boiling with good pumping. Select the lowest heat input to maintain steady pumping. With all the decades in the bridge at their maximum setting, obtain an approximate balance of the bridge with the coarse control. Increase the galvanometer sensitivity and make the final balance using the decades. Record the initial resistance (R_0) on the decades; the bridge is now calibrated for this particular volume of solvent; the initial balance control must not now be altered.

Switch off the heater, and when boiling has stopped add a weighed amount (*c.* 50 mg) of the solute in pellet form. Adjust the heater to its previous low setting to give steady pumping. As the boiling temperature has been increased there will be a decrease of the resistance of the thermistor, the bridge will be out of balance and in consequence there will be a deflection of the galvanometer. Continually balance the bridge using the decade resistance (galvanometer at maximum sensitivity), until all the solute has dissolved and equilibrium has been re-established, 2–3 minutes after the addition. Read the resistance on the decades (R). Switch off the heater and add a further 50 mg sample of solute and again balance the bridge when the solution is boiling steadily. Continue adding further amounts of solute until 0.2–0.3 g has been added.

It is necessary to calibrate the apparatus, ie. obtain an ebullioscopic constant for each solvent using a solute of known relative molecular mass.

In the experiment use naphthalene as the solute for the determination of the ebullioscopic constant for acetone and benzene. Finally repeat the experiment with each solvent in turn, now using benzoic acid as the solute.

Treatment of experimental data and discussion

1. Tabulate the change in resistance $\Delta R = R_0 - R$ for each addition of solute and plot ΔR against the total weight of solute added, w_C. The constant k/Ω mol^{-1} for the solvent and thermistor is given by:

$$k = \frac{\Delta R M_r(C)}{w_C} \qquad (3.21)$$

where $M_r(C)$ is the relative molecular mass of the calibrating solute C. Thus from the slope $\Delta R/w_C$ the value of k can be determined. Thereafter for the solute B

$$M_r(B) = \frac{k w_B}{\Delta R'} \qquad (3.22)$$

or

$$M_r(B) = \frac{\text{Slope of } \Delta R/w_C}{\text{Slope of } \Delta R'/w_B} \times M_r(C) \qquad (3.23)$$

where $\Delta R'$ is the change in resistance corresponding to the weight w_B. From a knowledge of the slopes of the two curves for each solvent calculate the value of M_r (benzoic acid).

2. Explain why each volume of solvent requires a separate calibration.
3. Why must the tip of the cold finger touch the side of the ebulliometer vessel?
4. What are the main sources of error? Discuss possible ways in which these could be minimized.
5. What general conclusions can be drawn from the experimental values of M_r in the two solvents? Assuming the benzoic acid dimerises in benzene, calculate the degree of association (equation (3.20)).

Experiment 3.12 Determine the relative molecular mass of polystyrene from viscosity measurements

Theory: The viscosity of a polymer solution is related to the relative molecular mass, M_r of the polymer by the equation:

$$\frac{\eta_{sp}}{c} = \frac{(\eta/\eta_0 - 1)}{c} = KM_r^{\alpha} \qquad (3.24)$$

where η_{sp} is the specific viscosity of the solution, η_0 and η the viscosities of solvent and solution respectively, c/g the weight of polymer in 100 cm^3 of solution and K and α constants. This equation is only valid for low concentrations (<0.5 to 1%) of a linear polymer, dissolved in a solvent in which there is no association. It is thus necessary to extrapolate the curve of c against η_{sp}/c to $c = 0$; the intercept is the intrinsic viscosity $|\eta|$. Thus, in the limit, equation (3.24) becomes:

$$|\eta| = KM_r^{\alpha} \qquad (3.25)$$

In general, the value of K depends on the type of polymer, the solvent and the temperature, while α is a function of the geometry of the molecule. The viscometric method of determining relative molecular masses is very convenient and gives accurate results if K and α are known. These are obtained from samples of polymer of known M_r determined by alternative methods (Table 3.2).

Table 3.2   The constants K and α for polystyrene in different solvents at 25 °C

Solvent	Relative molecular mass range	K	α
Benzene	$32 - 1300 \times 10^3$	1.03×10^{-4}	0.74
Methyl ethyl ketone	$2.5 - 1700 \times 10^3$	3.9×10^{-4}	0.58
Toluene	$30 - 1500 \times 10^3$	3.7×10^{-4}	0.62

Requirements: Viscometer (a suspended level viscometer, Fig. 2.3b, can be used with advantage), stop-watch, thermostat at 25.00 °C, polystyrene, benzene, toluene or methyl ethyl ketone. (See Expt. 2.4.)

Procedure: Prepare a solution containing about 0.5 g polystyrene in 25 cm^3 of the solvent.

Clean and dry the suspended level viscometer, transfer 20 cm^3 of solvent into the viscometer, and, when temperature equilibrium has been established, determine the time (p. 24) for the meniscus to move between the two etched marks (C to B). Take at least three timings on the solvent. Empty the viscometer and pipette in 20 cm^3 of the polystyrene solution and, following the standard procedure determine the flow time. Now add 5 cm^3 of the appropriate solvent to A, mix the solution well in A, and, when temperature equilibrium has again been established, determine the flow time (three readings). Repeat the determination after the addition of further 5 cm^3 aliquots of solvent.

If an Ostwald viscometer (p. 25) is used, prepare serial dilutions of the original solution and determine the flow time of each.

Determine the density of the solvent and the most concentrated solution (Expt. 2.1).

Treatment of experimental data and discussion

1.   From the tabulated data of times of flow at the different concentrations

calculate and tabulate the value of η/η_0 for each concentration (equation (2.6)), making the necessary correction for density. Tabulate the values of η_{sp} and η_{sp}/c, and plot η_{sp}/c against c and hence obtain the intrinsic viscosity. Finally calculate the relative molecular mass of the polystyrene (equation (3.25)).

2. Discuss the errors inherent in this method.
3. Is the relative molecular mass of a polymer obtained from viscosity measurements the same as that obtained from other methods, eg. osmotic measurements? Discuss the reasons for any differences.
4. What are the units of intrinsic viscosity?
5. Discuss the advantages of the viscosity method for the determination of relative molecular masses of polymers.

4
Heterogeneous equilibria

The Phase rule $p + f = c + 2$ gives the condition for equilibrium in heterogeneous systems, where c, is the number of components existing in p phases and f is the number of degrees of freedom (independent variables).

1 Solubility

(a) Solubility of gases in liquids

In this equilibrium $c = 2$, $p = 2$ and hence $f = 2$; in this case, it is necessary to define both temperature and pressure for a state of equilibrium. Henry's law, which states that the mass of gas (m), dissolved by a given volume of solvent at constant temperature, is proportional to the pressure of the gas (p) with which it is in equilibrium, is only valid for sparingly soluble gases. Mathematically:

$$m = kp \qquad (4.1)$$

All gases would conform in the range of low pressures, corresponding to dilute solutions, in the absence of chemical reaction. High gas solubilities, eg. hydrogen chloride, ammonia and sulphur dioxide in water, occur as a result of chemical reaction.

Gas solubilities may be expressed in various ways; for gases obeying Henry's law, the most convenient measure is the volume dissolved per unit volume of solvent at the specified temperature.

(b) Solubility of solids in liquids

When a solid is in contact with a liquid in which it is soluble, it passes into solution until saturation is attained. In solubility measurements it is necessary to establish the nature of the solid phase in equilibrium with the solution as well as the amount dissolved.

The solubility depends on the temperature, increasing or decreasing with rise in temperature depending on whether the act of dissolving is endo- or exo-thermic respectively. The van't Hoff isochore written:

$$\left(\frac{\partial \ln S}{\partial T} \right)_p = \frac{\Delta H}{RT^2} \qquad (4.2)$$

or in the integrated form:

$$\log S = \frac{-\Delta H}{2.303 \, RT} + C \qquad (4.3)$$

where S is the solubility (mol kg^{-1} solvent), T/K the temperature, ΔH/J mol^{-1} the heat of solution, can be used to determine ΔH from a knowledge of S at different temperatures. The integration assumes ΔH to be independent of temperature; this is not entirely true.

An abrupt change in the solubility–temperature curve indicates a change in the nature of the solid phase.

The time necessary for a solution to become saturated varies with the nature and state of subdivision of the substance as well as with the efficiency of agitation. It is advisable, where possible, to approach equilibrium from the molecular side (ie. allow a solution saturated at a higher temperature to cool to the required temperature) rather than from the crystal side.

Experiment 4.1   Determine the solubility of benzoic acid over a range of temperatures and hence calculate its heat of solution

Theory: The heat of solution to be determined is the heat absorbed when 1 mole of the solid is dissolved in a practically saturated solution. It differs from the heat of solution at infinite dilution (usually quoted in tables) by an amount equivalent to the heat of dilution.

Requirements: Large beaker as thermostat, boiling tubes, 10 cm^3 pipette connected to short extension tube, packed with glass wool, by rubber tubing, pure benzoic acid, standard 0.05 mol dm^{-3} sodium hydroxide solution.

Procedure: Prepare about 200 cm^3 of a solution of benzoic acid saturated at about 80 °C, and dispense 30–40 cm^3 into each boiling tube. Place a tube in the beaker maintained at the required temperature and stir well with a ring stirrer while maintaining the temperature constant for 15–20 min. Record the temperature to 0.1 °C; immediately withdraw a 10-cm^3 sample, drain into a tared bottle and reweigh. Under no circumstances must any small crystals, which have separated out on cooling, be withdrawn. This can be avoided by fitting to the pipette, with rubber tubing, a piece of drawn-out tube packed with glass wool; remove the endpiece before discharging into the bottle. At the higher temperatures, crystals of benzoic acid separate out in the pipette, this can be avoided by passing the pipette through a Bunsen flame before use. Carefully transfer the weighed contents to a conical flask together with washing water. Titrate with the standard alkali solution using phenolphthalein as indicator.

Making duplicate determinations of the solubility at 0, 20, 25, 30, 35 and 40 °C.

Treatment of experimental data and discussion

1.   From the titration figures determine the solubility S, expressed as mol of benzoic acid per kg water. Plot the graph of S against T and confirm that it is continuous. Plot log S against T^{-1} and draw a smooth curve through the points. Draw tangents to the curve at convenient temperatures and determine the slope and hence ΔH (equation (4.3)).

2.   Explain why the graph of log S against T^{-1} is not linear.

3.   Why is it advisable to approach the equilibrium from the molecular side?

Experiment 4.2   Determine the solubility of sodium sulphate over the temperature range 10–50 °C

Theory: This provides an example of a discontinuous solubility curve due to a change in the solid phase:

$$Na_2SO_4 . 10H_2O \rightleftharpoons Na_2SO_4 + 10H_2O$$

which occurs at the transition point.

Requirements: Large beaker as thermostat, boiling tubes, anion exchange column, (e.g. De-Acidite FF), solid sodium sulphate and standard 0.1 mol dm^{-3} hydrochloric acid.

Procedure: Prepare saturated solutions of sodium sulphate by the general method described in Expt. 4.1 at 10, 20, 25, 28, 30, 32, 35, 40 and 50 °C. The concentration of sodium sulphate in the aliquot removed at each temperature is most conveniently determined using an ion-exchange column. First pour approximately 50 cm^{-3} of 2 mol dm^{-3} sodium hydroxide solution through the column to regenerate it, and wash through with several 50-cm^3 aliquots of distilled water until there is no positive test for alkali in the emergent liquid. Transfer the weighed solution of sodium sulphate to a separating funnel, clamped above the column, well wash the bottle and add the washings to the liquid in the funnel. Allow this liquid to flow through the column and wash through with two 50-cm^3 amounts of water; collect all the liquid in a conical flask. As a result of anion exchange, this is a solution of sodium hydroxide equivalent to the sodium sulphate in the original solution. Titrate with the standard acid solution using methyl red as indicator. Continue washing the column and titrating the washings until there is no further titratable alkali. Regenerate the column after three to four determinations.

Remove the excess solid in contact with the liquid at 30–35 °C by rapid filtration in a sintered-glass crucible. Dry the solids by pressing between filter paper and determine the water of crystallization in each by heating a weighed sample to constant weight.

Treatment of experimental data and discussion

1.   From the total titration of hydrochloric acid calculate the solubility of anhydrous sodium sulphate (g per 100 g water) at each temperature. Plot the temperature–solubility curve.

2. Explain why there is a discontinuity in the solubility curve. What is the transition temperature?
3. Label the areas on the solubility curve and discuss the phase diagram in terms of the phase rule.
4. From the slopes of the curves, predict whether the solution of Na_2SO_4 and of $Na_2SO_4.10H_2O$ is exothermic or endothermic.

Experiment 4.3   **Determine the transition temperature of sodium sulphate by the thermometric method**

Theory: The method depends on the heat change occurring at the transition point of:

$$Na_2SO_4 . 10H_2O \rightleftharpoons Na_2SO_4 + 10H_2O$$

The application of the phase rule shows that there is only one temperature at which the anhydrous salt, hydrated salt, solution and vapour can exist in equilibrium.

Requirements: Apparatus (Fig. 4.1), 0–50 °C thermometer graduated in 0.1 °C, sodium sulphate crystals.

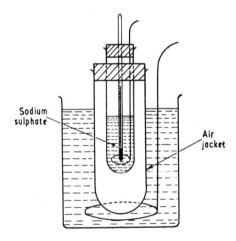

Sodium sulphate

Air jacket

Fig. 4.1   Apparatus for determining transition temperatures

Procedure: Recrystallize the sodium sulphate by preparing a saturated solution of the salt at 30 °C and cooling in an ice-bath. Filter the crystals on a Buchner funnel and press until dry.

Mix some of this hydrated sodium sulphate with a little anhydrous salt (prepared by allowing the hydrated salt to effloresce in the air). Fill the inner tube with sufficient of this mixture to cover the thermometer bulb and heat to 34 or 35 °C replace in the outer tube and set in a beaker of water at room temperature. Stir the resulting solution continuously with the ring stirrer, taking care not to break the thermometer, and record the temperature every minute. The temperature should fall steadily to about 32.4 °C and remain constant there for a considerable time. Finally the temperature falls steadily.

If the temperature falls below 32.4 °C without becoming constant then the solution has become supercooled; and a crystal of hydrated sodium sulphate should be added.

Repeat the experiment using 10 g of sodium sulphate to which is added four successive 0.1 g samples of either (a) urea or (b) ammonium sulphate. Determine the transition temperature after each addition.

Treatment of experimental data and discussion
1.   Plot the cooling curve (ie. *T* against time) and deduce the transition temperature. If there is a large part of the mercury column of the thermometer which is not at the temperature of the mixture, then an emergent stem correction must be made.
2.   Explain why the temperature remains constant at the transition point. Discuss the equilibrium at the transition point in terms of the phase rule. By reference to standard books, draw the complete phase diagram for the system $Na_2SO_4–H_2O$.
3.   Explain the changes in the transition temperature as a result of the addition of urea or ammonium sulphate. Assuming that a solution of urea in sodium sulphate is ideal, calculate the molar change of transition temperature. Hence determine the relative molecular mass of ammonium sulphate and comment on the state of ammonium sulphate dissolved in sodium sulphate. (An equation of the form of equation (3.18) will be of help.)

Experiment 4.4   Determine the transition temperature of sodium sulphate by the dilatometric method

Theory: This method depends on the change of volume which occurs at the transition temperature during the transformation.

Requirements: Dilatometer (Fig. 4.2), 0–50 °C thermometer graduated in 0.1°, hydrated sodium sulphate, xylene.

Procedure: Fix the thermometer in the ground-glass joint with sealing wax and ensure a tight fit. Alternatively, use a rubber bung and seal this to the tube with sealing wax. If not graduated, attach a millimetre scale to the capillary arm (20 cm long, 0.5 mm, bore).

Fill the lower half of the tube with hydrated sodium sulphate and completely fill the tube with an inert solvent, such as xylene, making sure that all the air bubbles are displaced from the crystals. Insert the stopper, carrying the thermometer, again ensuring that no air bubbles remain. The meniscus should now be in the capillary tube.

Place the dilatometer in a large beaker of water at about 31 °C with an efficient stirrer. Heat the water very slowly (1 °C in 10 min) to 34 °C and at minute intervals record both the meniscus level and the temperature. Keep the temperature constant at 34 °C until there is no further increase in level in the capillary tube and

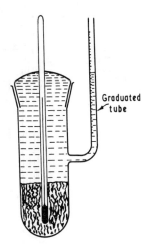

Graduated
tube

Fig. 4.2   Dilatometer

then remove the heat. During the cooling period, again take readings of the meniscus and temperature at minute intervals.

Treatment of experimental data and discussion

1.   Plot the graph of temperature against the meniscus level and hence determine the transition temperature. In addition to the change of volume of the solid at the transition point, there is a change of volume of the liquid with change of temperature; how are these two changes separated?
2.   Why is it necessary to displace all the air from the crystals?
3.   By reference to standard books, draw the complete phase diagram for the $Na_2SO_4-H_2O$ system and discuss it in terms of the phase rule.

Experiment 4.5   Determine the vapour pressure of sodium sulphate solutions over a range of temperatures

Theory: The most convenient method for the measurement of the vapour pressure is the dew point method. The dew point is the temperature at which condensation of water vapour in a system occurs, in this instance the water vapour is that in the equilibrium:

$$Na_2SO_4\ 10H_2O \rightleftharpoons Na_2SO_4 + 10H_2O$$

Requirements: Dew point apparatus with silvered boiling tube (Fig. 4.3), glass fronted thermostat adjustable from 25–45 °C. Hydrated sodium sulphate, chloroform.

Procedure: Grind about 10 g of $Na_2SO_4 . 10H_2O$ together with 0.1 g Na_2SO_4 to a fine powder in a mortar. Spread the powder over the bottom of the dry bottle of the dew point apparatus and assemble the apparatus as in Fig. 4.3. Evacuate the bottle on a water pump and close tap A. Clamp the apparatus in a thermostat at 25 °C and after about 30 minutes pour a small amount of chloroform into the silvered boiling tube. Slowly suck air through the chloroform (using a water pump)

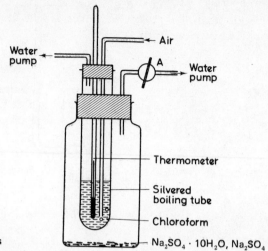

Fig. 4.3   Dew point apparatus

and note the first appearance of misting of the silvered surface, this is the dew point — record the temperature. Stop evaporating the chloroform, allow the temperature of the apparatus to rise and note the temperature at which the film of dew disappears. Repeat the experiment several times at 25 °C and also at 30, 35, 40 and 45 °C.

Treatment of experimental data and discussion

1.   From the measured dew point temperatures and tabulated values of the vapour pressure of water at different temperatures, obtain the vapour pressure of the $Na_2SO_4 . 10H_2O–Na_2SO_4–H_2O$ (vapour) system at the different temperatures. Plot a graph of p against T and determine the transition temperature.

2.   Plot the graph of log p against T^{-1} and calculate the heat of solution and the heat of transition. Explain why this graph consists of two linear portions. What equation is used in the calculation?

3.   From the vapour pressure of the solution at 25 °C and the known vapour pressure of water at 25 °C, calculate $\Delta G^{\ominus}$ for the reaction:

$$Na_2SO_4 + 10H_2O \rightleftharpoons Na_2SO_4 . 10H_2O$$

and hence ΔS for the transition. Compare your values with those obtained from tabulated values of $\Delta H_f^{\ominus}$, $\Delta G_f^{\ominus}$ and $S^{\ominus}$. Explain the significance of the values obtained.

(c) Solubility of sparingly soluble salts

The methods described previously are generally not sufficiently accurate or sensitive for the measurement of the solubility of such salts as silver chloride (1.95×10^{-3}g dm^{-3}) or lead sulphate (4.5×10^{-2}g dm^{-3}) in water. Electrical methods involving electrometric or conductometric measurements are used. The solubility of a sparingly soluble salt in water is a function of the ionic strength of the solution in which it is dissolved. This provides a method for the determination of mean ionic activities and activity coefficients.

Experiment 4.6   Determine the solubility of lead sulphate in water at 25 °C conductometrically

Theory: The conductivity method can be used to measure the concentration of a saturated solution provided that the solubility is not too high and that the salt is not hydrolysed. In a saturated solution of such a salt, the solution is so dilute that its molar conductivity can be assumed to be equal to the limiting value (Λ_0) obtained by extrapolation to infinite dilution. Thus Λ_0 may be taken as the sum of the molar conductivities of the ions; these must be known at the correct temperature.

The molar conductivity

$$\Lambda/\Omega^{-1}\ m^2\ mol^{-1} = \frac{\kappa/\Omega^{-1}\ m^{-1}}{c/mol\ m^{-3}} \tag{4.4}$$

but

$$\Lambda = \Lambda_0$$

Hence

$$c/mol\ m^{-3} = \frac{\kappa}{\Lambda_0} \tag{4.5}$$

Requirements: Conductivity bridge, conductivity cell (bottle type, Fig. 9.2), thermostat at 25 °C, conductivity water, solid lead sulphate.

Procedure: Following the instructions for the use of the bridge, determine the cell constant at 25 °C. Determine the conductivity of the water at 25 °C (p. 243).

Shake the lead sulphate repeatedly with conductivity water to remove any soluble impurities, which would otherwise affect the conductance measurement. Suspend this washed lead sulphate in conductivity water in a steamed-out Pyrex flask and bring to the boil. Filter the hot solution through a filter paper (well washed with conductivity water) into a clean Pyrex flask. Stopper and place in the thermostat at 25 °C, shake vigorously until temperature equilibrium is attained. Determine the conductance and hence the conductivity of this solution by repeated measurements on fresh samples until concordant results are obtained. (*NB.* For more sparingly soluble salts such as silver chloride, precautions must be taken to exclude carbon dioxide during the determination.)

Treatment of experimental data and discussion

1.   From the measured conductance, G/Ω^{-1}, and the cell constant, J/m^{-1}, determine the conductivity, κ ($= J\ G$), of the solution and of the water. Subtract the conductivity of water from that of the solution and calculate the molar conductivity for lead sulphate (equation (4.4)). Using the tabulated molar conductivities of the ions (p. 305), calculate the solubility of lead sulphate (equation (4.5)) and hence the solubility product.
2.   Estimate the magnitude of the error involved in assuming that $\Lambda = \Lambda_0$. Discuss an alternative method by which the correct values of Λ could be obtained.

3.   How does the solubility of lead sulphate depend on particle size? Comment on the difficulty of ensuring that the solution is saturated with lead sulphate. Will the presence of fine particles of lead sulphate in the final solution affect the measurement?

4.   Calculate the solubility product of $PbSO_4$ given the following data at 298 K:

Substance	$\Delta G_f^{\ominus}/kJ\ mol^{-1}$
Pb^{++} aq	-24.3
SO_4^{--} aq	-742.0
$PbSO_4$ s	-811.2

Clearly explain the theory underlying your calculation.

Experiment 4.7   Determine the solubility of silver chloride in water electrometrically

Theory: The cell:

$$Ag|AgCl(s), 0.01\ mol\ dm^{-3}\ KCl \vdots Satd.\ KNO_3 \vdots 0.01\ mol\ dm^{-3}\ AgNO_3|Ag$$
$$(a_{Ag^+})_1 \qquad\qquad\qquad\qquad (a_{Ag^+})_2$$

in which the solution in the left-hand compartment is obtained by adding one drop of silver nitrate solution to $0.01\ mol\ dm^{-3}$ potassium chloride solution, can be treated as a concentration cell without liquid junction. The emf. (E) is given by:

$$E = \frac{RT}{\mathscr{F}} \ln \frac{(a_{Ag^+})_2}{(a_{Ag^+})_1} \qquad (4.6)$$

Knowing the activity coefficient of Ag^+ in $0.01\ mol\ dm^{-3}$ silver nitrate solution, the activity of silver ions in the potassium chloride solution $(a_{Ag^+})_1$ can be determined, and hence knowing γ_{Cl^-} in $0.01\ mol\ dm^{-3}$ potassium chloride, the solubility and solubility product (K_s) of silver chloride can be calculated.

$$K_s = a_{Ag^+}\ a_{Cl^-} \qquad (4.7)$$

Requirements: Potentiometer and galvanometer, silver electrodes, saturated potassium nitrate bridge (p. 307), $0.01\ mol\ dm^{-3}$ potassium chloride and $0.01\ mol\ dm^{-3}$ silver nitrate solutions.

Procedure: Set up the cell either in two small beakers with a saturated potassium nitrate—agar bridge or as shown in Fig. 7.11 (make sure that the screw-clips are tightly closed to prevent syphoning of the liquid). To the $0.01\ mol\ dm^{-3}$ potassium chloride solution, add one drop of silver nitrate solution. Standardize the potentiometer and determine the emf. of the cell; record the polarity of the electrodes. Record the temperature and repeat the experiment using fresh electrolytes.

Repeat the experiment using solutions of potassium bromide or potassium iodide instead of the potassium chloride.

Treatment of experimental data and discussion

1. From the experimental value of E (with the correct sign) calculate $(a_{Ag^+})_1$ assuming the value of γ_{Ag^+} in 0.01 mol dm^{-3} silver nitrate solution (p. 316). Calculate the thermodynamic solubility product using the appropriate value of γ_{Cl^-}, and hence the solubility of silver chloride at the experimental temperature.
2. Briefly indicate the derivation of equation (4.6) from basic thermodynamic principles.
3. Explain the role of the salt bridge in this cell.
4. From tabulated values of $\Delta G_f^{\ominus}$ calculate the solubility product of silver chloride.

Experiment 4.8 **Study the variation of the solubility of silver bromate in potassium bromate solutions and hence determine the solubility product of silver bromate**

Requirements: Glass-stoppered bottles, thermostat at 25 °C, potassium bromate, potassium nitrate, silver bromate, standard 0.002 mol dm^{-3} potassium thiocyanate solution and ammonium iron (III) sulphate indicator.

Procedure: Prepare a series of solutions (total volume 100 cm^3) of potassium bromate and potassium nitrate in varying proportion but with constant ionic strength, I, of 0.05 mol dm^{-3}. ($I = \frac{1}{2}\Sigma\, c_i z_i^2$, where c_i and z_i are the ionic concentration and valency respectively.)

    Add solid silver bromate in excess to each solution in the bottles, stopper tightly and shake vigorously in a thermostat for several days. Filter each solution through a dry filter paper, rejecting the first 10 cm^3. Determine (in duplicate) the silver concentration in each solution by titrating a 25-cm^3 aliquot with the standard potassium thiocyanate solution, using iron (III) solution as indicator.

Treatment of experimental data and discussion

1. From the titration figures, calculate the solubility of $AgBrO_3$ in the various solutions and hence the solubility product. Plot a graph of c_{Ag^+} against $1/c_{BrO_3^-}$, determine the slope of this line and hence obtain a value for the solubility product.
2. The solubility product so calculated is not the true thermodynamic solubility product; suggest the method whereby this quantity could be calculated.

Experiment 4.9 **Study the variation of the solubility of silver acetate with ionic strength and hence determine K_s**

Theory: The thermodynamic solubility product, K_s, of silver acetate is given by:

$$K_s = a_{Ag^+}\, a_{Ac^-} = c_{Ag^+}\, c_{Ac^-}\, \gamma_{Ag^+}\, \gamma_{Ac^-} \tag{4.8}$$
$$= K_c \gamma_{\pm}^2$$

The classical solubility product, K_c, at a particular ionic strength, I/mol dm^{-3}, can be determined from solubility measurements, and K_s can be obtained using the Debye–Hückel equation (see also p. 316):

$$\log \gamma_\pm = \frac{-Az_+z_-\sqrt{I}}{1 + B\mathring{a}\sqrt{I}} \tag{4.9}$$

For the system under study equation (4.9) becomes:

$$\log \gamma_\pm = \frac{-0.5\sqrt{I}}{1 + 1.4\sqrt{I}} \tag{4.10}$$

Taking logs of equation (4.8) and combining with equation (4.10) gives on re-arrangement:

$$pK_c = pK_s - \sqrt{I}/(1 + 1.4\sqrt{I}) \tag{4.11}$$

Requirements: Stoppered boiling tubes, mechanical shaker, 1.0 mol dm^{-3} potassium nitrate solution, standard 0.1 mol dm^{-3} ammonium thiocyanate solution, iron (III) ammonium sulphate indicator, solid silver acetate.

Procedure: Prepare a series of solutions (total volume 40 cm^3) of potassium nitrate of concentrations 0, 0.05, 0.1, 0.2, 0.3, 0.5, 0.75 and 1.0 mol dm^{-3} in the boiling tubes. To each add about 1 g of silver acetate, stopper the tubes and shake mechanically for 1–2 hours. As the tubes are not thermostatted, record the room temperature. Filter each solution through a dry filter paper, rejecting the first 10 cm^3. Determine (in duplicate) the silver concentration in each solution by titrating a 10-cm^3 portion with the standard ammonium thiocyanate solution using iron (III) ammonium sulphate as indicator.

Treatment of experimental data and discussion
1.   From the titration figures, calculate the solubility of silver acetate and hence K_c in each solution. From the known ionic concentrations of all species calculate the ionic strength of each solution. Plot a graph of pK_c against $\sqrt{I}/(1 + 1.4\sqrt{I})$ and determine pK_s from the intercept, equation (4.11). Finally calculate $\gamma_\pm$ at each concentration, equation (4.8).
2.   Does the slope of the graph agree with the theoretical value?
3.   List the errors involved in the determination and comment on the likely error involved in not keeping the solutions in a thermostat.
4.   Explain, in terms of the ionic atmosphere, why a sparingly soluble salt is more soluble in the presence of an electrolyte that it is in water.

Experiment 4.10   Study the variation of the solubility of potassium hydrogen tartrate with ionic strength and hence determine the mean ionic activity coefficients

Theory: This experiment differs from 4.9 in that the electrolyte contains an ion in common with the saturating salt. For the uni-univalent potassium hydrogen tartrate

$$K_s = a_+a_- = c_+c_-\gamma_+\gamma_- = (c_\pm\gamma_\pm)^2 \tag{4.12}$$

Thus
$$\gamma_\pm = \frac{K_s^{1/2}}{c_\pm} \qquad (4.13)$$

The graph of log $c_\pm$ against $\sqrt{I}$ is linear (why?), the intercept at $I = 0$, $\gamma_\pm = 1$, is log $\sqrt{K_s}$, hence $\gamma_\pm$ can be calculated at any ionic strength (equation (4.13)).

Requirements: Stoppered bottles, thermostat at 25 °C, potassium chloride, sodium chloride, potassium hydrogen tartrate, standard 0.01 mol dm^{-3} sodium hydroxide solution.

Procedure: Prepare the following series of solutions (100 cm^3 of each):

(a) Potassium chloride solutions: 0, 0.025, 0.05, 0.1, 0.2 and 0.3 mol dm^{-3}.

(b) Sodium chloride, potassium chloride mixtures containing the same amount of potassium chloride as in (a) (1−5) but made up with sodium chloride so that the total ionic strength is 0.2 mol dm^{-3}.

Add excess solid potassium hydrogen tartrate to each solution in a glass-stoppered bottle and shake in a thermostat at 25 °C for one to two days. Filter each solution through a dry filter paper, rejecting the first 10 cm^3 of filtrate. Titrate (in duplicate) 10-cm^3 samples of each with the standard sodium hydroxide solution using phenolphthalein as indicator.

Treatment of experimental data and discussion

1. From the titrations, determine the concentration of tartrate, c_2/mol dm^{-3}, in the solution and hence $c_\pm = \sqrt{c_2 (c_1 + c_2)}$, where c_1/mol dm^{-3} is the concentration of KCl in solution. Plot log $c_\pm$ against $\sqrt{I}$ (taking into account also the potassium hydrogen tartrate in solution). Extrapolate the line, determine K_s and hence $\gamma_\pm$ at each ionic strength (equation (4.13)). For both series (a) and (b) plot log $\gamma_\pm$ against $\sqrt{I}$.

2. Consider briefly the error involved in plotting log $c_\pm$ against $\sqrt{I}$ instead of the more complex function of equation (4.11).

3. Explain why the solubility of a sparingly soluble salt is lower in an electrolyte containing a common ion than it is in an electrolyte of the same concentration but not containing a common ion.

4. Explain why a sparingly soluble salt is more soluble in the presence of a neutral electrolyte than it is in water.

5. Explain the differences in the results obtained for the two series (a) and (b).

2 Mutual solubility

Some liquids are miscible in all proportions (ethanol−water), but in some cases the solubility of each in the other is limited (ether−water, phenol−water). Generally both liquids become more soluble with increasing temperature and eventually a critical solution temperature, or consolute point, is reached above which the liquids are miscible in all proportions. Triethylamine and water exhibit a lower critical solution temperature, while nicotine and water give a closed solubility curve with upper and lower critical solution temperatures.

At any temperature below an upper critical solution temperature, the compositions of the two liquid layers in equilibrium are fixed and independent of the relative amounts of the two phases.

Two methods are available for the study of such systems.

(a) Analytical method. The two liquids are shaken together in a thermostat until equilibrium is established. The resulting two layers are analysed by an appropriate physical or chemical method.

(b) Synthetic method. Weighed amounts of the two substances are placed in small glass tubes (sealed if necessary) and the temperature determined at which a single homogeneous solution is obtained.

The mutual solubility of a pair of partially miscible liquids is generally markedly affected by the presence of a third component (p. 84).

Experiment 4.11   Determine the mutual solubility curve of phenol and water

Requirements: Large beaker, large tube as air jacket, 0–100 °C thermometer reading to 0.1°, 80% solution of phenol (add 125 cm³ of water to a 500-g bottle of phenol, shake until a homogeneous solution is obtained).

Procedure: Dispense into separate test tubes by means of a burette, 10, 9, 8 . . . 1 cm³ of phenol solution and 0, 1, 2 . . . 9 cm³ of water, to give a range of phenol concentrations of 8–80%.

Insert the thermometer and small glass ring stirrer into one of these tubes, fitted into the outer tube, which acts as an air jacket and reduces the rate of heating or cooling. Clamp this in the beaker of water and heat gently. Stir the phenol solu-

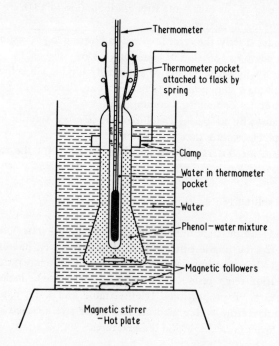

Fig. 4.4   Phenol tube

tion well and record the temperature at which the turbid liquid just clears. Remove the source of heat and allow the temperature to fall slowly, record the temperature at which the liquid becomes turbid and two layers separate. Determine the mean temperature for each tube at which the two phases disappear or appear.

Repeat the above experiment with the addition of a constant amount of (*a*) naphthalene (final concentration 0.2%), and (*b*) sodium chloride (1.0%).

It has been found that, in the interests of safety, it is advisable to prepare the diluted solutions in advance. These are most conveniently prepared and stored in long-necked flasks (Fig. 4.4) fitted with a joint and thermometer pocket; each flask contains a small magnetic follower. Each flask is clipped, in turn, on to a bar and immersed in a beaker of water (containing a large magnetic follower) and heated on a magnetic stirrer/hot plate. In this way both the water and the phenol solution are adequately stirred. The temperatures at which the solution becomes homogeneous on heating and separates into two layers on cooling are recorded. Solutions at concentrations suggested above, stored in the flasks, can be used repeatedly.

Treatment of experimental data and discussion
1. Plot graphs of temperature of complete miscibility against the composition of phenol in the different mixtures. Determine the critical solution temperature.
2. Account for the effect of the additives on the critical solution temperature.
3. State the phase rule and determine the number of degrees of freedom for the system which exists in two phases.
4. Explain what is meant by a 'tie line', draw a typical tie line on your phase diagram and, for a mixture defined by a point on it, determine the ratio of the amounts of the two phases which separate out.
5. Explain the physical cause of partial miscibility.
6. Suggest reasons why the analytical method of studying such systems is not always convenient.

3   Freezing point diagrams of binary mixtures

When a homogeneous liquid binary mixture (the components of which are completely miscible in the liquid phase) is cooled, a solid phase separates out; the composition of the solid depends on the nature of the two components. The equilibrium temperature, ie. the freezing point, depends only on the composition of the liquid and not on the relative amounts of liquid and solid. When the solid separating out has the same composition as the liquid phase then no change in the composition of the liquid occurs during freezing, and, in consequence, the freezing point remains constant during the change. This condition applies to the separation of solid from a pure liquid and the separation of a eutectic mixture or compound with a congruent melting point from a binary liquid mixture. If, on the other hand, the composition of the solid phase differs from that of the liquid, then the freezing temperature will change as more of the solid separates out. The method of thermal

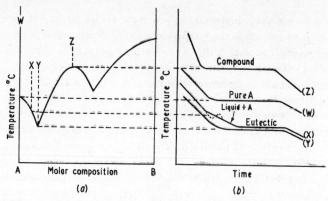

Fig. 4.5   (a) Phase diagram for binary mixture which forms a compound. (b) Corresponding cooling curves

analysis, ie. the study of the cooling curves obtained when a liquid mixture is cooled, depends on these principles (Fig. 4.5).

The cooling curve for mixture (X) shows a gradual fall until solid A starts to separate and then there is a reduced rate of cooling. The transition is, however, rarely sharp and often supercooling occurs (dotted line on X). Provided this supercooling is not excessive, the two parts of the curve can be extrapolated giving the temperature of the break. Further cooling results in the solution being saturated with B as well as A, these then crystallize out in the same proportions as present in the liquid; this is the eutectic.

Experiment 4.12   Determine the freezing point curve for mixtures of naphthalene and biphenyl

Requirements: Cooling apparatus (Fig. 4.1), 6 × 1 inch boiling tubes, 0—100 °C thermometers, naphthalene and biphenyl.

Procedure: Prepare boiling tubes containing the following weighed mixtures:

Wt. of naphthalene/g	15	15	15	15	10	7.5	5	2	0
Wt. of biphenyl/g	0	5	10	15	15	15	15	15	15

Fit a cork, with a thermometer and ring stirrer, to each tube. (*NB*. These tubes may be stored and used by other students.) Melt the sample by immersing the tube in a beaker of hot water, stir well to ensure complete mixing. Replace the tube in the air jacket and clamp this in a beaker packed with crushed ice to ensure uniform cooling. Stir thoroughly and record the temperature every half-minute until the temperature remains constant and the whole mass solidifies (a temperature range of 80—30 °C is satisfactory). Repeat with each sample.

Treatment of experimental data and discussion

1:    Plot the cooling curve for each sample, and, from each, determine the maximum temperature of the first arrest (corrected for supercooling) and also the eutectic temperature. Plot the complete phase diagram of these values of the freezing point and eutectic temperatures against x_A (the mole fraction of component A), and label all the areas on the diagram.

2. If such a binary system behaves ideally, then at temperatures above the eutectic temperature:

$$\log x_A = -\frac{\Delta H(A)}{2.303R}\left\{\frac{1}{T} - \frac{1}{T_0(A)}\right\} \tag{4.14}$$

where $T_0(A)/K$ is the freezing point of pure A and T/K is the temperature at which a solution containing mole fraction, x_A, deposits pure A, $\Delta H(A)$ is the enthalpy of fusion of A. Test the validity of this equation for the two components by plotting $\log x_A$ (or $\log x_B$) against T^{-1}; from the slopes determine $\Delta H(A)$ or $\Delta H(B)$.

3. Using the phase rule determine the number of degrees of freedom at the eutectic point. Explain what is meant by the terms 'eutectic point' and 'eutectic arrest'.

4. Explain why there is a break in a cooling curve when a solid separates out from a mixture of different composition, or when a pure component crystallizes.

5. Show how equation (4.14) may be rearranged to give equation (3.16b), a form which may be used to determine the relative molecular mass of the solute.

6. What factors decide whether a solution will behave ideally or not? What conclusions can be drawn about the naphthalene–biphenyl system?

Experiment 4.13 Determine the freezing point curve for mixtures of *o*-nitrophenol and *p*-toluidine

Requirements: Cooling apparatus (Fig. 4.1), 0–50 °C thermometer, *o*-nitrophenol and *p*-toluidine. (Once prepared the mixtures of different concentrations can be used repeatedly.)

Procedure: Fit the boiling tube with a cork carrying a thermometer and ring stirrer. (*a*) Nitrophenol half: Introduce 15 g of pure *o*-nitrophenol into the boiling tube and melt by immersing it, without the air jacket, in hot water. Replace the tube in the air jacket and clamp this in a beaker packed with crushed ice. Stir thoroughly and take temperature readings every half-minute, until the arrest in cooling shows that the freezing point of the pure component has been reached. The temperature should now remain constant until the whole mass has solidified. (Temperature range 50–12 °C.)

To the *o*-nitrophenol add 1 g *p*-toluidine, melt the contents of the tube and again determine the temperature at half-minute intervals, as described above, continuing until the final eutectic arrest is obtained, at about 15 °C. The time for which the eutectic temperature is maintained is proportional to the amount of minor constituent present. Add further 1, 2, 4 and 6 g of *p*-toluidine and determine the cooling curve for each mixture.

(*b*) Toluidine half: Determine the freezing point of *p*-toluidine alone and in the presence of increasing amounts of *o*-nitrophenol. The weights given previously are suitable.

Treatment of experimental data and discussion (see Expt. 4.12)

Experiment 4.14   Determine the freezing-point curve for mixtures of phenol and either α-naphthylamine or *p*-toluidine

Theory: These two components form a compound with a congruent mp. and two eutectic mixtures (Fig. 4.5).

Procedure: Determine cooling curves for the pure components and mixtures as described in Expt. 4.13 (temperature range 55–15 °C).

Treatment of experimental data and discussion
1.   From the cooling curves for each sample, determine the temperature of the first arrest (corrected for supercooling) and of the eutectic. Plot the complete phase diagram of these values of the fp. and eutectic temperature against the mole fraction of one component.
2.   Determine the formula of the molecular compound formed.
3.   Label all the areas in the phase diagram and discuss it in terms of the phase rule.

Experiment 4.15   Determine the phase diagram for mixtures of metallic tin and cadmium

Theory: The type of curve obtained indicates the formation of two limited regions of solid solutions, so that between definite concentration limits, depending on temperature, two conjugate solid solutions can exist. The cooling curves will be similar to those previously described (Fig. 4.5), with the exception of liquids approaching 0 and 100% cadmium. There will be two points of inflexion corresponding to the beginning and end of separation of solid solution, but no horizontal portion showing that there is no temperature at which a solid phase of invarying composition is deposited.

Requirements: Potentiometer and galvanometer or thermocouple millivoltmeter, copper–constantan thermocouple enclosed in silica sheath, furnace, cadmium–tin alloys containing 5, 10, 25, 30, 50, 70, 75 and 90% cadmium.

Procedure: Set up the thermocouple (Fig. 4.6) with one junction immersed in melting ice in a Dewar vessel; and calibrate it using appropriate secondary standards of temperature. Prepare samples of each of the alloys and also the pure metals in Pyrex boiling tubes, fitted with a cork carrying the thermocouple sheath and two tubes for gas inlet and outlet. Pass a slow stream of nitrogen over the metal suface (to prevent oxidation) and suspend each tube in turn in the furnace until the metal is molten. Transfer the tube to the air jacket, replace the thermocouple junction and allow the melt to cool. Record the potentiometer reading every half-minute;

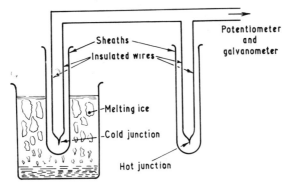

Fig. 4.6   Typical arrangement for thermocouples (junctions formed either by soldering or melting appropriate wires in an oxy-gas flame)

except for the pure metals, the temperature should be allowed to fall to at least 175 °C. After the mass has solidified, warm the tube, extract the thermocouple sheath and store the alloy for future use.

Treatment of experimental data and discussion

1.   Plot graphs showing the decrease of thermocouple emf. with time; read off the emf. values at the points of inflexion or horizontal portions. From the calibration curve for the thermocouple, convert these emf. readings to temperature/°C, and plot the phase diagram.

2.   Label all the areas of the diagram and discuss it in terms of the phase rule.

3.   Explain what is meant by the term alloy and distinguish clearly between an alloy, an eutectic mixture, a solid solution and a compound.

4   Liquid—vapour equilibria

(a) Completely miscible liquids

When two completely miscible liquids are mixed, the composition of the vapour differs from that of the liquid. In an ideal solution, each component exerts a partial pressure (p), given by Raoult's law:

$$p(A) = p^{\ominus}(A)x_A \qquad (4.15)$$

where $p^{\ominus}(A)$ is the vapour pressure of the pure component and x_A its mole fraction in solution. The graph of composition against the total vapour pressure (ie. the sum of the individual partial pressures, above liquid mixtures) should therefore be linear and join the vapour pressures of the pure components. The vapour pressure usually lies above (position deviation) or below (negative deviation) this straight line required by equation (4.15). If the deviation is small, the vapour pressure increases continuously as the composition changes from the pure component of low vapour pressure to that of high vapour pressure and shows no maximum or minimum.

Larger deviations from Raoult's law give rise to a maximum or minimum vapour pressure at a certain composition.

When the vapour pressure—composition curve is continuous, the bp.—c curve (now at constant pressure) is also continuous, the bp. of any mixture lying between the bp. of the two pure components. When such a mixture is distilled, the vapour (distillate) will be richer in the component of higher vapour pressure and the bp. will rise continuously as distillation proceeds. A complete separation of the two components in such a zeotropic mixture can be achieved by repeated distillation of such a mixture or by an efficient fractionating column.

If the vapour pressure—c curve passes through a maximum or minimum, then the bp.—c curve passes through a minimum or maximum respectively. Such mixtures cannot be separated into two pure components, distillation gives the pure component present in excess and the constant bp. mixture, the composition of which corresponds to the maximum or minimum on the bp.—c curve.

Experiment 4.16   Determine the boiling point—composition curves for the completely miscible binary system carbon tetrachloride—ethanol

Requirements: Distillation apparatus (Fig. 4.7) with microburner, refractometer, 2×1 cm^3 burettes, dropping tubes, small stoppered sample tubes, carbon tetrachloride and ethanol. (Other suitable mixtures are acetone—water, chloroform—acetone and methanol—benzene.)

Procedure: Prepare 1 cm^3 samples of mixtures containing $1:9$, $2:8$, $4:6$, $5:5$, $6:4$, $8:2$ and $9:1$ of the components. Determine the refractive index of each of these mixtures and also of the two pure solvents (Expt. 5.1).

With a small flame, heat about 10 cm^3 of carbon tetrachloride in the flask and reflux it with the three-way tap in the position which allows the condensed distillate to flow back into the flask by the lower arm. When equilibrium has been attained, record the boiling point and turn the tap to an 'off' position and collect a sample of the distillate. Remove the flame, run out some of the condensed distillate ($0.2-0.5$ cm^3 is sufficient) into a sample tube and immediately stopper the tube. Remove the stopper from the flask and by means of a dropping tube fitted with a teat transfer a sample ($0.2-0.5$ cm^3) of the residue to a stoppered tube. Measure the refractive index of both samples; they should agree to within 0.005. Repeat with the ethanol. It is essential that the liquid be refluxed for a minimum of 10 min so that equilibrium is established.

To the 10 cm^3 of ethanol in the flask add about 1 cm^3 of carbon tetrachloride. Reflux the mixture, as described previously, and when equilibrium has been attained (10 min) record the boiling point of the mixture, and collect samples of the distillate and residue. Determine the refractive index of each sample. Repeat the determinations with successive additions of about, 1, 1, 1, 1.5, 2, 2.5, 3.5, 4, 5 and 5 cm^3 of carbon tetrachloride.

Discard this solution and place about 10 cm^3 of carbon tetrachloride and 0.5 cm^3 of ethanol in the flask. Determine the boiling point of the mixture and the

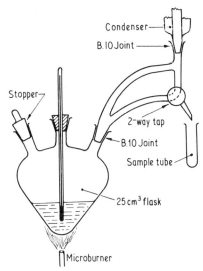

Fig. 4.7   Apparatus for determining the composition of the vapour in equilibrium with a binary liquid mixture

refractive index of a sample of distillate and a sample of residue. Repeat with six successive additions of 0.5 cm^3 of ethanol.

Treatment of experimental data and discussion

1.   Construct the calibration curve of refractive index against the known mole fraction of one component. From this calibration curve obtain the composition of each sample of distillate and residue in equilibrium at the measured temperature. Plot the phase diagram showing the bp. as a function of the composition of the vapour (vaporous curve) and of the residue (liquidus curve).

2.   Discuss the phase diagram in terms of the phase rule and state the composition and bp. of any azeotropic mixture formed. Give an account of the behaviour of the system during fractional distillation.

3.   Does the mixture obey Raoult's law or do your results suggest positive or negative deviations? Explain any deviations in terms of the energies of interaction between like and unlike molecules. Would you expect this system to exhibit a heat of mixing and a volume change on mixing?

4.   What is understood by the boiling point of a liquid mixture? Explain why the bp. passes through a minimum value if the vapour pressure passes through a maximum as the composition is varied.

5.   Why is it essential that only a very small volume of distillate is removed from the system?

(b) Immiscible liquids

When a mixture of two immiscible liquids is heated, each exerts its vapour pressure irrespective of the presence of the other. Thus when the sum of the two vapour pressures equals the atmospheric pressure both components distil over together, the composition of the distillate and the temperature of distillation remaining constant until one of the components is exhausted. The composition of the distillate is determined by the vapour pressures and relative molecular masses of the two components and is independent of the composition of the liquid.

Quantitative distillation in a current of steam, a dynamical process of vapour pressure measurement which is easily carried out, is a valuable method for determining vapour pressures of organic substances in the neighbourhood of 100 °C. The method can only be used when the organic compound is practically insoluble in water.

If x is the weight of water of relative molecular mass $M(A)$ and y the weight of organic substance of relative molecular mass $M(B)$ in the distillate, then by Avogadro's law:

$$\frac{\text{No. of mol of water}}{\text{No. of mol of organic substance}} = \frac{\text{vp. of water}}{\text{vp. of organic substance}} = \frac{p(A)}{p(B)}$$

Hence

$$\frac{x/M(A)}{y/M(B)} = \frac{p(A)}{p(B)}$$

or

$$\frac{x}{y} = \frac{p(A) \times M(A)}{p(B) \times M(B)} = \frac{[P - p(B)]M(A)}{p(B) \times M(B)} \tag{4.16}$$

where P is the barometric pressure.

Thus from a determination of x/y, the vapour pressure of the organic compound, $p(B)$, can be determined.

Experiment 4.17   Determine the vapour pressure of dekalin

Requirements: Steam generator, connected to two 500 cm³ round-bottom flasks in series (Fig. 4.8), measuring cylinder, dekalin (decahydronaphthalene), (other suitable liquids are chlorobenzene, nitrobenzene and aniline).

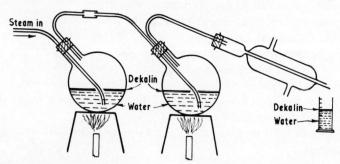

Fig. 4.8   Steam distillation apparatus

Procedure: Two distillation flasks are essential to ensure that equilibrium has been established. Place about 50 cm³ of water in each flask and cover with about 50 cm³ of dekalin. Heat the two flasks and at the same time pass steam through; this avoids the condensation of large amounts of steam during the initial stages. As soon as distillation commences, remove the source of heat from both flasks and lag well. Adjust the rate of passing in the steam until distillation proceeds at the rate of one drop every 2 s. When this rate has been established, collect a sample, of about 40 cm^3 in a 50 cm^3 measuring cylinder (or burette). Record the volumes of dekalin and water collected and record the barometric pressure.

Treatment of experimental data and discussion
1. From the volumes of dekalin and water calculate the proportions by weight of the two components, (ie. *x/y*), density of dekalin at 293 K = 0.895 × 10^3 kg m^{-3}. Use equation (4.16) to calculate the vapour pressure of dekalin; express your answer in mmHg and N m^{-2}.
2. Discuss the validity of the assumptions made in deriving equation (4.16).
3. What is the temperature of distillation?
4. Why is it essential that the distillation rate is low?
5. Why is water a good second component for this type of distillation?

5   Equilibria in three component liquid systems

For a system of three components, four degrees of freedom are possible since pressure, temperature and the concentration of two components are independently variable. To simplify graphical representation of the conditions of equilibrium, it is customary to consider condensed systems, in which the vapour phase is neglected, at constant temperature. Triangular diagrams are commonly used in which the pure

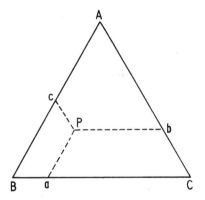

Fig. 4.9   Principle of a triangular diagram

components are shown at the apices of an equilateral triangle (Fig. 4.9). If from any point, P, within the triangle, lines are drawn parallel to the three sides of the triangle, then the sum is always equal to a side of the triangle; ie. Pa + Pb + Pc = AB. Thus, by taking the sides as unity (or 100%) and expressing the amounts of 3 components as mole fractions (or percentages of the whole) the composition of any ternary system can be represented by a single point on the triangle. Any point within the triangle represents 3 components; on a side, 2 components, and an apex, 1 component.

The addition of a third component to a pair of partially miscible liquids alters their mutual solubility. If the third component is soluble in only one of the other two components, the mutual solubility of the two liquids is decreased, but when it is soluble in both, then, their mutual solubility is increased. Thus when ethanol is added to a mixture of benzene and water the mutual solubility of these increases until a point is reached when the mixture becomes homogeneous.

Experiment 4.18   Determine the phase diagram for the system ethanol, benzene, water

Requirements: Glass-stoppered bottles (50 cm^3), thermostats at 25 and 40 °C, benzene, ethanol.

Procedure: Prepare six different mixtures of ethanol and benzene in the stoppered bottles with the following percentages (by weight) of ethanol: 10, 30, 50, 70, 90 and 95. It is sufficiently accurate to make up approximately 20 g of each mixture by running out the calculated volumes (accurately) from a burette (density of ethanol: 0.789×10^3, benzene: 0.879×10^3 kg m^{-3}). Thermostat at 25 °C and titrate each with water to the first appearance of turbidity due to the presence of a second phase. Add the water in small amounts only and shake vigorously after each addition.

Repeat the experiment at 40 °C.

Treatment of experimental data and discussion
1. From the titre of water when the second phase starts to separate out, calculate the percentage by weight of each component. Plot these points on triangular co-ordinate graph paper.
2. Discuss the phase diagram in terms of the phase rule, and explain the variation of the solubility curve with temperature.
3. Explain the meaning of the 'plait point'.
4. Show geometrically that the composition of the mixture, P, in Fig. 4.9, is given by $c_A : c_B : c_C = Pa : Pb : Pc$.

Experiment 4.19   Determine the complete phase diagram for the system chloroform, acetic acid, water

Theory: The first part of this experiment is the construction of a phase diagram similar to Expt. 4.18. Analysis of mutually saturated solutions for one component permits the construction of the tie lines across the miscibility gap.

Requirements: Glass-stoppered bottles (50 cm^3), chloroform, glacial acetic acid, standard 0.1 mol dm^{-3} sodium hydroxide solution.

Procedure: Prepare approximately 20 g mixtures of chloroform and acetic acid containing 10, 20, 40, 60, 80 and 90% by weight of chloroform. The calculated volumes may be run in from a burette (density of chloroform = 1.50×10^3 kg m^{-3}, acetic acid = 1.05×10^3 kg m^{-3}). Titrate with water, adding small amounts at a time with vigorous shaking, until the first appearance of turbidity. Record the temperature.

Prepare, in separating funnels, three different mixtures of chloroform, acetic acid and water from within the binodal curve, which are known to separate into two layers. Shake vigorously and allow the two layers to separate out. Carefully run off each layer and weigh. Remove a small aliquot, 1 or 2 cm^3, weigh, add 50 cm^3 of water and titrate with the standard sodium hydroxide solution using phenolphthalein as indicator.

Treatment of experimental data and discussion

1. Calculate and tabulate the percentage by weight of each component when the solution just becomes turbid. Plot these on triangular graph paper. Draw the complete phase diagram, at room temperature, a saturated solution of chloroform in water contains 0.8% (w/w) CHCl$_3$, and a saturated solution of water in chloroform contains 99.0% (w/w) CHCl$_3$.

2. From the alkali titration figures, the weight of sample titrated and the total weight of the layer, calculate the percentage by weight of acetic acid in each layer. For each pair of conjugate solutions plot the percentage of acetic acid on the binodal curve and join by a tie-line. This line will pass through the composition of the original mixture.

3. Explain why the binodal curve is not symmetrical and why the tie-lines are not parallel. Determine the 'plait point' and discuss its significance. Does this coincide with the maximum of the binodal curve?

4. Show, by means of the phase rule, that a condensed three-component system at constant temperature possesses a maximum of two degrees of freedom.

5

Optical and spectrophotometric measurements in chemistry

Optical measurements are of utmost importance in chemistry for analysis of mixtures, determination of purity of organic compounds and the determination of molecular structures. The infrared absorption spectrum of an organic compound is a better criterion of purity than is any other physical property of the compound.

1 Refractometry

The refractive index, n, a characteristic of each substance, is temperature dependent; for many liquids $dn/dT \sim -4 \times 10^{-4}$. The value of the refractive index depends on the wavelength of light used for measurement; in regions remote from an absorption band, n decreases gradually with increasing wavelength; this is known as normal dispersion. Within or near a region of absorption, n changes sharply, anomalous dispersion.

The molecular refraction $[R]$, of a compound, defined by the equation:

$$[R] = \frac{n^2 - 1}{n^2 + 2} \frac{M_r}{\rho} \tag{5.1}$$

where M_r and ρ are the relative molecular mass and density of the substance respectively, depends only on the wavelength of the light used to measure n. $[R]$ is a molecular volume (units: m^3), and, as such, is an additive and constitutive property. It is thus possible to determine refraction equivalents for various atoms, radicals and bonds.

The molar refraction $[R]_{A,B}$, of a mixture is the sum of the contributions of the separate components, ie.

$$[R]_{A,B} = x_A [R]_A + x_B [R]_B \tag{5.2}$$

where x_A and x_B and $[R]_A$ and $[R]_B$ are the mole fractions and molecular refractions respectively of the two components. In addition:

$$[R]_{A,B} = \frac{n_{A,B}^2 - 1}{n_{A,B}^2 + 2} \left(\frac{x_A M_A + x_B M_B}{\rho_{A,B}} \right) \tag{5.3}$$

where $n_{A,B}$ and $\rho_{A,B}$ refer to the mixture.

Critical angle refractometers

In this type of instrument light passes through the substance under investigation into a glass prism of higher refractive index; the light is observed on emerging from this prism. The two main types of critical angle refractometer are the Abbé and the Pulfrich.

The maximum accuracy of the refractive index for both these instruments is $\pm 1 \times 10^{-4}$ and the range 1.3–1.85. The Abbé refractometer has the advantage that it only requires 0.05 cm^3 of liquid in contrast to the Pulfrich, which requires at least 1 cm^3. In both instruments the prism surface should be cleaned with ethanol between readings and then rinsed with the liquid to be investigated.

The main disadvantages of critical angle refractometers are: (1) unsymmetrical field of view, (2) measurements can only be made on substances of refractive index less than that of the prism, and (3) the index measured is that of a surface layer which may or may not be the same as that in bulk. These disadvantages have been overcome in the non-critical angle, refractometers which have maximum accuracy of $\pm 1 \times 10^{-5}$, but require 2–3 cm^3 of liquid.

Experiment 5.1   Determine the variation of the refractive index with composition for mixtures of benzene and carbon tetrachloride

Requirements: Refractometer, sodium lamp, pure benzene and carbon tetrachloride.

Procedure: Prepare mixtures of benzene and carbon tetrachloride by mixing known volumes (1:9, 2:8, etc.). Determine the refractive indices of the pure components and each mixture, and the density of at least one of the mixtures.

Treatment of experimental data and discussion
1. Plot graphs of refractive index against (a) composition by volume, (b) composition by weight, and (c) mole fraction. Calculate the molecular refractions of two pure components and the mixture.
2. Is equation (5.3) valid for these liquids? Under what conditions would it not apply?
3. Calculate the polarizability of the liquids investigated. From your results would you conclude that carbon tetrachloride and benzene were spherical molecules?
4. Is it meaningful to use atomic refractions to calculate molecular refractions?

Experiment 5.2   Determine the values of the molecular refraction for a series of homologous alcohols and esters and hence deduce the refraction equivalents of the methylene group, carbon, hydrogen and oxygen

Requirements: Refractometer, sodium lamp, pure organic compounds: methanol, ethanol, propan-1-ol, propan-2-ol, butan-1-ol, butan-2-ol, methyl formate, ethyl formate, *n*-propyl formate, methyl acetate, ethyl acetate and *n*-hexane.

Procedure: Determine the refractive index of each of the above liquids.

Treatment of experimental data and discussion
1.   Calculate the molecular refraction of each liquid (equation (5.1)).
2.   Determine the contribution of the methylene group in each series.
3.   Is the contribution of the methylene group the same in both series?
4.   Calculate the refraction equivalent for C, H, and $-O-$ (in the alcohol series).

2   Polarimetry

Certain substances, optically anisotropic, have the property of rotating the plane of polarization of radiation in its passage through them. This change in direction is known as optical rotation; the magnitude and direction of rotation are measured with a polarimeter. Substances which rotate the plane of polarization in a clockwise direction (when looking towards the oncoming radiation) are said to be dextro-rotatory and in an anticlockwise direction laevo-rotatory.

A polarimeter consists of a radiation *source*, a *polarizer* to produce polarized radiation, a tube for holding the test liquid or solution, an *analyser* to measure the amount of rotation of the plane of polarization by the liquid, and a *detector* for investigating the radiation passing the analyser.

The optical rotation of a liquid or solid depends on the layer thickness, wavelength of radiation and temperature. For a solution, the rotation depends, in addition, on the concentration of the optically active substance and on the nature of the solvent. Often it is also a function of time, eg. mutarotation of glucose in water (Expt. 8.4). It is therefore essential to define the experimental conditions, and desirable to reduce the experimental rotations to standard values.

Experiment 5.3   Determine the specific and molecular optical rotatory power of sucrose

Theory: The specific optical rotatory power of a homogenous active substance is given by:

$$\alpha_m = \frac{\alpha V}{ml} \tag{5.4}$$

where α is the angle of optical rotation measured in radians for l m length of a solution containing m kg of substance dissolved in V m^3 of solution.

The molar optical rotatory power is defined by:

$$\alpha_n = \frac{\alpha V}{nl} = \frac{\alpha}{cl} \tag{5.5}$$

where c/mol m^{-3} is the concentration.

Requirements: Polarimeter, tubes of various lengths, sodium lamp, sucrose.

Procedure: Prepare a solution of 10 g sucrose (dried at 105 °C and cooled in a desiccator) in 100 cm³ water. Dilute this stock solution (*a*) 25 cm³ to 50 cm³ and (*b*) 10 cm³ to 50 cm³ with water. Fill the polarimeter tube with water and determine the mean zero position. Fill the tube with each of the sucrose solutions in turn and, from at least ten observations, determine the mean angle of rotation. Finally determine the angles of rotation of one solution in tubes of different lengths.

Treatment of experimental data and discussion

1. Determine the direction and the angle of rotation for each sugar solution.
2. Establish that the angle of rotation is proportional to both the concentration and the path length.
3. Determine the specific and molecular optical rotatory power of sucrose, equations (5.4) and (5.5).
4. Why is it important to specify both the solvent and concentration of solute when quoting specific optical rotatory power?
5. Do α_m and α_n vary with wavelength?
6. What is polarized light?
7. What type of substance exhibits optical anisotropy?
8. Is it important to specify the temperature at which the measurements were made?

3 Emission spectroscopy

A spectrometer will separate a beam of light into its constituent wavelengths. Each wavelength gives rise to an image of the entrance slit in the field of view. In prism instruments, the dispersion, $dS/d\lambda$ (where S is the distance of wavelength, λ, from a fixed position in the spectrum), decreases with increasing wavelength and depends on the material of the prism. The resolution, ie. the closest wavelength separation of two lines which can be recognized as two, depends also on the slit width. In grating instruments the dispersion is independent of the wavelength.

To set up a simple spectrometer for wavelength determination the scale must be calibrated or the drum correctly positioned. Illuminate the slit (very narrow) with a sodium source. View the yellow sodium doublet and adjust the eyepiece and the telescope until the spectral lines and either the illuminated scale or the pointed arrowhead are sharply in focus. Care must be taken that there is no parallax between the lines and the scale or pointer. Record the position of the doublet on the illuminated scale; this is the reference for subsequent observations. With the constant deviation spectrometer, set the drum reading to 589 nm and adjust the pointer to coincide with that line. Subsequent wavelengths can then be recorded directly.

Experiment 5.4   Observe the predominant lines in the flame spectra of the alkali and alkaline earth metals

Requirements: Spectrometer, eg. constant deviation, solid LiCl, NaCl, KCl, $CaCl_2$, $SrCl_2$, $BaCl_2$, TlCl.

Procedure: Set up and calibrate the spectrometer as described. Volatilize each salt moistened with concentrated hydrochloric acid on a clean platinum wire in a non-luminous flame. Place a sheet of asbestos, in which a hole has been cut, between the flame and the spectrometer to prevent excessive heat reaching and damaging the slit. Thoroughly clean the wire before each change of salt by dipping in concentrated hydrochloric acid and heating until no colour is produced. Observe and record the wavelengths of as many lines as possible, compare with the values quoted in the literature (p. 319). Difficulty may be encountered in locating the violet lines of potassium and calcium as they are faint and disappear soon after the salt is introduced into the flame. To observe the red potassium line, it may be necessary to open up the slit slightly.

Experiment 5.5   Observe the predominant lines in the hydrogen spectrum and hence determine the Rydberg constant

Theory: Balmer showed from the mathematical analysis of the hydrogen spectrum, that the wavenumber, $\bar{\nu}$, for the lines could be expressed by the equation:

$$\bar{\nu} = R_H \left\{ \frac{1}{2^2} - \frac{1}{n_1^2} \right\} \tag{5.6}$$

where n_1 has the values 3, 4, 5, etc., for successive lines and R_H is the Rydberg constant. The wavenumber (ie. the number of waves per cm) is the reciprocal of the wavelength expressed in cm.

Requirements: Spectrometer, eg. constant deviation, discharge tube containing hydrogen at low pressure, induction coil for exciting tube (a vacuum leak tester is very suitable, one end of the tube is earthed).

Procedure: Set up and calibrate the spectrometer as described. Arrange the hydrogen discharge tube immediately in front of the slit and observe the following lines in the hydrogen spectrum:

Line	n_1	Approx. λ/nm
Red H α	3	656
Green H β	4	486
Violet H γ	5	434
Violet H δ	6	410

Treatment of experimental data and discussion
1.   Record the wavelengths of the observed lines.
2.   Plot a graph of $\bar{\nu}$ against n_1^{-2} and hence calculate R_H from the slope and from the intercept.
3.   Describe the other spectral series observed in atomic hydrogen.
4.   Explain why every line in any given series can be represented as a difference of two terms one $R_H n_2^{-2}$ where n_2 has a fixed value for a given series, and a second $R_H n_1^{-2}$, where n_1 can take on a series of consecutive values beginning with $n_1 = n_2 + 1$.

Experiment 5.6   Observe and compare the spectra of hydrogen, helium, nitrogen and oxygen

Requirements: Spectrometer, gas discharge tubes, excitor unit.

Procedure: Set up and calibrate the spectrometer as described. Record the predominant lines in the spectra of these gases in discharge tubes. Observe the different types of spectra. If a camera attachment is available, photograph these spectra as a permanent record.

Treatment of experimental data and discussion
1. Discuss the difference between atomic and molecular spectra.
2. Explain why the spectrum of helium gas is more complicated than that of hydrogen.
3. Assign six of the emission lines in the nitrogen spectrum to transitions within the molecule.

4   Absorption spectroscopy of gases

Absorption of visible or ultraviolet radiation by a gas results in electronic, vibrational and rotational excitation. The observed spectrum for each electronic transition has a coarse structure due to vibrational changes and a fine structure due to rotational changes.

The potential energy curves of the two states of the iodine molecule are represented in Fig. 5.1, where for clarity only a few of the vibrational levels are shown and none of the rotational levels.

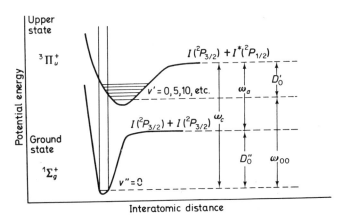

Fig. 5.1   Approximate potential energy curves for I_2

At room temperature the majority of the molecules in the ground electronic state will also be in the ground vibrational level. There are no restrictions on the change of vibrational quantum number in an electronic transition, apart from that imposed by the Franck—Condon principle; therefore a large number of transitions are observed.

The vibrational energy levels in the upper state, relative to the potential minimum, are given, to a first approximation by equation (5.7).

$$E(v') = (v' + \tfrac{1}{2})\omega'_e - (v' + \tfrac{1}{2})^2 x'_e \omega'_e \qquad (5.7)$$

Where v' is the vibrational quantum number, x'_e is the anharmonicity constant, and ω'_e is the equilibrium vibration frequency in the upper state, expressed in wavenumbers.

The energy separation of the vibrational levels in the excited electronic state is given by:

$$E(v' + 1) - E(v') = \omega'_e - 2x'_e \omega'_e (v' + 1) \qquad (5.8)$$

The levels gradually converge as v' approaches the dissociation level, the convergence limit corresponds to ω_c (cm^{-1}).

Experiment 5.7   Observe the visible absorption spectrum of iodine vapour

Requirements: Spectrometer, sealed bulb containing iodine with heater, source of white light, tungsten filament lamp, sodium lamp.

Procedure: If using a simple spectrometer (constant deviation) calibrate using the sodium lamp. If a recording spectrophotometer is used calibrate using a suitable filter.

Place the vessel containing the iodine between the slit of the spectrometer and the filament lamp. Adjust the pressure of the iodine vapour until the overlapping series of absorption bands can be clearly seen in the region 540—570 nm. Adjust the slits to give good resolution.

Measure the wavelength of the dark edge of each band starting from the first discernible band at the red end of the spectrum. Measure the bands to at least 510 nm. In the region 540—570 nm bands may be imagined to exist at regularly decreasing spacings so as to join the two sets of more clearly defined bands outside these limits.

Note the position and direction of degradation of the bands.

Treatment of experimental data and discussion
1. Convert all measurements to true wavelengths by means of your calibration curve, working to the nearest 0.1 nm.
2. Assign each band an arbitrary band number n, counting from the first visible band at the red end of the spectrum.
3. Calculate the wavenumber (ω) of each band observed (use five significant figures in the reciprocal).
4. For the region below 544 nm plot $\Delta\lambda/\Delta n$ against λ. Obtain a value for the convergence limit in nm and cm^{-1}.
5. From equation (5.8) it may be seen that from a graph of band separation

against n (or v') a value of ω_e' and x_e' may be obtained. Plot $\Delta\omega$ against n, and determine ω_e', x_e' and the convergence limit.

6. The bands at the extreme red end of the spectrum correspond to transitions from a higher vibrational level in the ground electronic state to relatively low-lying vibrational levels in the upper electronic state. From the separation of the observed bands calculate the lower limit of the value of ω_e' (cf. with the actual value of 128.0 cm^{-1}).

7. Given that the energy of the $^2P^{1/2}$ iodine atom is 7598 cm^{-1} relative to the ground state $^2P^{3/2}$, calculate D_0'' in cm^{-1} and J mol^{-1}. Explain why the $^2P^{3/2}$ state is of lower energy than the $^2P^{1/2}$ state, and show how the energy difference may be determined.

8. The wavenumber of the (0,0) band is 15,600 cm^{-1}, calculate the wavenumber of the hypothetical first line in the series with $v'' = 0$. Is it possible to calculate the wavenumber of the (0,0) band from your experimental data?

9. Calculate D_0' in cm^{-1} and J mol^{-1}.

10. Using the calculated value for ω_e' and given that ω_e'' is 214.6 cm^{-1}, calculate the force constants (k) for the two electronic states from the equation:

$$\omega_e = \frac{1}{2\pi c}\left(\frac{k}{\mu}\right)^{1/2} \qquad (5.9)$$

where μ is the reduced mass and c the velocity of light.

11. What determines the shape of the absorption bands, ie. the direction of degradation of the band?

12. From the values calculated for dissociation energies, vibration frequencies and force constants of the two electronic states and the shading of the bands, suggest what type of molecular orbital the excited electron occupies.

13. How may ω_e'' be determined?

14. Show why more than one series is observed in the iodine spectrum.

5   Absorptiometry

A great many salts and most organic compounds have absorption spectra which can be used to identify them or determine their concentration in solution. These absorption bands may be in the ultraviolet (185–400 nm), visible (400–760 nm) or infrared (0.76–25 μm) regions of the spectrum.

Absorptiometers are instruments used to compare the intensities of light transmitted by two solutions. Absorptiometers use either physical or visual detectors, the band width used in one measurement may vary from 1 nm in a high-class spectrophotometer to about 60 nm in a filter instrument.

The essential components of any photoelectric absorptiometer is a light source with filter to give an approximation to monochromatic light, glass cells to contain the solutions or solvent, and a photoelectric cell connected to a measuring device. The photoelectric current produced by the light after passing through the solvent or solution is thus measured. In some instruments two photocells are used;

one receives light which has passed through the solution (or solvent) and the other simultaneously receives light from the same source without passing through any liquid. The photocell outputs pass, in opposite directions, through the measuring device (galvanometer) and are adjusted to equality. This null-point instrument has the advantage that fluctuations of light intensity are balanced out, while errors arising from cell fatigue and temperature changes are largely eliminated.

Filters: The most suitable filter to use in a particular determination is that whose absorption is the greatest, ie. the filter which gives the greatest difference in absorbance between two concentrations of the unknown solution. In general, it is the filter of colour complementary to that of the test solution (p. 319).

Beer–Lambert law

The intensity of light (I) after passing through a solution of concentration c/mol m^{-3}, and a layer thickness l/m, is related to the incident intensity (I_0) by the Beer–Lambert law:

$$I = I_0 \, 10^{-\epsilon c l} \tag{5.10}$$

where ϵ/m^2 mol^{-1} is the molar extinction coefficient. The decadic absorbance A is given by the expression:

$$A = \log \frac{I}{I_0} = \log \frac{1}{T} \tag{5.11}$$

where T is the transmittance of the sample.

Thus
$$A = \epsilon c l \tag{5.12}$$

Experiment 5.8   Test the validity of the Beer–Lambert law using a photoelectric absorptiometer

Requirements: Photoelectric absorptiometer, filters, methylene blue solution (0.01 g dm^{-3}; ie. 10 ppm.).

Procedure: From the standard solution, prepare further solutions by dilution, containing 8, 6, 4, 2 and 1 ppm. The filter required is red; ie. the complementary colour to the blue of the solution. This may be checked as follows: determine the absorbance of one of the methylene blue solutions against a water blank using each of the set of filters in turn; the filter giving the highest absorbance is the correct one. Using this filter, determine the absorbance of each solution in turn against a water blank. Replicate readings should be concordant.

If a spectrophotometer is available, determine the absorption spectrum of one solution against a water blank at 10 nm intervals in the range 500–650 nm. Now determine the absorbance of each solution in turn at the wavelength of maximum absorption.

Treatment of experimental data and discussion
1.   Plot a graph of absorbance against concentration. Does the Beer–Lambert law hold for this system?

2.   Discuss the limitations of the combined law. Describe some causes for deviations from the law.
3.   Is it important to keep the temperature of the sample constant?
4.   What is the effect on A of changing the pH?
5.   How will suspended particles alter the true absorbance?
6.   Is it possible to analyse for a substance which itself does not absorb in the ultraviolet region?

Experiment 5.9   Construct a calibration curve for the determination of the concentration of creatinine in solution

Theory: Creatinine, in the presence of picric acid in alkaline solution, forms a red tautomer of creatinine picrate. This may be used for the absorptiometric determination of creatinine; other biological materials which react similarly may interfere.

Requirements: Photoelectric absorptiometer, 100 cm³ volumetric flasks, 1% solution of picric acid in water, 2.5 mol dm⁻³ sodium hydroxide solution, standard creatinine solution (1 g dm⁻³).

Requirements: Photoelectric absorptiometer, 100 cm^3 volumetric flasks, 1% solution of picric acid in water, 2.5 mol dm^{-3} sodium hydroxide solution, standard creatinine solution (1 g dm^{-3}).

Procedure: From the standard solution, prepare solutions containing 0.2, 0.4, 0.6 and 0.8 g dm^{-3} creatinine. Accurately transfer, to separate 100 cm^3 volumetric flasks, 1 cm^3 water (as reagent blank), 1 cm^3 of each of the standard solutions and 1 cm^3 of test solution (eg. 0.5 cm^3 urine + 0.5 cm^3 water). Add 20 cm^3 of picric acid solution to each flask, followed by 2 cm^3 of sodium hydroxide solution. Mix and allow to stand for exactly 15 min, dilute to 100 cm^3 with water and mix well. Measure the absorbance of each solution against the reagent blank at 520 nm (green filter).

Treatment of experimental data and discussion
1.   Plot a calibration curve of absorbance against concentration and hence determine the concentration of creatinine in the test solution.
2.   What is the purpose of the picric acid?
3.   List the advantages and disadvantages of this method.

Experiment 5.10   Construct a calibration curve for the determination of the concentration of inorganic phosphate in solution

Requirements: Photoelectric absorptiometer, red filter, 25 cm^3 volumetric flasks, 8.3% ammonium molybdate solution, add one drop of 0.880 ammonia solution per dm^3 to aid solution, 60% perchloric acid. Amidol reagent: dissolve 2 g amidol and 40 g sodium metabisulphite in 200 cm^3 distilled water, filter and store in dark bottle (this solution only keeps for a week). Stock phosphate solution: 1.0967 g KH_2PO_4 (dried at 100 °C) dissolved in 250 cm^3 distilled water; this contains 1 mg P per cm^3.

Procedure: From the standard solution, prepare further solutions by accurate dilution containing 10, 20, 30, 40 and 50 μg P per cm^3. Accurately transfer to separate 25 cm^3 volumetric flasks 2.0 or 2.5 cm^3 of (i) water (reagent blank), (ii) each of the above solutions, and (iii) the solution under test (suitably diluted to be in the concentration range). To each flask add 2 cm^3 60% perchloric acid, 2 cm^3 amidol reagent and 1 cm^3 ammonium molybdate solution, in that order, shaking gently after each addition. Dilute each solution to 25 cm^3 with water and shake well. The colour is fully developed in 5–30 min. Measure the absorbance of each solution against the reagent blank, using a red filter.

Treatment of experimental data and discussion
1.   Plot a calibration curve of absorbance against concentration and hence determine the concentration of phosphorus in the test solution.
2.   List the advantages and disadvantages of this method for determining the concentration of inorganic phosphates.

6   Spectrophotometry

Spectrophotometers consist essentially of a light source (hydrogen lamp for use in the ultraviolet, a tungsten lamp for the visible, Nernst filament in the infrared), a monochromator (either prism or diffraction grating) for selecting radiation of narrow-band width, and a suitable detector (photocell or themocouple).

(a) The use of ultraviolet and visible spectra in qualitative and quantitative analysis

Calibration: To be sure that a spectrophotometer is working satisfactorily it must be tested under conditions as stringent as those under which it will be used.

The wavelength calibration, of recording and non-recording ultraviolet and visible spectrophotometers, may be checked either with a didymium glass filter (sharp absorption maxima at 441, 475.5, 528.7, 585, 684.8, 743.5, 745, 808, 883 and 1067 nm) or a holmium filter (sharp absorption maxima at 279.39, 287.5, 333.7, 360.9, 418.7, 453.2, 460, 536.2 and 637.5 nm). The absorbance of the filter, supported on the movable cell carriage, is measured against air. Calibration curves for each filter are supplied by the manufacturer.

These glasses are not suited for checking the photometric scale of any spectrophotometer. Values of maximum absorption are too dependent on slit widths, while values on the steep parts of the curve are susceptible to slight wavelength errors. The best transmission standards are neutral glass for the visible region and thin deposits of rhodium on quartz for the ultraviolet region; these can be calibrated by the National Physical Laboratory. If two or more such glass discs are placed in series, then the total absorbance is the sum of the individual values. The rhodiumized-quartz filters are available in a limited density range; they have spectral transmission curves which vary only very slowly with wavelength.

The spectral transmittancies of certain stable solutions have been accepted, mainly in the United States, as suitable standards for checking the photometric scale of spectrophotometers (p. 318). The chromate solution is considered to be the most suitable solution in the ultraviolet region where the copper and cobalt solutions have little absorption. The values for the mercury and helium lines are the most reliable, with an uncertainty in absorbance not exceeding 0.001 for the particular chemical used.

A serious drawback to the use of alkaline chromate solution is that it damages glass and silica cells; in its place a 1% solution of potassium dichromate in 0.005 mol dm^{-3} sulphuric acid has been recommended. Since this solution has broad maxima and minima in the ultraviolet, the measurements are not greatly affected by slit width.

When using 'matched' cells in double-beam spectrophotometers it is important always to ensure that they are placed in the same position in the cell holder; that is, the same cell is used in the sample beam with the same surface nearest the source.

Solvents: In analytical work it is customary to use dilute solutions of substances in solvents whose absorption is low in the ultraviolet range. The choice of solvent is finally governed by solubility of solute and interaction between solvent and solute. In general polar solvents should be avoided. The most useful solvents and the lowest wavelength at which they transmit ultraviolet radiation are listed on p. 320.

Experiment 5.11   Plot some typical absorption spectra

Requirements: Ultraviolet spectrophotometer, a recording instrument, can be used to advantage, 1 cm quartz cells, pure materials and spectroscopically pure solvents.

Procedure: If not pure, the solutes should be purified by the appropriate means, eg. recrystallization, distillation, etc., and their purity checked after each purification stage until a constant ultraviolet spectrum is obtained.

Prepare solutions of the concentration indicated; these have been chosen to give a maximum absorbance not much in excess of 1.0. Owing to the high cost of spectroscopically pure solvents, the total volume used must be kept to a minimum; this is achieved by accurate serial dilution using standard flasks and pipettes, eg. for benzene.

(*a*)   Original solution 0.25 g benzene in 10 cm^3 cyclohexane.

(*b*)   Dilute solution (*a*) 1 in 10 with cyclohexane.

(*c*)   Dilute solution (*b*) 1 in 10 (concentration 0.25 g dm^{-3}; range of study: 220–280 nm).

(*d*)   Dilute solution (*c*) 1 in 10 (concentration 0.025 g dm^{-3}; range of study: 200–230 nm).

Determine the absorbance of each solution against the solvent blank over the range of wavelengths indicated. Make measurements at large increments (eg. 10 nm) at first, completing any maximum or minimum at intervals of 1 to 5 nm.

Solute	Solvent	Concentration /g dm^{-3}	Range of study /nm
Benzene	Cyclohexane	0.25	220–280
	Cyclohexane	0.025	200–230
Naphthalene	95% ethanol	0.025	240–350
	95% ethanol	0.001	200–260
Anthracene	Cyclohexane	0.028	290–400
	Cyclohexane	0.00056	220–310
o, m, p-Toluidines	Cyclohexane	0.065	250–310
Anthranilic acid	Cyclohexane	0.007	200–350
m and p-amino-benzoic acids	Cyclohexane	0.007	200–350
p-nitraniline	Hexane	0.014	220–500
	Dioxan	0.014	220–500
	Water	0.014	220–500
Aniline	Water	0.015	210–300
	0.5 mol dm^{-3} HCl	1.0	220–280
Benzoic acid	Cyclohexane	0.02	210–290
Benzylamine	Isooctane	0.68	220–280
Acetone	Hexane	3.6	200–300
Mesityl oxide	Hexane	2.4	280–350
	Hexane	0.0078	200–290

Treatment of experimental data and discussion

1. Calculate the molar extinction coefficient at selected wavelengths.
2. Why do certain compounds absorb ultraviolet radiation? What determines the position and intensity of the absorption?
3. Compare the spectra and attempt to explain changes in the wavelength and intensity of bands arising from substitution and change of solvent.
4. Would you expect geometric isomers to show differences in their absorption spectra?

Experiment 5.12   Determine the composition of a mixture of potassium permanganate and potassium dichromate spectrophotometrically

Theory: Spectrophotometric methods of quantitative analysis of mixtures are simplified if the Beer–Lambert law is valid and if the absorbance of various components in a mixture are additive.

Two particular cases will be cited by way of example:

1. Mixtures in which the absorption bands of the components do not overlap. This is the simplest condition for analysis since a set of wavelengths can be selected, at each of which one substance has high absorption, and, all the others, negligible absorption. This means that the absorption of each component can be measured independently of the others. A calibration curve must be constructed for each component to establish the validity of the Beer–Lambert law.

2. Mixtures in which the absorption bands of the components overlap. Analysis of n mutually interfering components involves the measurement of the absorbance of the mixture at n suitably chosen wavelengths and the setting

up of n simultaneous equations. The spectra of the pure components must be compared, and suitable wavelengths selected, at each of which one component shows stronger absorption than the others.

Figure 5.2 shows the spectra for solutions of pure A and B of known concentrations c_A and c_B respectively and the mixture of unknown concentrations c'_A and c'_B.

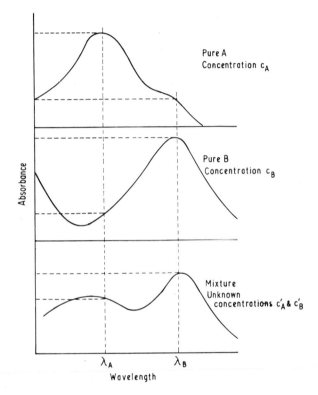

Fig. 5.2   The absorption spectra of two components A and B singly and as a mixture of unknown composition

The absorbance of the mixture at the two wavelengths λ_A and λ_B may be expressed as follows.

At λ_A $$A_A = \epsilon_A c'_A l + \epsilon'_B c'_B l \tag{5.13}$$

At λ_B $$A_B = \epsilon_B c'_B l + \epsilon'_A c'_A l \tag{5.14}$$

where ϵ_A is the extinction coefficient of A at λ_A, ϵ'_B is the extinction coefficient of B at λ_A, ϵ_B is the extinction coefficient of B at λ_B, and ϵ'_A is the extinction coefficient of A at λ_B. The molar extinction coefficients can be evaluated from measurements, at λ_A and λ_B, of individual standard solutions of A and B.

Requirements: Spectrophotometer, 1.0 cm glass cells. Potassium dichromate,

2×10^{-3} mol dm^{-3} in 0.5 mol dm^{-3} sulphuric acid, potassium permanganate 5.0×10^{-4} mol dm^{-3} in 0.5 mol dm^{-3} sulphuric acid. Mixture of these solutions. 0.5 mol dm^{-3} sulphuric acid.

Procedure: Measure the absorbance of the given permanganate and dichromate solutions in the region 400—610 nm (at 10 nm intervals if using a manual instrument). Use 0.5 mol dm^{-3} sulphuric acid in the reference beam. Select two wavelengths for the analysis (these will be near 440 and 520 nm). Prepare from each of the given solutions, by dilution, about five standard solutions. Measure the absorbance of each solution at the wavelengths chosen for the analysis. Measure the absorbance of the solution of unknown concentration at the chosen wavelengths.

Treatment of experimental data and discussion
1.   Plot graphs of absorbance against concentration for each component at the two wavelengths and calculate the best values of $\epsilon_A l$, $\epsilon_B l$, $\epsilon'_A l$, and $\epsilon'_B l$.
2.   Derive an expression for the concentration of the components in terms of the absorbance and extinction coefficients.
3.   Calculate the concentration of potassium permanganate and potassium dichromate in the given mixture.
4.   Comment on the validity of the Beer—Lambert law in the system.
5.   Why is it important to use the same cell as reference each time?

(b) Measurements in the mid-infrared region

Infrared radiation is selectively absorbed by substances, producing spectra which can be used qualitatively for the identification of the substance, and, quantitatively, for its estimation. With organic compounds, the spectra consist of a large number of fairly narrow bands. Each band corresponds to a particular vibration of the molecular structure, the frequency depending on the masses of the atoms, the strengths of the binding forces (bond energies), the configuration of the compound and the state of dispersion (ie. solid or solution). Thus, apart from optical enantiomorphs, in solution, no two compounds have the same infrared spectrum. This unique property of a compound has, in recent years, been widely used for the identification, and 'finger printing', of organic compounds and the assessment of their purity. It thus supplements the classical methods such as melting and boiling points, refractive index, etc., previously used.

Each observed band is due to the vibration of the molecule as a whole; this vibration may be largely localized in a particular bond or group when the movement of the atoms directly concerned is large in comparison with the movement of the remainder of the molecule. Thus certain absorption bands have, as the result of extensive work, been assigned to the relative vibration of two adjacent parts of a molecule, eg. C—H, C=C, C=O, O—H, N—H. The presence of such a band in the spectrum of an unknown compound is evidence of the presence of the corresponding group or bond in the molecule. Structural differences, eg. between isomers, are also revealed. In the $3n$ degrees of freedom of a non-linear molecule of n atoms, three are for translational, three for rotational motion and the remaining $3n - 6$ are the possible modes of vibration. These modes of vibration can be generally classified as those along a bond (stretching vibrations), and those perpendicular to the bond (bending or deformation vibrations). The frequency associated with a particular bond or group will be affected by the rest of the molecule; the values given in correlation charts in reference books are therefore only approximate values for the given bond or group.

For detailed correlation charts, reference should be made to such books as *The Infra-Red Spectra of Complex Molecules*, by L. J. Bellamy (Methuen), 1958, and *Introduction to Practical Infra-red Spectroscopy*, by A. D. Cross and R. A. Jones 3rd edn. (Butterworth's Scientific Publications), 1969.

In addition to the fundamental frequencies, overtone bands occur at multiples of the fundamental frequencies and combination bands at frequencies which are the sum or difference of two or more fundamental frequencies. Thus, if the fundamental frequencies are y and z, overtones will occur at $2y$ and $2z$ (first), $3y$ and $3z$ (second), etc., and combination bands can occur at $y + z$ and $y - z$. Overtones and combination bands are usually of lower intensity than fundamentals.

Calibration: For high accuracy the wavelength scale should be calibrated using ammonia gas or the water vapour spectrum or a liquid sample of pure indene. (Tables of wavenumbers for the calibration of infrared spectrometers, IUPAC (Butterworths) London 1961.) A thin film of polystyrene is, however, commonly used for checking calibration as polystyrene possess a large number of sharp absorption bands whose wavelengths are accurately known.

The linearity of the transmittance scale should also be checked regularly using rotating discs.

It is essential that any calibrations are performed under identical conditions to those for which the spectrometer will be used to make subsequent measurements, ie. slit width, scanning speed, period, scale expansion, etc.

Experiment 5.13 Plot the infrared absorption spectra of various organic compounds as mulls and as KBr discs

Requirements: Infrared spectrophotometer, NaCl plates and holder, 'mini-press', agate pestel and mortar, pure Nujol, hexachlorobutadiene. High purity KBr powder. Benzoic acid, sodium benzoate, acetanilide, *p*-nitrotoluene (or other substance showing characteristic functional group absorptions).

Procedure: Following the instructions for the instrument, plot the spectrum with no sample in either the sample or reference beam from 4000–650 cm^{-1}. The I_0 or 100% transmission line should be constant to better than ±2% over this range. Plot a calibration spectrum of a polystyrene film; the most useful bands for this purpose occur at the following wavenumbers: 2924 cm^{-1}, 1603 cm^{-1}, 1029 cm^{-1} and 907 cm^{-1}.

Mull: Place 2–5 mg of the substance in a dry clean agate mortar and grind the dry powder *well*, add one drop of Nujol and again grind well to give a uniform dispersion of the fine solid particles. Using a microspatula, transfer a portion of the slurry to a clean, dry NaCl plate and immediately cover with a second plate thereby forcing the mull to spread as a thin film. Clamp the plates in a cell holder and view through transmitted light; a brownish colour indicates satisfactory dispersion of the solid. Place the holder in the sample beam of the spectrophotometer. Following the instrument instructions, adjust the 100% transmission control to give minimum background absorption throughout the spectrum. The salt plates should give a constant background absorption throughout the range, this level of absorption is adjusted by the 100% transmission control. If there is still an unduly high background absorption, insert a similar salt plate in the reference beam.

Plot the spectra of Nujol and hexachlorobutadiene (or Kel F-oil) alone and mulls of the various compounds in Nujol, and then in hexachlorobutadiene (or Kel F-oil) in the region 4000–650 cm^{-1}.

Discs: The halide disc technique consists of grinding a few milligrams of the solid with approximately 500 mg of alkali halide (usually potassium bromide) and placing the mixture in a special die and compressing the powder into a small disc.

The 'mini-press' consists of two highly polished stainless steel bolts turned against each other in a rugged steel barrel which also serves as the disc holder. Insert one of the bolts into the cylinder and place about 70–100 mg of mixed halide and sample in the barrel. Insert the second bolt and tighten. Fix the lower bolt in a vice and apply pressure to the upper bolt using a torque wrench set for about 12 lb ft^{-1}. Remove bolts, leaving the disc inside the steel barrel. Place this in a suitable holder in the sample beam of the spectrometer. (The disc should exhibit a high visible clarity with no visible non-uniformity.) Mask the reference beam so that it is of the same size as the disc. Record the spectrum from 4000–650 cm^{-1}. If poor quality discs are obtained dry the KBr powder at 200 °C and store it at 110 °C or in a desiccator over P_2O_5.

Treatment of experimental data and discussion
1. With the aid of correlation charts identify and label the main absorption bands.
2. Why are liquids of high viscosity and high refractive index used for preparing mulls?
3. Is it valid to compare spectra of mulls and discs?
4. Why are better discs obtained if the press is evacuated?
5. Suggest reasons for changes in the 100% line and asymmetry of absorption bands, sometimes observed in the spectra of solids.

Experiment 5.14   Plot the infrared absorption spectra of various organic liquids using the liquid film technique

Requirements: Infrared spectrophotometer, demountable cell and fixed path length cell (0.1 mm path length) with NaCl windows, hypodermic needle and syringe, pure toluene, *o*, *m* and *p*-xylenes, mesitylene, benzaldehyde, cinnamaldehyde, acetonitrile, tert-butyl bromide, aniline and methyl benzoate.

Procedure: Calibrate the instrument as in Expt. 5.13. Prepare capillary films by sandwiching one drop of the appropriate liquid between two NaCl plates, record their spectra in the region 4000–650 cm^{-1}.

To show aromatic substitution patterns in the series toluene, *o*, *m* and *p*-xylenes, and mesitylene, introduce the liquid using the syringe and needle into the 0.1 mm cell. Ensure that all the air bubbles have been removed and stopper the holes with teflon plugs. Place the cell in the sample beam and plot the spectrum in the region 2000–1600 cm^{-1}.

Treatment of experimental data and discussion
1. Using correlation charts, identify and label the main absorption bands in the spectra.
2. Discuss the advantages of liquid phase measurements.

Experiment 5.15   Plot the infrared absorption spectra of solutions of benzene in carbon tetrachloride and hence test the validity of the Beer–Lambert law

Theory: In the study of solution spectra, it is essential to use matched solvent/ solution cells, thereby eliminating interference by solvent absorption. It is advisable to use a fixed path length cell for the solution and a variable path length cell for the solvent, rather than to use two fixed thickness cells of the same length for solvent and solution. In this way, it is possible to match the absorption of the cells completely.

Although a wide range of solvents are available in the different regions of the spectrum (p. 322) the commonest are chloroform, carbon tetrachloride and carbon disulphide. Apart from hydrogen bonding, there is little interaction when dilute solutions are prepared in non-polar solvents.

Requirements: Infrared spectrophotometer, fixed path length cell (0.1 mm) and variable path length cell with NaCl windows, hypodermic needle and syringe, pure benzene and carbon tetrachloride. Solution of benzene in carbon tetrachloride, of unknown concentration.

Procedure: Check the calibration of the instrument as indicated in Expt. 5.13. If possible check the linearity of the transmittance scale using sectored discs.

Plot the spectrum (using 0.1 mm cell) of pure CCl$_4$ and of pure benzene in the region 4000–850 cm^{-1}.

Prepare four or five solutions of benzene in carbon tetrachloride (concentrations ranging from 0.5–2.0 mol dm^{-3}). Plot the spectra of each of these solutions in the 0.1 mm cell using the variable path cell filled with carbon tetrachloride in the reference beam. Set the wavelength of the spectrophotometer so that it coincides with a strong absorption peak of the solvent but no solute absorption. Adjust the thickness of the variable path cell until 100% transmission is obtained. Plot the spectrum of the solution from 4000–850 cm^{-1}. Repeat the procedure with the other standard solutions and with the solution of unknown concentration.

Treatment of experimental data and discussion
1. Select three or four regions of maximum absorption of the solute which are

removed from the absorption peaks of the solvent. Measure the absorbance of each of the selected solute absorption bands by the 'base-line' method (*Chemical Infrared Spectroscopy*, Vol. 1., W. J. Potts, Jun., John Wiley & Sons (1963), pp. 165—70).

2. Plot absorbance against concentration for each selected band. Does the Beer—Lambert law apply (equation (5.12))?

3. From each graph determine the concentration of the unknown solution. Calculate the mean value for the concentration.

4. Why are quantitative measurements in the infrared region more difficult than those made in the ultraviolet or visible region?

5. What type of component mixtures are likely to show deviations from the Beer—Lambert law?

6. How would you accurately determine the path length of the empty fixed path length cell?

Experiment 5.16   Investigate the hydrogen bonding in cyclohexanol

Theory: Cyclohexanol forms intermolecular hydrogen bonds. Investigate the change in bonding on dilution with carbon tetrachloride.

Requirements: Infrared spectrophotometer. Liquid cell with spacers ranging from 0.1—3.0 mm, and variable path cell with NaCl windows. Pure dry cyclohexanol and pure dry carbon tetrachloride.

Procedure: Check the accuracy of the transmission scale of the spectrophotometer.
    Prepare a 1 mol dm^{-3} solution of cyclohexanol in carbon tetrachloride. Fill the 0.1 mm path cell with the solution and adjust the variable path cell to balance the solvent absorption (Expt. 5.15). Record the spectrum in the region 4000—3000 cm^{-1}. Note the presence of a broad absorption due to bonded —OH and a sharp absorption due to free —OH. Prepare a series of solutions by dilution. Record the spectra (4000—3000 cm^{-1}) of the solutions choosing path lengths such that there are a constant number of molecules of solute in the beam (ie. for 0.5 mol dm^{-3} solution use 0.2 mm cell). Ensure that you record the spectrum of one solution which shows monomer absorbance only (ie. about 0.005 mol dm^{-3}).

Treatment of experimental data and discussion
1. Calculate the extinction coefficient of the monomer.
2. Calculate the absorbance of monomer and dimer at each concentration.
3. Is the bonded —OH absorption due to dimer or are there several states of association?
4. Would you expect the sum of the absorbance of free and bonded —OH to be constant?

Experiment 5.17   Plot the infrared spectrum of simple molecules in the gas phase

Theory: High resolution infrared spectroscopy of simple molecules in the gas phase affords a powerful technique for the determination of molecular structure (bond lengths, bond angles), vibrational frequencies and force constants.

The vibrational energy (joules) of a harmonic oscillator is given by

$$E_v = \frac{h}{2\pi}\left(\frac{k}{\mu}\right)^{\frac{1}{2}}(v + \tfrac{1}{2}) \tag{5.15}$$

and the rotational energy (joules) of a rigid rotor is

$$E_J = \frac{h^2}{8\pi^2 \mu r^2} J(J + 1) \tag{5.16}$$

To a first approximation these two energies are additive. If infrared radiation is absorbed by the molecule then v can change by +1 and J by ±1. Since most molecules at room temperature have vibrational energy corresponding to $v'' = 0$,

$$\Delta E(v = 0 \rightarrow v = 1) = \frac{h}{2\pi}\left(\frac{k}{\mu}\right)^{\frac{1}{2}}$$

or

$$\Delta \bar{E} \text{ (cm}^{-1}) = \frac{1}{2\pi c}\left(\frac{k}{\mu}\right)^{\frac{1}{2}} = \bar{\nu}_0 \tag{5.17}$$

Superimposed on the vibrational transitions will be transitions between the rotational levels of the two vibrational states.

$$\Delta \bar{E}(J'' \longrightarrow J') = B'J'(J' + 1) - B''J''(J'' + 1) \tag{5.18}$$

The rotational constants B' and B'' equal $h/8\pi^2 \mu r^2 c$ using the r appropriate to the particular state.

Combining equations (5.17) and (5.18) and considering $\Delta J = \pm 1$; $\Delta J = +1$ (R branch)

$$\bar{\nu}_R = \bar{\nu}_0 + 2B' + (3B' - B'')J + (B' - B'')J^2 \tag{5.19}$$

J(Rotational quantum of the ground vibrational state) = 0, 1, 2 . . . J; $\Delta J = -1$ (P branch)

$$\bar{\nu}_P = \bar{\nu}_0 - (B' + B'')J + (B' - B'')J^2 \tag{5.20}$$

The vibration–rotation spectrum consists of two series of absorption bands (P and R branches) each side of $\bar{\nu}_0$, the components of each branch are almost equally spaced, approximately $2B$ cm^{-1} apart with a $4B$ cm^{-1} gap at the centre.

Requirements: Infrared spectrophotometer, 10 cm gas cell fitted with polished NaCl windows. Carbon monoxide or hydrogen chloride gas.

Procedure: Calibrate the instrument in the region in which you wish to record the spectra of the gases. Set the instrument to give high resolution (narrow slits) and scan at a slow speed. Make sure the instrument settings for calibration are the same as those to be used for recording the gas phase spectra.

Record the vibration–rotation spectrum of carbon monoxide in the region 2300–2000 cm^{-1} or hydrogen chloride in the region 3200–2590 cm^{-1}.

Treatment of experimental data and discussion

1. Measure the wavenumber of each component in the P and R branches. Record this value and the J value of the component (numbering from the centre gap, remember that the P branch starts with $J = 1$ and the R branch with $J = 0$).
2. Plot $\bar{\nu}_R(J) - \bar{\nu}_P(J)$ against $(2J + 1)$; the slope of this line gives B_1 (ic. B' in equation (5.18)); prove this.
3. Plot $\bar{\nu}_R(J) - \bar{\nu}_P(J + 2)$ against $(2J + 3)$; the slope of this line gives B_0 (ie. B'' in equation (5.18)); prove this.
4. Determine B_e using the relationship: $B_v = B_e - \alpha(v + \frac{1}{2})$.
5. Calculate r_1, r_0 and r_e from the calculated B_1, B_0 and B_e.
6. From the value of $\bar{\nu}_0$, calculate the force constant.
7. Why does the moment of inertia differ for the different vibrational states?
8. In the HCl spectrum, why is each component split into a doublet of unequal intensity?
9. Is it possible to calculate the intensity of the components of the P and R branches?
10. Would the accuracy be considerably improved if the anharmonic oscillator and non-rigid rotor expressions had been used?

(c) Uses of Raman spectroscopy

Raman scattering is essentially a single-beam emission experiment. A monochromatic line source is required for excitation of the Raman line. In the past a mercury arc was used with filter systems required for isolating the exciting lines (435.8 or 253.7 nm); He/Ne or Argon ion, lasers are now commonly used. A multiple monochromator is used to minimize stray light and a photomultiplier detector.

In Raman spectroscopy the lower frequency limit is largely a function of the sample; it depends on the ratio of Raman to Rayleigh and/or Tyndall scattering. Typical values are $50-100$ cm^{-1}.

The intensity of the Raman shift, though linear with concentration, is not a function of the sample alone. Intensity information is therefore not often used.

Both infrared and Raman spectroscopy are highly suitable for identification procedures by the technique of matching spectra. For very symmetric molecules both techniques are required for complete analysis. The Raman technique frequently has the marked advantage of permitting polarization measurements which usually allow immediate identification of those bands which are due to the totally symmetric modes of vibrations.

Experiment 5.18   Plot the Raman and infrared spectra of some liquids

Requirements: Raman spectrophotometer. Liquid cell. Infrared spectrophotometer, fixed path cell (0.1 mm) with sodium chloride windows. (i) Dimethyl acetylene, acetonitrile, α-chloroacetonitrile. (ii) A series of methyl-substituted aryl compounds. Solution (10%) of series (i) compounds in carbon tetrachloride or carbon disulphide.

Procedure: Record the Raman and infrared spectra of the liquids in group (i) from $200-4000$ cm^{-1}, and of the liquids in group (ii) from $1300-1400$ cm^{-1}.

Treatment of experimental data and discussion

1. Compare the spectra of the liquids in group (i) with the infrared spectra of solutions of these liquids in carbon tetrachloride or carbon disulphide.
2. Compare the intensities of the band near 1380 cm^{-1} shown by the group (ii) compounds. What type of vibration causes this Raman shift?
3. What are the advantages of using Laser Raman sources as compared with a mercury arc?
4. What determines the position and intensity of a Raman shift?
5. Why are Stokes lines more intense than anti-Stokes lines?

7   Flame photometry

Characteristic emission lines can be stimulated in many elements by excitation (usually in solution) in a suitable flame. The measurement of the intensity of such lines is the basis of flame photometry, a rapid and accurate analytical method for the determination of a wide range of elements.

The essential requirements of a flame photometer are that the flame intensity should not fluctuate, that the atomizer and sampling system should be reliable and easily cleaned and that the radiation detecting system should be stable and very sensitive. In general, the test solution is atomized and introduced into a non-luminous flame burning under carefully controlled conditions. The flame then becomes coloured and the intensity of the light emitted is measured by means of a photocell. The various regions of the spectrum appropriate to the different elements are isolated by passing the light through an optical filter or monochromator. The intensity of the light emitted with the sample is then compared with that emitted with a standard solution of known concentration burning under standard conditions.

A much greater range of elements can be estimated by a monochromator instrument than by a filter instrument; the elements which can be estimated include: Ag, Al, Au, B, Ba, Ca, Cd, Co, Cs, Cu, Fe, Ga, Hg, In, K, La, Li, Mg, Mn, Mo, Na, Ni, Pb, Pd, Ru, Sn, Sr, Ti, Tl and V. The filter instruments while limited to Ca, K, Li and Na nevertheless find extensive use for rapid, routine estimations of these elements (eg. biochemistry, agricultural chemistry, cement industry).

The accuracy obtainable depends on the element to be determined, its concentration and on the other elements in solution. Optical interference between elements can usually be avoided by a suitable choice of spectral line and slit width. The glass prism, with a greater dispersion than the silica prism, is of great use when close lines must be separated. There are interferences in the flame itself (a feature common to all instruments) and standard solutions should be prepared to correspond as nearly as possible to the test sample in respect of all elements.

Experiment 5.19   Plot calibration curves for the determination of sodium and potassium in solution using a flame photometer

Theory: The liquid under examination is sucked up by a stream of air and atomized, producing a fine mist. The mist is mixed with coal gas under carefully controlled conditions and burnt in a non-luminous flame. The intensity at the particular wavelength (isolated by filters) is then compared with that produced by standard solutions, using a photoelectric cell.

Requirements: Flame photometer, standard sodium chloride solution (127.1 mg dry sodium chloride in 500 cm³ deionized water), ie. containing 10 mg Na per 100 cm³; standard potassium chloride solution (382.0 mg dry potassium chloride in 500 cm³ deionized water), ie. containing 40 mg K per 100 cm³.

Procedure: The standard solutions above are merely stock solutions; these have to be diluted to come within the working range of the instrument. Careful dilution is necessary, particularly of high concentrations in unknown solutions. Dilutions must always be made stepwise, using standard bulb pipettes and volumetric flasks. Prepare the working substandard sodium (0.5 mg Na per 100 cm³) and potassium (0.8 mg per 100 cm³) solutions. From these substandard solutions, prepare serial dilutions for calibration of the instrument (eg. for sodium, 0.4, 0.3, 0.2, 0.1 and 0.05 mg Na per 100 cm³).

   Set up the instrument and make the necessary adjustments to air pressure and the flame. Now insert a beaker of water under the inlet tube, and adjust the galvanometer deflection to zero. Replace the water with the substandard solution, and adjust the galvanometer to 100%. Repeat these operations until no further adjustments are necessary. Now spray each of the dilutions in turn, and record the galvanometer deflection. Always spray water between the solutions, and make determinations on duplicate dilutions.

   Determine the concentration of these metal ions in tap water, or in a biological fluid, eg. blood serum or urine.

Treatment of experimental data and discussion
1.   Plot calibration curves of galvanometer deflection against concentration for each metal.
2.   Knowing the dilution of the unknown fluid, determine the concentration of each metal ion.
3.   What is the effect of the presence of other metal ions on the calibration curves?

8   Atomic Absorption

Atomic Absorption has as its basis, the tendency of a population of unexcited atoms to absorb strongly the radiation emitted by excited atoms of the same element.

   The source of radiation is a hollow cathode lamp the cathode of which incorporates the element being determined. An atomizer—burner is used to aspirate a fine mist of the sample into a flame (air/acetylene or nitrous oxide/acetylene). Most of the atoms of the sample remain in their ground state in the inner-cone and are thus available to absorb selected wavelengths of radiation from the source. One of these wavelengths is isolated with a grating monochromator. A photomultiplier tube tuned to respond to the modulated signal from the hollow cathode tube is used as detector.

The absorption due to a transition from the ground electronic state to a higher energy level is virtually an absolute measure of the number of atoms in the flame and hence the concentration of the element in a sample.

Experiment 5.20   Estimate the nickel content of iron and steel

Requirements: Atomic Absorption spectrometer, orthophosphoric acid ($1.75 \ \text{g cm}^{-3}$), sulphuric acid ($1.84 \ \text{g cm}^{-3}$), nitric acid ($1.42 \ \text{g cm}^{-3}$). Ni solution special for atomic absorption (or pure sample of Ni). Standard steel sample containing (0.5—5% Ni).

Procedure: Prepare an acid solution containing 15% (v/v) orthophosphoric acid and 15% (v/v) sulphuric acid.

Prepare four nickel standard solutions containing, 50, 40, 30 and 20 $\mu g \ \text{cm}^{-3}$ of Ni respectively. Each solution must contain 4.5% (v/v) orthophosphoric acid and 4.5% (v/v) sulphuric acid.

To prepare the sample for analysis, accurately weigh and transfer about 0.1 g of iron or steel sample to a 125 cm^3 beaker. Add 30 cm^3 of the acid solution and heat on a hot plate. When the acid simmers, add nitric acid dropwise. (In some cases some aqua regia may be required for dissolution.) Evaporate to fumes and fume gently for 1 min. Extract the cooled residue with 30 cm^3 of water and digest for 5 min to ensure dissolution of all soluble salts. Filter if necessary through a Whatman 541 filter paper into a 100 cm^3 volumetric flask. Rinse the beaker and filter paper several times with hot 2% (v/v) sulphuric acid. Cool and make up to the mark.

Set up the instrument as described in the manual. Use the 341.5 nm line for analysis. Set the zero using distilled water. Aspirate one of the standard solutions into the flame and note the absorbance. Aspirate distilled water until zero is regained. Repeat the procedure with the standard solutions and finally aspirate the unknown solution.

Treatment of experimental data and discussion

1. Plot a calibration curve of absorption against concentration for the standard nickel solutions and hence calculate the nickel content of the original sample.

2. Discuss other methods for the analysis of nickel, indicating clearly advantages and disadvantages.

3. What are the main limitations of Atomic Absorption analysis?

9   Nuclear magnetic resonance

Nuclei possessing a non-zero spin exhibit nuclear magnetic resonance.

Nuclei with an odd mass number have half integral spin, eg. ^{1}H, ^{15}N, ^{19}F.

Nuclei with an even mass number and an even charge number have zero spin eg. ^{12}C and ^{16}O.

Nuclei with an even mass number and an odd charge number have integral spin eg. ^{2}H, ^{14}N.

For a nucleus having nuclear spin quantum number I_z there are $(2I_z + 1)$ spin states. These will have the same energy unless the nucleus is placed in a strong magnetic field when the energies will be different.

^{1}H has two spin states $+\frac{1}{2}$ and $-\frac{1}{2}$ which have different energies (ΔE) when placed in an applied magnetic field of B tesla.

$$\Delta E = h\nu = \frac{\gamma h}{2\pi} B \tag{5.21}$$

For a proton, resonance is observed at about 60 MHz in a magnetic field of 1.5 T.

Chemical Shift: The exact frequency of the absorption depends on the chemical environment of the nucleus. Differences in resonance conditions are referred to as chemical shifts. Therefore the resonance condition becomes

$$\nu_j = \left| \frac{\gamma}{2\pi} \right| B_0 (1 - \sigma_j) \tag{5.22}$$

ν_j and σ_j refer to the resonance frequency and shielding constant for nucleus j.

Chemical shifts are usually quoted relative to an internal or external standard; tetramethylsilane (TMS) is usually used for proton magnetic resonance spectroscopy. Chemical shifts in ppm relative to TMS are based on the tau (τ) or delta (δ) scale

δ scale TMS at 0.00, τ scale TMS at 10.00.

Integration: The signal areas are a direct measure of the amount of absorption (number of nuclei) provided the spectrometer controls are not altered.

Spin-spin coupling: When there are two or more non-equivalent magnetic nuclei in a molecule then interaction of spins may take place, causing a splitting of some of the resonance bands. Coupling constants like chemical shift depend on chemical environment and are therefore of great importance in structure determination: Coupling constants are independent of operating frequency and are therefore expressed in Hz.

Experiment 5.21   Measure the nuclear magnetic resonance spectra of a number of organic molecules

Requirements: 60 MHz nmr. spectrometer. Carbon tetrachloride, tetramethylsilane, formaldehyde, acetone, methyl ethyl ether, toluene, acetaldehyde, 1,1,2 trichloroethane, N,N dimethyl formamide, ethyl acetate.

Procedure: Prepare 10 cm^3 of 7% solution of the above compounds in CCl$_4$, add a drop of TMS to each solution.

Follow the instrument instructions and record the nmr. spectrum of each solution. If possible measure the integral of each spectrum.

Treatment of experimental data and discussion
1.   With the aid of correlation tables assign the bands observed in the spectra.
2.   How could you confirm the assignment?
3.   Explain the intensity distribution within a multiplet.
4.   Calculate the energy separation between the two states of the hydrogen nucleus in a 1.4 T field. Calculate the ratio of the number of nuclei in the upper state to the number in the lower energy state at 300 K.

Experiment 5.22   Investigate the nmr. spectrum of ethanol

Requirements: 60 MHz nmr. spectrometer. Absolute ethanol, dried absolute ethanol (distilled from freshly ignited quicklime), dilute hydrochloric acid.

Procedure: Record the nmr. spectra of absolute ethanol, dried absolute ethanol, a mixture of 25 cm^3 absolute ethanol and 5 cm^3 dilute HCl, a mixture of 25 cm^3 absolute ethanol and 5 cm^3 water.

Treatment of experimental data and discussion
1.   Determine the chemical shifts relative to the methyl protons for each sample.
2.   Measure the coupling constants.
3.   Explain the changes in the spectrum of ethanol on adding acid and water.
4.   How could you measure the rate of exchange of the hydroxyl protons?

Experiment 5.23   Determine the concentration of sodium trichloroethylphosphate in a pharmaceutical formulation, from the nmr. spectrum

Theory: The measurement of the relative areas of signals in proton nmr. spectra may be used as a quantitive analytical method. The intensity of each signal (the integral) is proportional to the number of protons giving rise to it and independent of the chemical nature of the proton.

$$\text{Integral} = \text{Number of protons} \times \text{concentration} \qquad (5.23)$$

If one signal from a component can be identified in the spectrum of a mixture, that compound can be assayed without being required pure as a standard. The area of the signal is measured relative to that of a reference compound added in known concentration.

Requirements: 60 MHz nmr. spectrometer. Sedative 'triclofos', acetonitrile, D$_2$O (other mixtures can be chosen so long as the reference and sample signals are well separated).

Procedure: Dilute the triclofos syrup with D$_2$O so as to obtain a well resolved spectrum. The signal of the CH$_2$ group of CCl$_3$ CH$_2$ OPO$_3$ HNa appears 0.4 ppm to high field of the water signal. Add a known amount of acetonitrile to the triclofos and then dilute with D_2O, record the nmr. spectrum (acetonitrile signal about 3 ppm to high field of the water signal). Scale expand so as to obtain maximum accuracy in the integral. Record the integrals of the CH_2 and acetonitrile signals. Repeat the integration about five times.

Treatment of experimental data and discussion

1.   Measure the integrated area of the CH_2 peaks of the triclofos and the CH_3 peak of acetonitrile from the recorded spectrum integrals; take a mean of your readings. From the known concentration of the acetonitrile and the ratios of the integrals of the two components use equation (5.23) to calculate the concentration of triclofos.
2.   Why is the CH_2 signal a doublet?
3.   Discuss means of optimising the accuracy of quantitative nmr. measurements.

10   Electron spin resonance

Substances containing one or more unpaired electrons are paramagnetic and may be studied by electron spin resonance spectroscopy.

The spin energy levels of an electron are separated in an applied magnetic field, B, by an amount:

$$\Delta E = h\nu = g\beta B \qquad (5.24)$$

where g = Landé splitting factor and β = Bohr magneton.

For a free electron with g = 2.0023, resonance occurs at 8 388.225 MHz in a field of 0.3 tesla.

Multiplet structure of esr. absorptions arise for two different reasons. Fine structure is due to the presence of more than one electron of unpaired spin. Hyperfine structure arises through coupling of the spins of the unpaired electron with a neighbouring nuclear spin. In general n nuclei with spin I will split the resonance line of an electron into a multiplet with $(2In + 1)$ components.

Experiment 5.24   Investigate the esr. spectra of some organic free radicals

Requirements: esr. spectrometer, cells suitable for organic liquids and for aqueous solutions. Triphenylmethylchloride, benzene, hydroquinone, methylhydroquinone, 50% aqueous ethanol, 0.05 mol dm^{-3} NaOH. Nitrogen cylinder. Zn dust.

Procedure: Dissolve about 20 mg of triphenylmethylchloride in 5 cm^3 of air-free benzene under an atmosphere of nitrogen, add Zn dust. Decant the liquid into an esr. tube. Record the esr. spectrum. Dissolve 20 mg of hydroquinone in 50% aqueous ethanol, add a drop of 0.05 mol dm^{-3} NaOH. Shake the solution and transfer it to an esr. aqueous cell. Record the esr. spectrum. Repeat the procedure using methylhydroquinone instead of hydroquinone.

Treatment of experimental data and discussion

1.   From the recorded spectra deduce the structure of the radicals.
2.   What factors influence the magnitude of the electron/nucleus coupling?

3. Why are the coupling constants observed in esr. spectra about 10^6 larger than nuclear coupling constants?

4. When an electron couples with a proton, four possible energy levels are involved but the esr. spectrum only shows two absorption bands. Explain this observation.

5. Calculate the relative population of the two spin states of a free electron.

6

Methods of titration

A titration is performed to determine the chemical equivalence of one reagent for another in a well characterized reaction, by observing the change in some property of the solution. The equilibrium of the reaction must lie far toward completion so that the equivalence point will be accompanied by a large and sudden change in the concentration of one of the reactants. When this change has proceeded sufficiently far to actuate the means of observation, the 'end-point' is obtained.

At the stoichiometric or equivalence point the reaction has occurred according to a known stoichiometric equation and neither reagent is present in excess. The concentration of the unknown solution may be calculated from the volumes of the reacting solutions and the concentration of the known solution using the equation for the reaction.

On account of the possible variability of the equivalent weight of a substance in different reactions, the concepts of equivalent weight and normality have been abandoned in favour of the mole concept. According to this, a solution containing 1 g molecular (formula) weight per dm^3 of solution is uniquely defined irrespective of the reaction in which it is participating. The stoichiometric equation of the reaction is the basis for all calculations. Thus for the general reaction:

$$aA + bB \longrightarrow \text{Products}$$

if V_A cm^3 of solution A of concentration c_A mol dm^{-3} are equivalent to V_B cm^3 of solution B of concentration c_B mol dm^{-3} then:

$$\frac{\text{No. of moles of A}}{\text{No. of moles of B}} = \frac{c_A V_A/1000}{c_B V_B/1000} = \frac{a}{b}$$

or

$$b(c_A V_A) = a(c_B V_B) \tag{6.1}$$

As an example, during the standardization of a solution of iron (II) ammonium sulphate, 25.00 cm^3 were equivalent to 32.45 cm^3 of a solution of potassium dichromate of concentration 0.015 mol dm^{-3}. The equation for this reaction is:

$$6Fe^{++} + Cr_2O_7^{--} + 14H_3O^+ \longrightarrow 6Fe^{+++} + 2Cr^{+++} + 21H_2O$$

Thus from equation (6.1):

$$1 \times 25.00 \times c_{Fe^{++}} = 6 \times 32.45 \times 0.015$$

or

$$c_{Fe^{++}} - 0.1169 \text{ mol dm}^{-3}$$

The equivalence point is often located visually with chemical indicators which change colour with change of pH or oxidation potential. To overcome some of the limitations of visual indicators, such as in highly coloured solutions, instrumental methods have been developed to locate the end-point. Potentiometric, conducto-metric and amperometric titrations are now extensively used. In addition, these methods are attractive in extending the advantages of titrimetric analysis to titra-tions generally accepted as not feasible because of the values of their equilibrium constants. An instrumental end-point need not be sharp to be useful. Other proper-ties of solutions, which may be used to follow the progress of a titration, include the absorbance at a fixed wavelength: spectrophotometric titrations, and the enthalpy: thermal titrations.

A further type is the coulometric titration, so named because a definite amount of one of the reagents is generated in the solution electrically, this amount being determined from the number of coulombs. The equivalence point is detected by any of the usual ways (potentiometric, conductometric, amperometric or with indicators). It thus differs from other instrumental methods in that these refer to the different ways of determining an end-point, whereas coulometric, like volu-metric, refers to the method of measuring the quantity of reagent.

A general feature of all titration methods, which should be stressed, is that they are unaffected by many changes in experimental conditions which would seriously disturb analyses by the corresponding direct methods. Titration methods are inherently more accurate than the corresponding direct method, since many observations are used to locate the end-point.

1 Titrations using indicators

(a) Acid-base titrations

The accepted definition of the pH value of a solution is:

$$\text{pH} = -\log a_{H^+} \tag{6.2}$$

or

$$a_{H^+} = 10^{-\text{pH}}$$

where a_{H^+} or $a_{H_3O^+}$ is the activity of the hydrogen ion in solution.

Hydrogen ion activities, however, except in strongly acid solutions, are nearly equal to the hydrogen ion concentration, and pH is generally expressed in the nota-tion first suggested by Sørensen:

$$\text{pH} = -\log c_{H^+} \tag{6.3}$$

or

$$c_{H^+} = 10^{-\text{pH}}$$

At 25 °C, in dilute aqueous solution, the product of the activities of the hydrogen and hydroxyl ions is 10^{-14}|($K_w = a_{H^+}a_{OH^-} = 10^{-14}$ mol^2 dm^{-6}). At the neutral point, $c_{H^+} = c_{OH^-} = 10^{-7}$ mol dm^{-3}, thus, the pH of a neutral solution is 7.0. In acid solutions $c_{H^+} > 10^{-7}$ mol dm^{-3} and pH < 7, whilst in alkaline solutions $c_{H^+} < 10^{-7}$ mol dm^{-3} and pH > 7. The pH range is 0 (for 1 mol dm^{-3} solution of a strong acid) to 14 for 1 mol dm^{-3} solution of a strong alkali. For a weak acid such as acetic acid ($K_a = 1.75 \times 10^{-5}$ mol dm^{-3} the hydrogen ion concentration is a function of both K_a and the acid concentration [$c_{H^+} = \sqrt{(K_a c)}$], hence, the pH of 0.1 mol dm^{-3} acetic acid is 2.88.

The pH of the exact equivalence point in acid-base titrations depends on the relative strengths of the acid and base. For strong acid–strong base, it is at pH 7.0, while it is on the acid or alkaline side of neutrality if the base or acid respectively is weak, owing to the hydrolysis of the salt (equations (6.4), (6.5)).

$$pH = \tfrac{1}{2}pK_w - \tfrac{1}{2}pK_b - \tfrac{1}{2}\log c \qquad (6.4)$$

$$pH = \tfrac{1}{2}pK_w + \tfrac{1}{2}pK_a + \tfrac{1}{2}\log c \qquad (6.5)$$

For weak acid–weak base, the pH of equivalence is given by

$$pH = \tfrac{1}{2}pK_w + \tfrac{1}{2}pK_a - \tfrac{1}{2}pK_b \qquad (6.6)$$

where K_a and K_b are the acidic and basic dissociation constants of the weak acid and weak base respectively, K_w, the ionization constant of water and c/mol dm^{-3} the concentration of salt.

Although indicators are now only used for approximate measurements, a brief account of their use is relevant. Acid-base indicators are weak acids or bases which exhibit different colours depending on their ionic state:

$$HIn_A + H_2O \rightleftharpoons In_B^- + H_3O^+$$

Acidic form   Basic form
Colour A   Colour B

The dissociation (or indicator) constant, K, using concentrations, is given by:

$$K = \frac{c_{In_B^-}\, c_{H^+}}{c_{HIn_A}} \qquad (6.7)$$

whence

$$pH = pK + \log \frac{c_{In_B^-}}{c_{HIn_A}} \qquad (6.8a)$$

$$= pK + \log \frac{\text{Intensity B}}{\text{Intensity A}} \qquad (6.8b)$$

assuming that the concentrations of the two forms are proportional to the intensities of light transmitted by solutions of colours A and B (Intensity A and Intensity B respectively). Thus, when the indicator is added to a solution of given pH, the equilibrium adjusts itself so that equation (6.8a) is obeyed. Since the two forms have different colours, the colour of the indicator is therefore adjusted. Thus, from the ratio of the intensities of the two colours the pH of the solution may be determined, provided K is known (p. 314).

The useful range of an indicator is limited to:

$$pH = pK_a \pm 1 \qquad (6.9)$$

(b) Precipitation titrations

These are generally limited to those employing silver nitrate as one reagent. Three different types of indicator may be used; those resulting in:

(i)   the formation of coloured precipitate, eg. the use of potassium chromate in the titration of a neutral solution of a chloride, with the formation of the red sparingly soluble silver chromate (solubility product 1.7×10^{-12} mol^2 dm^{-6}) at the end-point (solubility product of silver chloride 1.2×10^{-10} mol^2 dm^{-6});

(ii) the formation of a soluble coloured compound, eg. the Volhard titration of silver in the presence of free nitric acid with potassium thiocyanate using iron (III) ammonium sulphate as indicator;

(iii) the formation of a coloured precipitate using an adsorption indicator; at the equivalence point the indicator, eg. fluorescein or eosin, is adsorbed on the precipitate, and, at the same time the colour changes.

(c) Oxidation-reduction titrations

In this type of titration, the end-point can be determined either by a change in the colour of the reagent at the end-point (eg. permanganate titrations), or by using external or internal indicators. The ideal oxidation-reduction indicator (p. 314) is that with oxidation potential midway between that of the two reagents in the titration and which exhibits a sharp colour change. These indicators, generally organic dyestuffs, have different colours in the oxidized and reduced forms. Like acid-base indicators, their range is limited, in this case to ± 0.05 V from $E^{\ominus}$. Diphenylamine, in the presence of phosphoric acid to complex the Fe^{+++} ions, may be used in the titration of a iron (II) salt with potassium dichromate.

(d) Complexometric titrations

The introduction of chelons (the general name for a class of reagents including the polyaminocarboxylic acids and polyamines) which form stable, soluble and usually 1:1 complexes with metal ions, has opened up a new approach to metal ion analysis. The most common reagent is the disodium salt of ethylenediaminetetr-aacetic acid (EDTA), which can be represented by $Na_2H_2Y \cdot 2H_2O$. This forms complexes with practically all the metals with the exception of the alkali metals, eg.

$$H_2Y^{--} + Ca^{++} \longrightarrow CaY^{--} + 2H^+$$

$$H_2Y^{--} + Bi^{+++} \longrightarrow BiY^- + 2H^+$$

Titration of the liberated hydrogen ions with a standard base suffers from many disadvantages and is, in consequence, seldom used. The most popular approach consists of adding a trace quantity of a complexing dye, 'metallochromic indicator', which exhibits different colours in the 'metallized' and 'unmetallized' forms.

$$M^{++} + HIn^{(m-1)} \longrightarrow MIn^{(m+1)} + H^+$$

eg. blue   eg. red

During the major part of a direct titration, the indicator exists entirely in the metallized form, only the free metal ion remaining in solution being titrated with the chelate. At the equivalence point, the chelate removes the metal ion from the metal-dye complex (lower stability constant than that of the metal chelate) with a resulting colour change. The metallochromic indicator is a multidentate chelating agent containing for example polyaminocarboxylate, hydroxy, basic nitrogen and/or azo ligands.

The following conditions are required for such an indicator:

1. large molar extinction coefficient in the visible region, making possible their use in very dilute solutions;
2. the stability constant of the metal-chelate (p. 315) must be greater than that of the metal indicator (p. 315);
3. the indicator must be sufficiently stable to exist as the metallized complex in extremely dilute solutions.

Few substances fulfil all these conditions. Thus, a given indicator is useful only for a limited number of metals under specified conditions of pH and buffer (type and concentration). The position is still further complicated by the acid-base properties of the indicator itself and by the fact that many are blocked by traces of metal ions, eg. copper, cobalt, nickel, iron.

2   Potentiometric titrations

The variation of the potential of an electrode in equilibrium with its ions, with the concentration of the ions, may be used as an indicator in volumetric analysis. This method is applicable to a wide range of titrations, including those on a microscale, provided that an appropriate electrode is available. It can be used in the titration of highly coloured solutions where indicators are useless. The precise location of the end-point in a potentiometric titration is obtained from a series of independent observations, rather than from one estimate of the end-point, as in the indicator method.

(a) Indicator electrodes

The electrode potential, $E(M^+, M)$, existing between a metal and a solution of one of its salts:

$$M^{n+} + n\epsilon \rightleftharpoons M$$

is given by:

$$E(M^{n+}, M) = E^{\ominus}(M^{n+}, M) + \frac{RT}{n\mathscr{F}} \ln a_+ \qquad (6.10)$$

where $E^{\ominus}(M^{n^+}, M)$ is the standard electrode potential, n the valence of the ions and a_+ their activity. For most purposes in quantitative analysis, the activity can be replaced by c_+, the ionic concentration; thus:

$$E(M^{n+}, M) = E^{\circ}(M^{n+}, M) + \frac{RT}{n\mathscr{F}} \ln c_+ \tag{6.11}$$

For an electrode in equilibrium with anions:

$$E(A, A^{n-}) = E^{\ominus}(A, A^{n-}) - \frac{RT}{n\mathscr{F}} \ln c_- \tag{6.12}$$

The electrode potential of the oxidation-reduction system (using platinum as the electrode):

Oxidized form $+ n\epsilon \rightleftharpoons$ Reduced form

given by:
$$E(O, R) = E'(O, R) + \frac{RT}{n\mathscr{F}} \ln \frac{c_{ox}}{c_{red}} \tag{6.13}$$

depends on the concentrations of both the oxidized and reduced forms in solution. E' known as the 'formal' reduction potential, is the electrode potential when $c_{ox} = c_{red}$. It is not a true constant, but is effectively constant for a given ionic strength, such as is used in a redox titration.

The main essential of any potentiometric titration is the determination of E since this reflects changes in either c_{M^+}, c_{A^-} or c_{ox}/c_{red}. As it is impossible to determine E on its own, the indicating electrode must be used in conjunction with a reference electrode, the potential of which does not change during the course of a titration.

(b) Reference electrodes

(i) *Calomel electrode:* This reference electrode consists essentially of mercury, mercury(I)chloride and potassium chloride solution of specified concentration, ie.

$$Hg \cdot Hg_2Cl_2 | KCl$$

The electrode potential is given by:

$$E(cal) = E^{\ominus}(cal) + \frac{RT}{2\mathscr{F}} \ln K_s - \frac{RT}{\mathscr{F}} \ln a_{Cl^-} \qquad (K_s = a_{Hg_2^{+}} \cdot a_{Cl^-}^2)$$

$$= E' - \frac{RT}{\mathscr{F}} \ln a_{Cl^-} \tag{6.14}$$

Thus the electrode behaves like a reversible chlorine electrode, the standard potential $E'(=E^{\ominus}(cal) + (RT/2\mathscr{F}) \ln K_s)$ depends on the concentration of the potassium chloride solution and on the temperature.

0.1 mol dm^{-3} KCl	$E' = 0.3335 - 0.0007(t - 25)$
1.0 mol dm^{-3} KCl	$E' = 0.2810 - 0.00024(t - 25)$
Saturated KCl	$E' = 0.2420 - 0.00076(t - 25)$

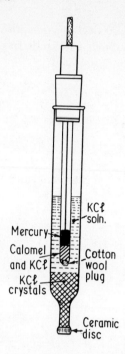

Fig. 6.1   Calomel electrode

The 0.1 mol dm^{-3} electrode is preferred for accurate work as it has the lowest temperature coefficient; the saturated electrode is the most convenient owing to the ease of replacing the solution.

Various types of calomel electrode are available commercially (Fig. 6.1), while others can be easily prepared in the laboratory using a salt bridge to join the half cell to the solution under test. In the commercial electrodes, the liquid junction is generally made by leakage of potassium chloride through a ceramic disc.

(ii) *Silver–silver chloride electrode:* This electrode consists of a strip or disc of silver on which is deposited a film of silver chloride. It behaves as a reversible chlorine electrode with a potential given by:

$$E(\text{AgCl}, \text{Cl}^-) = E^{\ominus}(\text{AgCl}, \text{Cl}^-) - \frac{RT}{\mathscr{F}} \ln a_{\text{Cl}^-} \qquad (6.15)$$

$$E^{\ominus}(\text{AgCl}, \text{Cl}^-) = 0.22239 - 645.52 \times 10^{-6}(t - 25) - 3.284 \times 10^{-6}(t - 25)^2 +$$
$$+ 9.948 \times 10^{-9}(t - 25)^3.$$

(c) Salt bridges

Salt bridges are used to connect the reference electrode with the test solution. The salt bridge (Fig. 6.2) consists of tubing plugged at the ends either with sintered discs of fine porosity or tightly rolled spirals of filter paper or asbestos. The tube is filled with the required bridge solution and the bridge stored with the legs dipping in vessels containing the same solution. Bridges of this type have the advantage over agar bridges since they last indefinitely and do not suffer from syneresis of the gel. Where contamination of the test solution with the bridge solution must be kept to a minimum an agar bridge must be used. (p. 307).

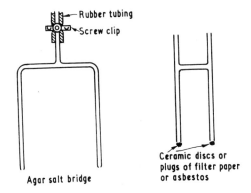

Fig. 6.2   Salt bridge   Agar salt bridge

(d) Cells

An electrochemical cell is prepared from a reference electrode and an indicating electrode, dipping into a solution containing an ion to which it is reversible, either directly or through a salt bridge. Depending on the nature of the electrodes, they are connected to a potentiometer, a digital voltmeter or a pH meter. Thus for an acid-base titration a typical cell would be:

$$\ominus \quad Pt \cdot H_2 | H^+ \text{ Satd. } KCl | Hg_2Cl_2 | Hg \quad \oplus$$

The emf. of such a cell at 25 °C given by:

$$E = E(\text{cal}) - E(H^+, H_2) = E' + 0.0591 \text{ pH}$$

where E' is a constant, gives a measure of the pH of the solution. Change of pH is reflected in a change of E. This principle applies to all potentiometric titrations, the general form of the titration curve being the same for all.

Figure 6.3*a* shows a typical potentiometric titration curve. The initial addition of a small amount of one standard reagent to another produces very little change in the emf. of the cell, since this depends on the fraction of a particular ion removed. Towards the equivalence point, however, the fraction of the ion removed by a constant amount of the standard reagent increases rapidly; this is reflected in a rapid change of the emf. Above the equivalence point, the curve again flattens out. The exact location of the equivalence point is not always so readily apparent. It can be precisely determined from the differential plot in which $\Delta E/\Delta V$ or $\Delta \text{pH}/\Delta V$ is plotted against V (Fig. 6.3*b*). From the original graph (Fig. 6.3*a*), the change of E or pH resulting from successive uniform increments (eg. 0.2 or 0.4 cm^3) is plotted against the mid-volume of the titration interval. The abscissa, corresponding to the maximum, is the equivalence point.

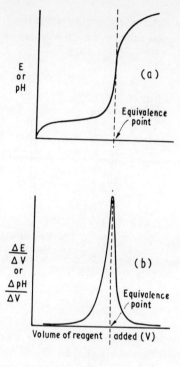

Fig. 6.3   *(a)* Potentiometric titration curve. *(b)* Differential titration curve

(a) Acid-base titrations

Experiment 6.1   Titrate sodium hydroxide with hydrochloric acid using the hydrogen electrode

Theory: The hydrogen electrode, in which hydrogen gas is in equilibrium with hydrogen ions in solution at a platinum black electrode, behaves like a metallic electrode:

$$H^+ + \epsilon \longrightarrow \tfrac{1}{2}H_2(g)$$

for which:

$$E(H^+, H_2) = E^{\ominus}(H^+, H_2) + \frac{RT}{\mathscr{F}} \ln \frac{a_{H^+}}{p_{H_2}^{1/2}} \tag{6.16}$$

It is the recognized convention to take the standard hydrogen electrode, in which $p_{H_2} = 1$ atmosphere (760 mmHg) and $a_{H^+} = 1$, as the arbitrary zero of electrode potential. Thus $E^{\ominus}(H^+, H_2) = 0$ and hence:

$$E(H^+, H_2) = \frac{RT}{\mathscr{F}} \ln \frac{a_{H^+}}{p_{H_2}^{1/2}} \tag{6.17}$$

Thus for the cell:

$$\ominus \ \ Pt . H_2(g) \mid HCl \ \vdots \ Satd. \ KCl \mid Hg_2Cl_2 . Hg \ \ \oplus$$

$$E = E(cal) - E(H^+, H_2) = E(cal) - \frac{RT}{\mathscr{F}} \ln a_{H^+}$$

$$= E(cal) + \frac{2.303RT}{\mathscr{F}} \, pH \qquad\qquad (6.18)$$

when p_{H_2} = 1 atmos. Rearranging equation (6.18) (T = 298 K) gives:

$$pH = \frac{E - E(cal)}{0.0591} \qquad\qquad (6.19)$$

Requirements: Potentiometer, galvanometer, hydrogen (Fig. 6.4) and calomel elec-trodes, supply of pure hydrogen, 0.1 mol dm^{-3} solutions of sodium hydroxide and hydrochloric acid.

Procedure: First titrate, in duplicate, 25 cm^3 of the hydrochloric acid with sodium hydroxide solution using the correct indicator. It is now possible to select the correct distribution of readings (ie. volumes of sodium hydroxide at which the emf. should be determined) during the potentiometric titration, particularly in the region of the end-point. Set up the above cell in a beaker containing 25 cm^3 of hydrochloric acid.

The hydrogen, from a cylinder, must be passed through alkaline pyrogallol (2 g pyrogallol in 40 cm^3 of 16% w/v NaOH), dilute sulphuric acid and water before passing to the electrode and solution. The calomel electrode may dip directly into the solution, or be joined by a bridge to a beaker containing saturated potassium

Fig. 6.4   Hydrogen electrode

chloride solution. Mechanical stirring is unnecessary. Connect the electrodes to a standardized potentiometer or digital voltmeter. Bubble a steady stream of hydrogen through the solution until the cell assumes a steady emf. This may take 10–15 min, thereafter the emf. is independent of the rate of bubbling. Run in the sodium hydroxide solution, initially in large amounts, gradually decreasing the amount towards the end-point, and determine the emf. after each addition.

Treatment of experimental data and discussion

1. Tabulate the emf. of the cell after each addition of NaOH and calculate the corresponding pH of the solution (equation (6.19)). Plot the potentiometric (E or pH) and differential titration curves (Fig. 6.3a, b) and determine the end-point.
2. List the advantages and disadvantages of the hydrogen electrode.
3. Why is it necessary to pretreat the hydrogen gas?
4. Explain why it takes so long before a steady emf. can be obtained for the cell.
5. From the known volume of acid and added alkali calculate the pH of each solution and hence plot the theoretical titration curve.

Experiment 6.2   Determine titration curves for acids and bases using the glass electrode assembly

Theory: The shape of the potentiometric titration curves depends on the relative strengths of acid and base. The acid dissociation constant of a weak acid can be determined from the titration curve of the acid with a strong base. Thus for the weak acid:

$$HA + H_2O \rightleftharpoons H_3O^+ + A^-$$

$$K_a = \frac{a_{H_3O^+} a_{A^-}}{a_{HA}} \tag{6.20}$$

which on rearrangement becomes:

$$pH = pK_a + \log \frac{a_{A^-}}{a_{HA}} \tag{6.21}$$

Expansion of equation (6.21) gives:

$$pH = pK_a + \log \frac{c_{A^-}}{c_{HA}} + \log \gamma_{A^-} \tag{6.22}$$

assuming the activity coefficient of the undissociated acid is unity. The activity coefficient of A^-, is given by the Debye–Hückel relationship:

$$\log \gamma_{A^-} = \frac{-0.505 z_{A^-}^2 \sqrt{(I/\text{mol dm}^{-3})}}{1 + 3.3 \times 10^7 \overset{\circ}{a} \sqrt{(I/\text{mol dm}^{-3})}} \tag{6.23a}$$

where z_{A^-} is the valence of the ion, and $\overset{\circ}{a}$ is the average 'effective diameter' of the A^- ions in solution (p. 315). For very dilute solutions, $\sqrt{I} < 0.1$ mol dm^{-3}, the second term of the denominator is negligible, the equation becomes:

$$\log \gamma_{A^-} = -0.505 z_{A^-}^2 \sqrt{I} \tag{6.23b}$$

In dilute solution, assuming the activity coefficients are unity, equation (6.21) becomes:

$$K = \frac{c_{H_3O^+} c_{A^-}}{c_{HA}}$$

(6.24)

or

$$pH = pK + \log \frac{c_{A^-}}{c_{HA}}$$

(6.25)

where c_{A^-} is the concentration of the salt and c_{HA} is the concentration of free acid. During a titration, the concentration of the salt is proportional to the volume of added base, b, while the concentration of free acid is proportional to $(a - b)$, where a is the volume of base required for complete neutralization. Thus equation (6.25) becomes:

$$pH = pK + \log \frac{b}{(a - b)}$$

(6.26)

This experiment requires the use of a glass electrode.

The emf. of the cell:

$$\underbrace{\text{Ag . AgCl} \mid 0.1 \text{ mol dm}^{-3} \text{ HCl} \mid \text{Glass}}_{\text{Glass electrode}} \mid \text{Solution} \mid \underset{\text{(Satd.)}}{\text{KCl}} \mid \text{Hg}_2\text{Cl}_2 \text{ . Hg}$$

varies with the hydrogen ion concentration of the test solution. The glass electrode may therefore be used for the determination of pH; its electrode potential may be represented by the equation:

$$E_G = E'_G + \frac{RT}{\mathscr{F}} \ln a_{H^+}$$

(6.27)

A glass electrode consists of a bulb of special glass (72% SiO_2, 8% CaO, 20% Na_2O) blown on the end of ordinary glass tubing (Fig. 6.5). The bulb, when supplied, contains 0.1 mol dm^{-3} hydrochloric acid with a silver–silver chloride electrode dipping in it. The bulb must be soaked in 0.1 mol dm^{-3} hydrochloric acid for 24 hours before use. These electrodes should not be used in solutions of pH > 11 as the electrode tends to behave like a sodium electrode. Glass electrodes must never be wiped or allowed to dry; they should be stored in distilled water.

The glass electrode, with a resistance of 10^7 to 10^8 ohms, cannot be used with simple potentiometer circuits; special potentiometers with a valve circuit, ie. pH meters, are necessary. In use, the glass and its reference electrode must be calibrated using solutions of known pH values (p. 309) to eliminate asymmetry potentials in the glass.

Requirements: pH meter, glass and calomel electrodes, standard 0.1 mol dm^{-3} sodium hydroxide solutions, 0.1 mol dm^{-3} hydrochloric acid, 0.1 mol dm^{-3} acetic acid and 0.05 mol dm^{-3} oxalic acid.

Procedure: First titrate 25 cm^3 of each acid with the sodium hydroxide solution

HCl—

Fig. 6.5   Glass electrode   Ag.AgCl—

using the correct indicator. It is now possible to select the correct distribution of readings during the potentiometric titrations.

Set up and standardize the pH meter as described in the manual. Pipette 25 cm³ of the acid into a 250-cm³ beaker and add about 50 cm³ of water. Immerse the electrodes and determine the pH of the solution. Run in the sodium hydroxide solution, from a burette, stir well (preferably with a mechanical stirrer) and determine the pH after each addition. At first, run in 2 cm³ at a time; towards the equivalence point, add smaller and smaller amounts (about 0.05–0.1 cm³ at the end). Continue the titration now adding larger volumes of alkali until the pH ceases to rise markedly.

Treatment of experimental data and discussion

1. From the tabulated pH values after the addition of different amounts of NaOH, plot the potentiometric and differential titration curves (as Fig. 6.3) and locate each equivalence point. Compare the curves for the three acids and discuss their shapes in terms of the relative strengths of the acids.

2. From the titration curve for acetic acid (and also oxalic acid) determine a and b at known pH values and calculate and tabulate values of $b/(a - b)$ and $\log b/(a - b)$. Plot the graph of pH against $\log b/(a - b)$ and determine the intercept; this is the pK of the acid at this concentration (equation (6.25)). Calculate the ionic strength (I/mol dm⁻³) for the solution at the point where $c_{A^-} = c_{HA}$, and hence using equations (6.22) and (6.23b) obtain an estimate of K_a.

3. List the relative advantages and disadvantages of the glass electrode.

4. Explain the meaning of the terms 'buffer solution' and 'buffer action' and indicate on the potentiometric titration curves where the solution may be

said to be a buffer solution. Show how equation (6.25) may be used in the preparation of a buffer solution. Why is the useful range of a buffer solution limited to pH = pK ± 1?

5. Discuss the relative advantages and disadvantages of the potentiometric and conductometric titrations in the determination of the equivalence point during the titration of (a) a strong acid with a strong base, (b) a weak acid with a strong base and (c) a weak acid with a weak base.

Experiment 6.3 **Titrate acetic acid with sodium hydroxide potentiometrically using the quinhydrone electrode**

Theory: In the cell:

$\ominus$ Hg|Hg$_2$Cl$_2$|Satd. KCl$\vdots$Acetic acid, Quinhydrone|Pt $\oplus$

the quinhydrone electrode is used as a H$^+$ indicating electrode, while the calomel electrode is the reference electrode.

The quinhydrone electrode consists essentially of a shiny platinum electrode dipping in the test solution which must be saturated with quinhydrone (ie. equimolar amounts of quinone, Q, and hydroquinone, QH$_2$).

Hydroquinone behaves as a weak dibasic acid:

$$QH_2 \rightleftharpoons QH^- + H^+, \quad K_1 = 1.75 \times 10^{-10} \text{ mol dm}^{-3}$$
$$QH^- \rightleftharpoons Q^{--} + H^+, \quad K_2 = 4.0 \times 10^{-12} \text{ mol dm}^{-3}$$

The quinhydrone electrode is a reversible oxidation-reduction electrode:

$$Q + 2H^+ + 2\epsilon \rightleftharpoons QH_2$$

and as such its electrode potential is given by:

$$E(Q, QH_2) = E^{\ominus}(Q, QH_2) + \frac{RT}{2\mathscr{F}} \ln \frac{[Q]}{[Q^{--}]} \tag{6.28}$$

where the terms in square brackets respresent the concentration of the appropriate species. Taking into account the dissociations of hydroquinone, equation (6.28) becomes:

$$E(Q, QH_2) = E^{\ominus}(Q, QH_2) + \frac{RT}{2\mathscr{F}} \ln \frac{[Q]}{[QH_2]} + \frac{RT}{2\mathscr{F}} \ln \frac{1}{K_1 K_2} + \frac{RT}{\mathscr{F}} \ln a_{H^+} \tag{6.29}$$

The ratio [Q]/[QH$_2$] is constant and equal to unity provided that the equilibrium is not disturbed by the presence of other oxidation-reduction systems. Further K_1 and K_2 are constants, hence equation (6.29) becomes

$$E(Q, QH_2) = E' + \frac{RT}{\mathscr{F}} \ln a_{H^+} \tag{6.30}$$

The quinhydrone electrode therefore acts as a hydrogen ion indicating electrode for pH values up to 8.

Thus for the cell, $E = E(Q, QH_2) - E(cal)$

$$= E' - E(cal) + \frac{RT}{\mathscr{F}} \ln a_{H^+} \qquad (6.31)$$

Requirements: Potentiometer and galvanometer, or digital voltmeter, platinum and dip-type calomel electrodes, quinhydrone (p. 307), standard 0.5 mol dm^{-3} sodium hydroxide solution, approx. 0.5 mol dm^{-3} acetic acid solution.

Procedure: First titrate, in duplicate, 25 cm^3 of the acetic acid solution with sodium hydroxide solution using the correct indicator.

Clean the platinum electrode by washing in alcohol and igniting in a flame. Set up the cell by placing the electrodes in a small beaker containing 25 cm^3 of acetic acid with sufficient quinhydrone to saturate the solution; 0.5–1.0 g in 100 cm^3 solution. Connect the electrodes to the standardized potentiometer (platinum is positive in acid solution). Add the sodium hydroxide in 2-cm^3 portions, stir well (mechanically) and determine the emf. after each addition. Towards the end-point, reduce the amount added each time to 0.1–0.2 cm^3.

Using at least two buffer solutions of known pH calibrate the quinhydrone electrode, ie. determine E' (equation (6.30)).

Treatment of experimental data and discussion
1.   From the tabulated data plot E against V and $\Delta E/\Delta V$ against V to locate the equivalence point (as Fig. 6.3). From the known value of E', calculate the pH after each addition and using the theory of Expt. 6.2, calculate the pK of acetic acid.
2.   List the advantages and disadvantages of the quinhydrone electrode.

Experiment 6.4   Titrate acetic acid with sodium hydroxide potentiometrically using an antimony-antimony oxide electrode

Theory: This metal-metal oxide electrode, used for the determination of the pH value of a solution, consists of the pure metal dipping in the solution under test. A skin of antimony oxide rapidly forms on the surface of the metal and this is in equilibrium with the antimony ion in solution:

$$Sb_2O_3 + 3H_2O \rightleftharpoons 2Sb^{+++} + 6OH^-$$

The potential of the antimony electrode is:

$$E(Sb) = E^{\ominus}(Sb) + \frac{RT}{3\mathscr{F}} \ln a_{Sb^{+++}} \qquad (6.32)$$

Since $a_{Sb^{+++}} = \dfrac{\sqrt{K_s}}{a_{OH^-}^3}$ and $a_{OH^-} = \dfrac{K_w}{a_{H^+}}$

it follows that

$$E(Sb) = E' + \frac{RT}{\mathscr{F}} \ln a_{H^+} \qquad (6.33)$$

where E', a constant incorporating $E^{\ominus}(Sb)$, K_s, and K_w, must be determined experimentally for each electrode.

For the cell:

$$\ominus \quad Sb|Solution \ Satd. \ KCl|Hg_2Cl_2 \cdot Hg \quad \oplus$$

$$E = E(cal) - E(Sb) = E' + 0.059 \ pH \ (T = 298 \ K) \qquad (6.34)$$

Requirements: Antimony electrode (3 cm of antimony metal soldered onto 15 cm of stout copper wire, held in a glass tube with sealing wax so that liquid cannot come into contact with the Cu-Sb junction), dip-type calomel electrode, potentiometer and galvanometer, 0.2 mol dm^{-3} disodium hydrogen phosphate solution, 0.1 mol dm^{-3} citric acid solution (McIlvaine's buffer, p. 311), 0.1 mol dm^{-3} solutions of acetic acid and sodium hydroxide.

Procedure: Polish the antimony with fine emery paper and set up the above cell. Standardize the potentiometer and measure the emf. of at least six buffer solutions in the range pH 2–8.

Set up the electrodes in a beaker containing 25 cm^3 of acetic acid and determine emf. Add the sodium hydroxide solution in 2 cm^3 portions, stir well and determine the emf. after each addition. Towards the end-point, reduce the amount added each time to 0.1–0.2 cm^3.

Treatment of experimental data and discussion
1. Plot a graph of E against the pH of the different buffer solutions. Draw the best straight line through the points, determine E' and compare the slope with the calculated value.
2. Convert the emf. values, obtained during the titration, into pH values and plot the potentiometric titration curve. Using the theory of Expt. 6.2 calculate the pK of acetic acid.
3. Discuss the advantages and limitations of the antimony electrode.

(b) Precipitation titrations

Experiment 6.5   Titrate halide solutions with silver nitrate solution potentiometrically, and hence determine the solubility products of the silver halides

Theory: The cell used in this titration is:

$$\ominus \ Hg \cdot Hg_2Cl_2 \ \left| \ Satd. \ KCl \ \right| \ \begin{array}{c} NH_4NO_3 \\ bridge \end{array} \ \left| \ \begin{array}{c} Halide \\ solution \end{array} \ \right| \ Ag \ \oplus$$

At the equivalence point of the KCl − $AgNO_3$ titration, $a_{Ag^+} = a_{Cl^-}$ and thus $K_s = a_{Ag^+} a_{Cl^-} = a_{Ag^+}^2$.

The electrode potential of the silver electrode:

$$E(Ag^+, Ag) = E^{\ominus}(Ag^+, Ag) + \frac{RT}{\mathscr{F}} \ln a_{Ag^+} \qquad (6.35)$$

at the equivalence point may be written

$$E(Ag^+, Ag) = E^{\ominus}(Ag^+, Ag) + \frac{RT}{2\mathscr{F}} \ln K_s \qquad (6.36)$$

and the emf. of the cell at the equivalence point is:

$$E = E(Ag^+, Ag) - E(cal) = E^{\ominus}(Ag^+, Ag) - E(cal) + \frac{RT}{2\mathscr{F}} \ln K_s \qquad (6.37)$$

Requirements: Potentiometer, galvanometer, silver and saturated calomel electrodes, ammonium nitrate agar bridge. Standard 0.1 mol dm^{-3} silver nitrate solution, standard 0.1 mol dm^{-3} potassium chloride solution, solution of potassium chloride, bromide and iodide of unknown concentration (total salt concentration 0.1–0.2 mol dm^{-3}), 0.25 mol dm^{-3} potassium chromate solution, and fluorescein solution (0.1 g sodium fluoresceinate in 100 cm^3 water).

Procedure: Titrate 25 cm^3 of potassium chloride with silver nitrate determining the end-point:

(i)   using 1 cm^3 of potassium chromate as indicator, indicator blank is required;
(ii) using fluorescein as indicator.

Set up and standardize the potentiometer. Set up the cell and pipette 25 cm^3 of the potassium chloride solution into a beaker and add 2–3 cm^3 of 2 mol dm^{-3} nitric acid solution. Immerse the cleaned silver electrode in this solution and the bridge connection to a beaker containing saturated potassium chloride and a calomel electrode. Connect the electrodes to the potentiometer and determine the emf. of the cell. Record the polarity of the electrodes. Run in the silver nitrate solution from a burette, stir well and determine the emf. after each addition. The volume added should be decreased towards the equivalence point, to 0.05–0.1 cm^3.

Repeat the titration using the mixed halide solution, observing the three end-points where the emf. of the cell suddenly increases due to complete removal of iodide, bromide and chloride ions respectively.

Treatment of experimental data and discussion

1.   Plot the titration and differential curves and hence determine the individual concentrations of the three halide ions in the solution. Comment on the advantages of the method. Is the method applicable to other anions?
2.   From the equivalence point in the KCl–AgNO$_3$ titration calculate the value of K_s for silver chloride (equation (6.37)). Calculate the value of K_s from tabulated values of $\Delta G_f^{\ominus}$ at 298 K, explaining the principles underlying the theory.
3.   From the value of K_s calculate the full potentiometric titration curve for KCl–AgNO$_3$, use equations (6.35) and (6.36) in combination with equation (6.37).

Experiment 6.6   Standardize a potassium chloride solution with silver nitrate by a differential potentiometric titration

Theory: In this method, the differential curve (Fig. 6.3b) is determined using the

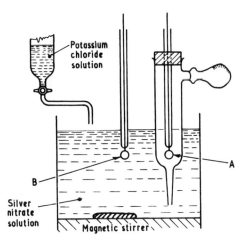

Fig. 6.6   Apparatus for differential potentiometric titration

apparatus (Fig. 6.6) and plotted directly, instead of being calculated from the E against V curve. A and B are silver electrodes connected to a potentiometer; A is enclosed in a tube which temporarily holds back a portion of the silver nitrate solution. Initially the emf. of the cell is zero; upon the addition of potassium chloride solution, the potential of B is changed, due to precipitation of silver chloride, whilst that of A is not changed, since the trapped solution has not mixed with the bulk. The emf. is recorded and the sheltered solution mixed with the bulk to give again a zero emf. This is repeated for further additions of potassium chloride; the maximum emf. recorded corresponds to the end-point.

Requirements: Potentiometer, galvanometer, magnetic stirrer, two silver electrodes, apparatus (Fig. 6.6), standard 0.1 mol dm^{-3} silver nitrate solution, approx. 0.1 mol dm^{-3} potassium chloride solution.

Procedure: Set up and standardize the potentiometer. Pipette 25 cm^3 of silver nitrate solution into the beaker, add 2–3 cm^3 of 2 mol dm^{-3} nitric acid solution and about 50 cm^3 water. Immerse the free and sheltered electrodes in the solution and connect to the potentiometer. Determine the emf. of the cell; this should be zero. Add a small amount of potassium chloride from the burette, stirring well; again determine the potential. Force the sheltered solution into the main volume and allow a sample to be trapped. Measure the emf. (zero). Continue adding further small amounts of potassium chloride, measuring the emf. after each addition.

Treatment of experimental data and discussion
1.   Plot the differential titration curve, ie. $\Delta E/\Delta V$ (the emf./KCl added) against V; the maximum in the curve corresponds to the end-point of the titration.
2.   Discuss the advantages of this method of determining the end-point.
3.   How could the method be adapted for the determination of other halides either singly or in mixtures?
4.   List any other applications.

Experiment 6.7   Determine the concentration of chloride in solution by the 'bottled end-point' method

Theory: In this method the reference calomel electrode (Expt. 6.5) is replaced by a compensation electrode, the potential of which is exactly equal to that of the indicator electrode in the solution under titration, at the end-point. The electrodes are coupled in such a way that the end-point is indicated by a sudden reversal of polarity. The most useful compensating electrode in the titration of a chloride is a silver electrode immersed in a solution of the same composition as that which the titration has at the end-point.

Requirements: Two silver wire electrodes (Fig. 6.7), sensitive galvanometer (16 mm per μA), standard 0.1 mol dm^{-3} silver nitrate solution, approx. 0.1 mol dm^{-3} sodium chloride solution, saturated potassium sulphate solution for bridge.

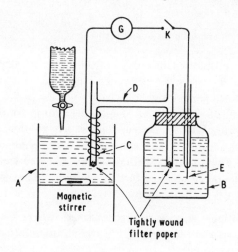

Fig. 6.7   Apparatus for bottled end-point titration

Procedure: Thoroughly clean the electrodes and wind C on to the arm of the bridge to prevent fouling the stirrer. Fill the H-shaped bridge, D, with saturated potassium sulphate solution; the ends are plugged with tightly wound strips of filter paper. Fill the reference half-cell B (amber glass bottle) with the following solution: 25.0 cm^3 of 0.1 mol dm^{-3} sodium chloride solution, 25.05 cm^3 of 0.1 mol dm^{-3} silver nitrate solution, 10.0 cm^3 of 1 mol dm^{-3} nitric acid and 40 cm^3 of water. Replace the rubber bung carrying one arm of the salt bridge and the indicator silver wire. Pipette 25 cm^3 of the sodium chloride solution into the beaker A and add 50 cm^3 of water. Ensure that the levels of liquids in A and B are the same to prevent siphoning. Run in the silver nitrate solution from the burette with continual mechanical stirring. Momentarily depress tapping-key, K, after the addition of a few cm^3 of the solution; towards the end-point, add the silver nitrate dropwise. Record the deflections of the galvanometer; these will initially be large, gradually decrease to zero at the end-point and then increase in the opposite direction.

Treatment of experimental data and discussion
1. Explain the theory underlying this cell.
2. Discuss the advantages of this method of locating the end-point in a titration and suggest other applications.

(c) Oxidation-reduction titrations

Experiment 6.8 **Titrate iron (II) ammonium sulphate solution with cerium (IV) sulphate solution potentiometrically and hence determine the formal redox potentials**

Theory: In the cell:

$$\ominus \quad Hg \; . \; Hg_2Cl_2|Satd. \; KCl \; Fe^{++} \; . \; Fe^{+++}|Pt \quad \oplus$$

the platinum electrode has a potential which depends on the formal redox potential of the system and $\log c_{Fe^{+++}}/c_{Fe^{++}}$ (equation (6.13)). Initially this ratio is small; on the addition of an oxidizing agent, this increases rapidly at the end-point giving a step in the $E-V$ curve. Beyond this point, the emf. is determined by the redox potential of the oxidizing system.

Requirements: Potentiometer and galvanometer, or digital voltmeter, bright platinum and dip type calomel electrodes, standard 0.1 mol dm^{-3} iron (II) ammonium sulphate and approximately 0.1 mol dm^{-3} cerium (IV) sulphate solutions both made up in 1 mol dm^{-3} sulphuric acid. N-phenylanthranilic acid (dissolve 1.07 g in 20 cm^3 of 5% sodium carbonate solution, dilute to 1 dm^3), 0.2% aqueous solution of sodium diphenylamine sulphonate, syrupy phosphoric acid.

Procedure: Titrate 25 cm^3 of the iron (II) salt solution, in the presence of 25 cm^3 1 mol dm^{-3} sulphuric acid, against the cerium (IV) sulphate solution, using either 0.5 cm^3 N-phenylanthranilic acid, or 0.5 cm^3 of sodium diphenylamine sulphonate and 5 cm^3 phosphoric acid.

Set up the cell by placing the electrodes in a beaker containing 25 cm^3 of iron (II) ammonium sulphate solution. Connect the electrodes to the standardized potentiometer or digital voltmeter (platinum positive) and determine the emf. of the cell. Titrate with cerium (IV) sulphate solution, stirring mechanically, and determine the emf. after each addition. Take numerous readings near the end-point.

Repeat the titration with cerium (IV) sulphate in the beaker.

Treatment of experimental data and discussion
1. Plot the potentiometric and differential titration curves (Fig. 6.3a, b) for both titrations, and determine the equivalence points. From the results determine $c_{Fe^{+++}}/c_{Fe^{++}}$ and plot E against $\log c_{Fe^{+++}}/c_{Fe^{++}}$; hence determine the formal redox potential, E', where $c_{Fe^{+++}} = c_{Fe^{++}}$; $E(cal) = 0.2420$ V for saturated electrode. Calculate the slope of the line and determine the number of electrons involved in the oxidation-reduction system. Repeat the calculation for the cerium system.
2. Explain the significance and usefulness of the 'formal electrode potential' for

a redox electrode. Why does E' differ from $E^{\ominus}(O, R)$? Is it possible to obtain an approximate value of $E^{\ominus}(Fe^{+++}, Fe^{++})$ from your experimental data?

3. Suggest a better method for the determination of $E^{\ominus}(Fe^{+++}, Fe^{++})$?

4. From the values of $E^{\ominus}(O, R)$ for the two electrode systems, calculate the value of E at the equivalence point.

$$E(\text{equiv}) = \frac{n_1 E_1^{\ominus} + n_2 E_2^{\ominus}}{n_1 + n_2}$$

where n_1 and n_2 are the numbers of electrons exchanged by couples 1 and 2 of standard potentials $E_1^{\ominus}$ and $E_2^{\ominus}$ respectively. Explain why diphenylamine $(E^{\ominus} = 0.76 \ V)$ is a suitable indicator for the titration.

5. $\Delta G_f^{\ominus}(Fe^{+++}aq, \ 298 \ K) = -10.6 \ kJ \ mol^{-1}$, and $\Delta G_f^{\ominus}(Fe^{++}aq, \ 298 \ K) = -84.9 \ kJ \ mol^{-1}$; calculate $E^{\ominus}(Fe^{+++}, Fe^{++})$, explaining the theory.

6. What additional precautions are necessary if the titrant is very easily oxidized?

7. Calculate the value of the equilibrium constant for the equilibrium:

$$Fe^{++} + Ce^{++++} \rightleftharpoons Fe^{+++} + Ce^{+++}$$

Experiment 6.9   Titrate iron (II) ammonium sulphate solution with potassium dichromate solution potentiometrically using a bimetallic electrode pair

Theory: The use of an attackable internal reference electrode paired with platinum as the indicator electrode, known as a bimetallic electrode pair, provides a very simple assembly for oxidation-reduction titrations. A tungsten electrode, which is slowly attacked by most oxidizing agents, is suitable as a reference electrode in such titrations. The potential at the tungsten surface increases slowly during the addition of an oxidizing agent, and, in concentrated solutions, the potentiometric end-point coincides with the true equivalence point. At the start of the titration (ie. in the reducing solution) there is usually a large emf. between the electrodes; on the addition of titrant, equilibrium is rapidly reached between the oxidized and reduced states and the emf. falls to a very low value (near 0). The emf. remains at this low value during the titration until to within 0.5 cm³ of the equivalence point. A slight increase indicates the nearness of the end-point at which there is an abrupt change of emf. which may be 100–200 mV.

Requirements: Potentiometer, galvanometer, bright platinum, tungsten and calomel electrodes, standard 0.1 mol dm⁻³ iron (II) ammonium sulphate and approximately 0.02 mol dm⁻³ potassium dichromate solutions both made up in 1 mol dm⁻³ sulphuric acid.

Procedure: Clean the tungsten electrode either by polishing with emery paper or by dipping into *just* molten sodium nitrite for a few seconds; wash well with distilled water. Set up and standardize the potentiometer circuit so arranged that the emf. is easily measured between the platinum (positive) and calomel, and tungsten (positive) and calomel. Clamp the three electrodes in a 250-cm³ beaker, containing

25 cm³ iron (II) ammonium sulphate solution, 5 cm³ concentrated sulphuric acid and 100 cm³ water. Measure the emf. between each of the two pairs of electrodes. Titrate with potassium dichromate solution, stirring the solution mechanically and make the same emf. determinations after each addition. Take numerous readings near the end-point.

Wash the electrodes and repeat the titration, using the metallic electrodes only (platinum positive).

Treatment of experimental data and discussion
1. Plot graphs of (*a*) the emf. of the platinum—calomel electrode system and (*b*) the emf. of the tungsten—calomel electrode system, against the volume of potassium dichromate. Also calculate and plot the difference curve (ie. the difference of emf. for the two systems against the volume of dichromate). Finally, plot the titration curve for the platinum—tungsten system; does this agree with the difference curve?
2. Suggest a theoretical basis for the results obtained with the tungsten electrode.
3. Mention any advantages and limitations of a bimetallic electrode system.

Experiment 6.10   Titrate iodine solution with sodium thiosulphate by the polarization or dead stop end-point method

Theory: This method does not require a reference electrode or a potentiometer. The principle of the method is to balance a small applied potential between two identical platinum electrodes with the back emf. due to polarization, so that no current flows through the galvanometer. The end-point is recorded when there is a sudden change from polarization of at least one electrode to depolarization of both, or vice versa. Polarization is assumed to be the result of adsorbed films of hydrogen and oxygen at the cathode and anode respectively. These may be depolarized by suitable oxidizing or reducing agents respectively.

Two types must be considered.

(*a*) Only one electrode is depolarized, the added reagent when present in excess causing depolarization of the other. In the titration of sodium thiosulphate with iodine solution, the thiosulphate acts as an anode depolarizer. On the addition of iodine, the iodide formed also acts as an anode depolarizer, until the end-point is reached, when the free iodine acts as a cathode depolarizer. This is indicated by a sudden flow of current through the solution resulting in a galvanometer deflection.

(*b*) Both electrodes are depolarized until the end-point, when one becomes polarized. In the reverse titration of iodine, dissolved in an iodide solution, with sodium thiosulphate, both electrodes are depolarized. At the end-point, the complete removal of the iodine from solution results in the cathode becoming polarized, and, in consequence, no current flows. Thus, during the titration the galvanometer records a current flowing; at the end-point and beyond there is no current flow.

A momentarily high local concentration round the electrode on each addition of reagent causes a movement of the galvanometer needle. This effect, depending

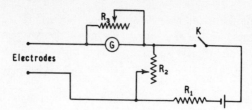

Fig. 6.8   Electrical circuit for polarization end-point titration

on the rate of titration and efficiency of stirring, is due to partial polarization or depolarization of an electrode. The size of the galvanometer deflections gives a warning of the approach of the end-point.

Requirements: Two bright platinum electrodes, galvanometer of sensitivity 50–200 mm per mA, mechanical stirrer, circuit (Fig. 6.8), standard 0.1 mol dm^{-3} sodium thiosulphate solution, approx. 0.05 mol dm^{-3} iodine in 0.1 mol dm^{-3} potassium iodide solution. Values of R_1 and R_2 are 100 kΩ and 1 kΩ respectively.

Procedure: Support the two clean electrodes in 50 cm^3 of sodium thiosulphate solution in a beaker and stir well (mechanically). Depress K and zero the galvanometer with R_2. Add the iodine solution in small amounts, recording the position of the galvanometer spot. This should remain on zero until the end-point is reached when it will be permanently displaced.

   Repeat the experiment, now adding the thiosulphate to the iodine solution. First, adjust the zero position of the galvanometer with sodium thiosulphate in the beaker using R_2. Wash the beaker and electrodes and introduce 50 cm^3 of iodine solution. Depress K momentarily; the galvanometer spot will be off the scale; bring this just on to the scale by adjusting the sensitivity control R_3. Run in the sodium thiosulphate solution, with continual stirring, and record the decrease in deflection. Proceed with small additions until the galvanometer spot is no longer deflected.

Treatment of experimental data and discussion
1. Plot the galvanometer deflection (μA) against the volume of sodium thiosulphate solution added, and estimate the end-point. Compare the titre with that obtained by the direct titration using starch as an indicator.
2. Discuss the usefulness of this technique and suggest other applications.

3   Conductometric titrations

The addition of one electrolyte solution (A^+B^-) to another (C^+D^-) will result in a change of conductance on account of volume changes and possible ionic reactions. If the addition is made so that there is no appreciable volume change, and there is no chemical reaction, the conductance of CD gradually increases on the addition of AB. On the other hand if there is an ionic reaction:

$$A^+B^- + C^+D^- \longrightarrow A^+D^- + CB$$

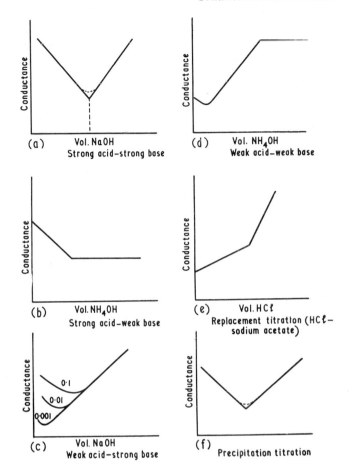

Fig. 6.9   Some typical diagrammatic conductometric titration curves

in which one of the products, CB, is either only slightly ionized or is insoluble, then a marked change of conductance occurs at the equivalence point. During the titration, A^+ replaces C^+ and the conductance may increase or decrease, depending on the relative mobility values of A^+ and C^+.

The addition of an alkali to a strong acid is accompanied by a decrease in conductance owing to the replacement of H^+ by the less mobile cation (Fig. 6.9). When neutralization is complete, ie. all the H^+ removed, further addition of base results in an increase in conductance owing to the presence of excess OH^-. The graph of volume of added alkali against conductance thus consists of two straight lines intersecting at the equivalence point.

The more acute the angle at which the lines intersect, the greater is the accuracy of location of the end-point. There should be no appreciable change of volume; this is best achieved using a titrating solution 20–100 times stronger than that under titration, and making additions from a microburette.

Conductometric titrations are applicable where potentiometric or visual methods fail, eg. where there is considerable hydrolysis at the end-point, weak acid—weak base and displacement titrations. The method can be used in very dilute or coloured solutions.

Typical conductance titration curves are shown in Fig. 6.9. In precipitation titrations, it has been found that the smaller the mobility of the ion replacing the reacting ion, the more accurate the result. Thus it is preferable to titrate a silver salt with lithium chloride rather than with hydrochloric acid. The time of the titration is affected by the rate of precipitation. The solubility and absorptive properties of the precipitate can give rise to inaccuracies in the location of the end-point.

Experiment 6.11   Determine the end-point in some typical titrations by the conductometric method

Requirements: Conductance bridge, dip-type conductance cell, microburette (1 or 2 cm^3), standard solutions as listed later.

Procedure: Pipette 50 cm^3 of the solution to be titrated (0.01 mol dm^{-3}) into a 100-cm^{-3} beaker. Immerse the electrodes in this solution and determine the conductance or resistance of the solution. It is not necessary to thermostat the solution unless the end-point is particularly indistinct. Add a small amount, 0.02–0.04 cm^3, of the titrating solution (approx. 1 mol dm^{-3}), stir well and again determine the conductance or resistance of the solution. Repeat, stirring well after each addition, for at least eight additions after the break in the curve has been observed.

Titrate 50 cm^3 of each of the following solutions:

(*a*)   0.01 mol dm^{-3} NaOH with 1 mol dm^{-3} HCl,

(*b*)   0.005 mol dm^{-3} H$_2$SO$_4$ with 1 mol dm^{-3} NaOH,

(*c*)   0.01 mol dm^{-3} HCl with 1 mol dm^{-3} NH$_4$OH,

(*d*)   0.01 mol dm^{-3} acetic acid with 1 mol dm^{-3} NaOH,

(*e*)   0.01 mol dm^{-3} acetic acid with 1 mol dm^{-3} NH$_4$OH,

(*f*)   0.01 mol dm^{-3} acetic acid (40 cm^3) mixed with 0.01 mol dm^{-3} HCl (10 cm^3) with (i) 1 mol dm^{-3} NaOH and (ii) 1 mol dm^{-3} NH$_4$OH,

(*g*)   0.01 mol dm^{-3} sodium acetate with 1 mol dm^{-3} HCl,

(*h*)   0.01 mol dm^{-3} AgNO$_3$ with 1 mol dm^{-3} solutions of HCl, KCl and LiCl,

(*i*)   0.001 mol dm^{-3} MgSO$_4$ with 0.1 mol dm^{-3} Ba(OH)$_2$.

Treatment of experimental data and discussion

1.   Construct titration curves of conductance (or R^{-1}) against the volume of added reagent, extrapolate the straight line portions of the graphs to give the end-point(s).

2.   Using the values of the molar conductances of the ions at infinite dilution (p. 305), explain the variation of the conductance as the titrations are performed.

3.   Explain why it is not necessary to know the cell constant of the conductance cell.

4. Why should readings near the end-point be ignored, particularly in titrations *d*, *e*, *f* and *g*?

5. Why is it preferable for the titrating solution to be of a much higher concentration than that of the solution to be titrated? Show that the correction for volume change is given by:

$$G' = GV/(V + v)$$

where G and G' are the measured and corrected conductances respectively, V the original volume of solution and v the volume of titrating solution added. For titration *a*, calculate G' for several readings of G near the end-point and hence decide whether it is reasonable to neglect the volume change.

6. Why must large amounts of non-participating electrolytes be excluded from conductometric titrations?

7. Compare and contrast the various methods of locating end-points, eg. potentiometric, conductometric, amperometric, spectrophotometric and thermal.

4   Amperometric titrations

The diffusion current (p. 162), ie. the difference between the limiting and residual currents, is directly proportional to the concentration of electro-active material in solution. Removal of some of this material with another reagent causes a lowering of the diffusion current. This is the fundamental principle of amperometric titrations involving the use of a polarograph. The diffusion current at a suitable applied voltage is measured and plotted as a function of the volume of added reagent. The end-point is located graphically, as in conductometric titrations.

The reagent added and/or the solution to be titrated must have a diffusion current at the selected applied voltage. Three types can be distinguished (Fig. 6.10), each giving a characteristic current-volume titration curve when the applied voltage lies between A and B:

(i) only the solute, ie. material being titrated gives a diffusion current (in region AB); when this material is removed the diffusion current falls to zero (eg. lead titrated with oxalic acid);

(ii) reagent gives diffusion current while solute does not. Thus, while the excess of electroactive material remains, the diffusion current (in region AB) is zero; in the presence of excess reagent beyond the end point, the diffusion current increases (eg. sulphate titrated with lead or barium salt);

(iii) both substances are electroactive in region AB, resulting in a V-shaped titration curve, one arm representing the diffusion current of the reagent and the other the solute (eg. lead titrated with potassium dichromate).

The dotted lines are plots of the actual readings; these are curves since the total volume in the titration cell is continually increasing. The continuous lines, required to locate the end-point, are obtained by plotting v against the diffusion current multiplied by $(V + v)/V$ (where V is the initial volume and v the volume of added reagent). This correction is considerably reduced, if not eliminated, by

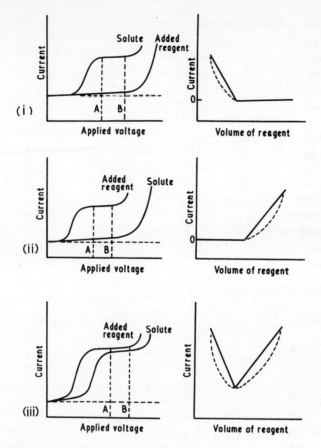

Fig. 6.10   Typical amperometric titration curves

adding the reagent from a microburette in a concentration at least twenty times that of the solution. The use of a more concentrated reagent has the further advantage that relatively little dissolved oxygen is introduced with each small addition; this reduces the degassing time after each addition.

Experiment 6.12   Determine the concentration of lead in solution by amperometric titration with potassium dichromate solution

Requirements: Polarograph, 100 cm^3 titration cell complete with electrodes and capillary assembly (Fig. 6.11), agar potassium nitrate salt bridge (p. 307), source of oxygen-free nitrogen, standard 0.005 mol dm^{-3} potassium dichromate solution, approx. 0.001 mol dm^{-3} lead nitrate, 0.01 mol dm^{-3} potassium nitrate solution as supporting electrolyte.

Procedure: Set up and adjust the polarograph (follow the instruction manual) with the mercury dropping freely into water. Record the polarograms (Expt. 7.5) of lead nitrate and potassium dichromate separately in the presence of 0.01 mol dm^{-3}

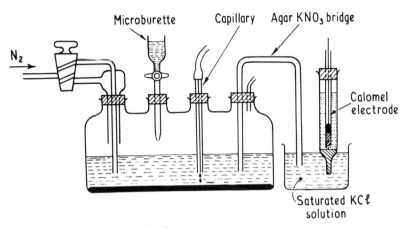

Fig. 6.11   Cell for amperometric titrations

potassium nitrate as supporting electrolyte. Polarograms similar to Fig. 6.10 (iii) will be obtained, both lead and dichromate ions give a diffusion current at an applied potential of -1.0 V against the saturated calomel electrode.

Pipette 25 cm^3 of the lead nitrate solution and 25 cm^3 of 0.01 mol dm^{-3} potassium nitrate solution into the cell and assemble the salt bridge, capillary and bubbling tube as shown (Fig. 6.11). Connect the electrodes to the polarograph, dropping mercury cathode, calomel anode. Bubble a slow stream of nitrogen (oxygen free) through the solution for 20 min to displace all dissolved oxygen; turn the two-way tap so that the nitrogen passes across the surface maintaining an inert atmosphere. Apply the selected voltage, -1.0 V, to the electrodes and record the deflection and sensitivity of the galvanometer; from this, the initial diffusion current is calculated. Now add 0.4 cm^3 of the standard potassium dichromate solution from the microburette, bubble nitrogen through the solution for 2 min to ensure good mixing, and remove dissolved oxygen from the added reagent. Turn the tap so that nitrogen now passes across the surface; apply the selected voltage and again determine the diffusion current. Make further additions (20) of dichromate solution until sufficient readings are available to plot a titration curve.

Treatment of experimental data and discussion
1.   For each observation calculate the true diffusion current, ie. observed current $\times (V + v)/V$. Plot the titration curve of diffusion current against the volume of dichromate solution added, draw the best straight lines through the points and locate the end-point. Hence calculate the concentration of lead in the original solution.
2.   List the advantages of the amperometric method of locating the end-point of a titration, with special reference to the concentration of the solution to be titrated, the presence of foreign electrolytes, and the titration of substances which can neither be oxidized nor reduced.
3.   Compare and contrast this method with other instrumental methods of locating end-points.

Experiment 6.13   Determine the concentration of sulphate in solution by ampero-metric titration with lead nitrate solution

Theory: Sulphate ions are not reducible, while lead ions give a diffusion current at −1.2 V (against saturated calomel electrode); thus, the curve obtained is as shown in Fig. 6.10 (ii). A supporting electrolyte is not required.

This titration can be carried out with reasonable accuracy in solutions as dilute as 0.001 mol dm^{-3} with respect to sulphate, in the presence of about 30% ethanol. The alcohol is added to reduce the solubility of the lead sulphate and thus reduce the rounded portion of the titration curve in the vicinity of the equivalence point.

Requirements: Polarograph, 100-cm^3 titration cell complete with electrodes and capillary assembly (Fig. 7.10), agar potassium nitrate salt bridge, source of oxygen-free nitrogen, standard 0.1 mol dm^{-3} lead nitrate solution, approx. 0.01 mol dm^{-3} potassium sulphate solution.

Procedure: Set up and adjust the polarograph and allow the mercury to drop into distilled water for 5 min. Record the polarograms of potassium sulphate and lead nitrate (Expt. 7.5) and hence select the required applied voltage for the titration (−1.2 V). Pipette 25 cm^3 of the potassium sulphate solution into the cell, add 2−3 drops of thymol blue, followed by a few drops of concentrated nitric acid until the colour is just red (pH 1.2); add 25 cm^3 of ethanol. Following the instructions of Expt. 6.12, titrate the solution with the standard lead nitrate solution at an applied voltage of −1.2 V.

Treatment of experimental data and discussion
1.   Apply the volume correction to the observed diffusion currents and plot a graph of the diffusion current against the volume of lead nitrate solution. Locate the end-point and hence determine the concentration of sulphate in solution.
2.   List the advantages of the amperometric method of locating the end-point of a titration, and compare the method with other instrumental methods of locating the end-point.

5   Spectrophotometric titrations

Although the theory and instrumentation have been long available, the practice of spectrophotometric (photometric) titrations has not been widely applied. In photo-metric titrations, the absorbance of a solution, at one or more fixed wavelengths, is used for this purpose. When small amounts of a concentrated solution of potassium permanganate are added to a large volume of water, so that the volume change is negligible, there is a progressive increase in the absorbance of the solution, at wavelengths where permanganate absorption is strong. A plot of absorbance against volume gives the normal Beer's law curve (p. 94) shown in Fig. 6.12a. If the titration is repeated, adding the permanganate, not to water but to a solution of an iron (II)

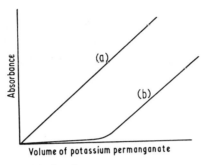

Fig. 6.12   Principle of photometric titrations

salt, a different curve is obtained (Fig. 6.12*b*). The permanganate colour is destroyed so long as there is some Fe^{++} present, and in consequence Beer's law plot is
delayed until the reducing agent has been destroyed. This simple experiment illustrates the principle underlying all photometric titrations.

The apparatus required for performing photometric titrations consists of
absorptiometer or spectrophotometer adapted to take a titration cell. The simplest
type of cell is illustrated in Fig. 6.13; the exact dimensions obviously depend on the
photometric instrument available. The cell consists of a 150 cm^3 beaker sealed on
to a suitable cell (either cylindrical or rectangular), covered to prevent entry of
stray light. (This design is very suitable for the Unicam SP.1400 spectrophotometer.) The inside of the cover should be painted with flat black, as also should the
burette tip and stirrer where these pass through the cover. The holes through which
the burette and stirrer pass must be protected by felt gaskets; the assembly is then
completely light tight. The tip of the burette should dip beneath the level of the
liquid, thus small amounts may be added without fear of the drop remaining on the
tip. Diffusion is negligible. In spectrophotometric titrations, the path length is

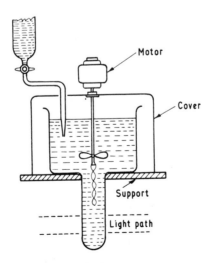

Fig. 6.13   Titration cell for photometric titrations

immaterial, provided it remains constant during a titration, it is thus essential when using cylindrical cells (or test tubes) that the cell is not moved during a titration. The range of use of photometric titrations can be extended into the ultraviolet region if the cell is made of quartz. A commercial filter instrument, using a built-in magnetic stirrer, is available from Evans Electroselenium Ltd.

Photometric titrations share with conductometric and amperometric titrations the advantage that the measured property (absorbance) is directly proportional to the concentration of the substance being followed. The end-point is determined (as in conductometric and amperometric titrations) by the intersection of two lines, each of which is determined from several observations. Reasonable extrapolation is permissible when the lines deviate from linearity in the vicinity of the end-point, since absorbance values in this region are no more significant than values far displaced on either side.

Experiment 6.14   Plot a spectrophotometric titration curve for the neutralization of hydrochloric acid with sodium hydroxide

Requirements: Spectrophotometer, titration cell (Fig. 6.13), 2 cm^3 microburette, standard 0.1 mol dm^{-3} sodium hydroxide solution, approximately 0.1 mol dm^{-3} hydrochloric acid, phenolphthalein.

Procedure: Plot the absorption spectrum of two drops of an alcoholic solution of phenolphthalein in 50 cm^3 0.1 mol dm^{-3} sodium hydroxide against a blank of sodium hydroxide, over the range 480–600 nm (maximum absorption: 545 nm). Wash the cell. Pipette 1 cm^3 of the hydrochloric acid solution into the titration cell and add 100 cm^3 of water and four drops of phenolphthalein. Set up the stirrer, cover and microburette and adjust the spectrophotometer to give zero absorbance at the wavelength of maximum absorption. Add the sodium hydroxide in 0.05 cm^3 aliquots, allow sufficient time for the solution to become homogeneous, and record the constant absorbance reading. Continue adding the alkali until the absorbance reaches 1.0.

Treatment of experimental data and discussion
1. Plot a graph of absorbance against the volume of alkali added and locate the end-point. Hence calculate the concentration of the acid.
2. Why is the precision of a photometric titration greater than that of a direct visual titration with an indicator?
3. Discuss the advantages and disadvantages of the photometric method of locating the end-point in a titration.

Experiment 6.15   Determine the concentration of iron (II) in solution by spectrophotometric titration

Requirements: Spectrophotometer, titration cell (Fig. 6.13), 2 cm^3 microburette, standard 0.02 mol dm^{-3} potassium permanganate solution, solution of iron (II) ammonium sulphate containing approx. 1 mg cm^{-3} Fe^{++} (ie. 7.0 g dm^{-3}

$FeSO_4 . (NH_4)_2SO_4 . 6H_2O$) made up in 1 mol dm^{-3} sulphuric acid, solution of iron (III) ammonium sulphate, approx. 1 mg cm^{-3} Fe^{+++}.

Procedure: Using a spectrophotometer, plot the absorption spectra of a 0.002 mol dm^{-3} potassium permanganate solution and the iron (II) and the iron (III) ammonium sulphate solutions over the range 440–640 nm. Potassium permanganate has two absorption bands at about 525 and 550 nm; select one of these wavelengths (or the appropriate filter with an absorptiometer) for the subsequent titrations. First, fill the cell with a known volume (50 or 100 cm^3) of 1 mol dm^{-3} sulphuric acid, adjust the spectrophotometer to give zero reading for the absorbance, and add the potassium permanganate solution in 0.05 cm^3 aliquots. After each addition allow sufficient time for the solution to become homogeneous on stirring, and record the absorbance. The solution is not adequately mixed until the absorbance remains constant. The graph of absorbance against the volume of permanganate should be linear — Beer's law (Fig. 6.12).

Wash the cell, and carefully introduce a 2-cm^3 sample of the iron (II) ammonium sulphate followed by 50 or 100 cm^3 of 1 mol dm^{-3} sulphuric acid. Adjust the initial absorbance to zero, and add 0.05 cm^3 aliquots of 0.02 mol dm^{-3} permanganate solution, recording the constant value of the absorbance after each addition. Continue adding the permanganate until there is a large absorbance reading.

Repeat the titration using 5 and 10-cm^3 samples of iron (II) ammonium sulphate solution.

Treatment of experimental data and discussion

1. Plot graphs of the absorbance of the solution against the volume of $KMnO_4$ added. Draw the best lines through the points and locate the end-points; hence calculate the concentration of Fe^{++} in the original solution.

2. It is necessary to correct the measured absorbance if relatively large volumes of permanganate solution have been added. Explain this statement and derive the appropriate correction factor.

3. Why is it necessary to plot the absorption spectra of the solutions of iron (II) and iron (III) ammonium sulphate?

Experiment 6.16 Determine the concentration of iron (III), copper or nickel in solution by photometric titration with EDTA solution

Theory: Photometric titrations of metal solutions with EDTA eliminates the use of specific indicators. The complex between salicylic acid and Fe^{+++} with a maximum absorption at about 525 nm is used as the basis for the titration with EDTA. At pH 2.4, the iron–EDTA complex is stronger than the iron–salicylic complex, and so in the titration there is a gradual disappearance of the iron–salicylic acid colour, as the end-point is reached. This change is not sharp enough for a good visual end-point, but, using the spectrophotometric method at 525 nm it is very sharp.

At a wavelength of 745 nm the copper–EDTA complex has a molecular extinction coefficient considerably greater than that of the copper solution alone.

A deepening of the blue copper colour accompanies the formation of the complex with EDTA. The end-point would be sufficiently sharp for visual work except that it is difficult to estimate when maximum colour saturation is reached. Similarly the spectrophotometric titration of nickel solutions is possible at 1000 nm where the nickel–EDTA shows characteristic absorption.

Requirements: Spectrophotometer, titration cell (Fig. 6.13), standard solution of copper (dissolve 2.5 g pure copper in 10 cm^3 concentrated nitric acid and dilute to 1 dm^3). Standard 0.05 mol dm^{-3} solution of iron (III) ammonium sulphate (standardize with cerium (IV) sulphate after reduction with Jones reductor), 0.05 mol dm^{-3} nickel sulphate solution (standardize gravimetrically with dimethyl-glyoxime). Standard 0.1 mol dm^{-3} EDTA, sodium acetate–hydrochloric acid buffer solution (add 1 mol dm^{-3} hydrochloric acid to 1 mol dm^{-3} sodium acetate to pH 2.4), sodium acetate–acetic acid buffer solution (0.2 mol dm^{-3} salt, 0.8 mol dm^{-3} acid pH 4.0), 6% salicylic acid in acetone.

Procedure: (a) *Iron* (III). Plot the absorption spectrum of a dilute solution of the iron (III)-salicylic acid complex at pH 2.0; maximum absorption 525 nm. Show that EDTA does not absorb in this region. Transfer an aliquot of iron (III) solution (5–10 cm^3) to the titration cell; add 20 cm^3 of acetate buffer solution (pH 2.4) and dilute to about 90 cm^3. This solution should have a pH in the range 1.7–2.3. If the iron (III) solution is very acid, neutralize with ammonia before adding the buffer solution. Set the wavelength to 525 nm and adjust the instrument (by varying the slit width) to give an absorbance of about 0.2. Add 1 cm^3 of the salicylic acid solution and stir well; the absorbance will now be very high (possibly off the scale). Add the EDTA solution, in 0.5-cm^3 portions, with continual stirring until the absorbance reading reaches 1.8; thereafter, make smaller additions (0.1–0.2 cm^3) until the reading is constant at about 0.2 after three additions.

(b) *Copper.* Plot the absorption spectra of the copper solution and the copper–EDTA complex (maximum absorption at 745 nm). Transfer an aliquot of the copper solution to the titration cell; add 20 cm^3 of buffer solution and water to make up to 90 cm^3 (pH 2.4–2.8). Set wavelength to 745 nm and adjust spectrophotometer to give zero absorbance. Add EDTA solution with constant stirring in 0.3–0.5-cm^3 portions, until at least three readings have been taken beyond the end-point.

(c) *Nickel.* Plot the absorption spectra of the nickel solution and the nickel–EDTA complex (maximum at 1000 nm). The procedure is as for copper except that the pH of the final solution is 4.0.

Investigate the effect of titrating at different pH values, in the presence and absence of foreign metal ions (eg. Zn, Al, Cr, Cd, Mn), on the accuracy of the estimation.

Treatment of experimental data and discussion

1.   Plot the absorbance against the titre of EDTA, corrected for volume change if necessary. Draw the best lines through the experimental points and hence calculate the concentration of metal ion in the original solution.

2. Explain the effect of the variation of pH or the presence of interfering ions on the titration.

Experiment 6.17 Determine copper and iron (III) simultaneously in solution by photometric titration with EDTA solution

Theory: The stability constants for iron (III) and copper (p. 315) are large and sufficiently different to give two consecutive end-points when a mixture of the two is titrated with EDTA solution. Further, the copper complex absorbs strongly at 745 nm where the iron (III) complex exhibits no absorption. Since the iron (III) complex is the more stable, the formation of the copper complex, resulting in increased absorption, serves as an indicator for iron. The copper end-point is indicated by a plateau indicating maximum absorption (Expt. 6.16).

Requirements: Spectrophotometer, titration cell (Fig. 6.13), standard 0.1 mol dm^{-3} EDTA solution, 0.01 mol dm^{-3} copper nitrate solution, 0.01 mol dm^{-3} iron (III) ammonium sulphate in 1 mol dm^{-3} sulphuric acid (standardized with cerium (IV) sulphate after passage through Jones reductor), buffer solution: 1 mol dm^{-3} monochloracetic acid adjusted to pH 2.0 with sodium hydroxide.

Procedure: Pipette known volumes of the copper and iron (III) salt solutions (eg. 10 and 5 cm^3 respectively) into the titration cell, followed by 25 cm^3 of the monochloracetate solution. Dilute to 50 or 100 cm^3 and adjust the pH to 2.0 if necessary. Allow the solution to stand for at least one hour before titrating. With this solution in the light path, and, at a wavelength of 745 nm, adjust the absorbance to zero. Titrate with small amounts of EDTA solution, with continual stirring, as described in the previous experiments.

Treatment of experimental data and discussion
1. Plot the titration curve of absorbance against the added volume of EDTA (corrected for volume change if necessary), locate the two end-points and hence determine the concentration of each ion in the mixed solution.
2. Suggest other mixtures that could be titrated spectrophotometrically.

6 Coulometric titrations

Coulometric methods of analysis are based on Faraday's original work on electrochemical equivalence. The physical methods already described are merely methods of determining the end-point of a titration, in which one of the reagents is a standard. In contrast, coulometric titrations require no standard solutions, because a definite amount of reagent is generated in the solution; this amount is directly determined by the number of coulombs. The equivalence point is then detected by any of the methods discussed.

Thus, in the titration of a base, water is used for the generation of H_3O^+ in solution, which immediately reacts with the base. At the end-point, detected by an indicator, the total amount of electricity passed is determined, and, hence the amount of H_3O^+. An accurate knowledge of the current and the elapsed time gives the number of coulombs; a simpler procedure is to pass a current, kept as constant as possible, through the titration cell and silver coulometer in series.

This method has many distinct advantages. It requires no standard solution; a single electrical standard replaces the large number of chemical standards generally employed. Unstable reagents, eg. chlorine, can be generated and used immediately without fear of decomposition or change of composition. As there is no minimum to the magnitude of currents that can be used to generate the reagents electrically, micro-additions of reagent can be made near the end-point: the difficulty is the accurate detection of the end-point. Since the titrant is actually the electrons themselves, there is no dilution of the sample. Since the amount of titrant added and the detection of the end-point may both be electrical, the actual titration can be operated by remote control more readily than with normal volumetric procedures.

Table 6.1 illustrates the versatility of the coulometric technique.

Table 6.1   Coulometric titrations

Substance to be estimated	Titrant generated	Substance used for generation of titrant
(a) *Anodic generation*		
Base	H_3O^+	H_2O
As^{+++}, SO_3^-	Cl_2	Cl^-
As^{+++}, Sb^{+++}, I^-, SO_2, Cu^+, NH_2OH, Aniline, resorcinol	Br_2	Br^-
As^{+++}, $S_2O_3^-$	I_2	I^-
Fe^{++}, hydroquinone	Ce^{++++}	Ce^{+++}
Cl^-, Br^-, I^-	Ag^+	Ag
(b) *Cathodic generation*		
Acid	OH^-	H_2O
MnO_4^-, CrO_4^{--}, Br_2, Cl_2	Fe^{++}	Fe^{+++}
CrO_4^{--}, VO_3^-, Br_2	Cu^+	Cu^{++} (in presence of Cl^- or Br^-)

Experiment 6.18   Standardize a solution of potassium hydrogen phthalate by a coulometric titration

Theory: A silver coulometer is used in this experiment; the measurements are made directly in terms of the weight of silver deposited.

Requirements: Apparatus (Fig. 6.14), silver coulometer, A (p. 307), titration cell (250-cm^3 beaker), platinum disc electrode, B (15—17 mm diameter), platinum wire anode, C, supported in sintered-glass disc filter stick, D, filled with sodium sulphate solution, source of variable d.c. up to 90 V, milliammeter (0—250 mA), 0.1 mol dm^{-3} solution of potassium hydrogen phthalate, sodium sulphate solution (25 g anhydrous salt in 100 cm^3 water adjusted to pH 7.0), 15% silver nitrate solution.

Procedure: Thoroughly clean the cylindrical platinum foil cathode and crucible (from coulometer) and dry to constant weight. Immerse this cathode and the silver anode in the silver nitrate solution in the coulometer. Pipette 10 cm³ of potassium hydrogen phthalate into the beaker; add 75 cm³ of sodium sulphate solution and three drops of phenolphthalein. Fill the filter stick with sodium sulphate solution; introduce the platinum wire anode, C, and assemble the apparatus (Fig. 6.14). Switch on the current and adjust it to about 50 mA; record the time. Check the ammeter reading periodically and keep it as constant as possible. The titration now proceeds, hydroxyl ions liberated by electrolysis react immediately with the acid. When the phenolphthalein changes to a permanent pink, switch off the current and record the time; the time for the titration is about 30 min. Carefully remove the cathode and crucible, and taking care not to lose any of the deposited silver, wash, and dry to constant weight.

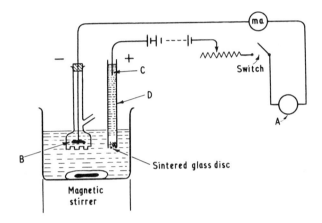

Fig. 6.14   Apparatus for coulometric titration

Treatment of experimental data and discussion
1. From the weight of silver deposited calculate the number of moles of silver deposited and hence the number of hydroxyl ions required for neutralization. This gives directly the number of moles of hydrogen ions in 10 cm³ of acid solution. Hence calculate the concentration of potassium hydrogen phthalate in solution.
2. Discuss the advantages of using an instrumental method to locate the end-point.
3. Why is it preferable to use a coulometer rather than to rely on the measurement of the current flowing for a known time?

Experiment 6.19   Standardize a solution of arsenic trioxide by a coulometric titration

Theory: In this experiment iodine, generated anodically from a solution of potassium iodide, is used to titrate the arsenic trioxide solution. When the current passes through the cell, the anode reaction is:

$$2I^- \longrightarrow I_2 + 2\epsilon$$

and the cathode reaction:

$$2H_2O + 2\epsilon \longrightarrow H_2 + 2OH^-$$

The products of the cathode reaction will only slowly diffuse into the bulk of the solution, and will, in any case, have a relatively small effect.

Requirements: Platinum foil anode (2 × 3 cm) fused on to platinum wire, held in a glass tube filled with mercury, for electrical connection, platinum wire cathode immersed in sodium sulphate solution in filter stick; 150-cm³ beaker mounted on magnetic stirrer, milliammeter (0–250 mA), silver coulometer; solution of arsenic trioxide (3 g dm⁻³) slightly acid for stability, 1 mol dm⁻³ sodium sulphate solution, buffered potassium iodide solution (60 g potassium iodide, 10 g sodium bicarbonate in 1 dm³ of water) to which is added 0.2 cm³ arsenic trioxide solution; starch solution.

Procedure: Set up the apparatus in a similar way to Fig. 6.14. This time, the cathode must be separated from the bulk of the solution so that the electrically generated iodine, formed at the anode, will not be stirred into direct contact with the cathode, and thus be reduced.

Clean, dry and weigh the platinum cathode and crucible and replace, with the silver anode, in the coulometer. The potassium iodide solution, stabilized with arsenic trioxide, must first be titrated. Transfer 50 cm³ of the potassium iodide solution to the beaker, add 2–3 drops of starch solution, and, with the electrodes in place switch on the current, and adjust to about 50 mA. At the first sign of a blue colour, switch off the current, carefully wash, dry and weigh the platinum cathode and crucible, taking care not to lose any deposited silver.

Rinse the beaker, and, now introduce 50 cm³ of potassium iodide solution, 10 cm³ arsenic trioxide solution and 2–3 drops of starch solution. Switch on the current (50 mA) and proceed with the titration as previously described; turn off the current when the solution just becomes blue. Dry and weigh the cathode and crucible.

Treatment of experimental data and discussion

1. From the increase in weight of the cathode, calculate the number of moles of silver deposited and hence iodine liberated. Thus, making allowance for the arsenic trioxide originally present as a stabilizer in the potassium iodide solution, calculate the concentration of the arsenic trioxide in solution.

2. Explain why a small amount of arsenic trioxide added to the potassium iodide solution stabilizes the solution.

3. List the main factors which limit the accuracy of the method.

7   Non-aqueous titrations

The poor end-point obtained in the titration of a very weak base (acid) in aqueous solution with a strong acid (base) arises from the basic (acidic) character of the water which is present in large excess. To minimize the competition of the water for the acid (base) and to improve the sharpness of the end-point, a solvent must be used which is a much weaker base (acid) than the solute, but which has a sufficiently high relative permittivity to permit potentiometric measurements. Convenient solvents are glacial acetic acid for the titration of weak bases and acetonitrile or pyridine for weak acids.

Titrations in non-aqueous solution may be carried out either with a suitable indicator or potentiometrically using a glass electrode and calomel electrode system. The operation of the glass electrode in a non-aqueous solvent is believed to depend on the presence of a thin film of water on the glass surface; provided the titration is carried out relatively rapidly this layer can be maintained. The electrode must be stored in water between titrations to regenerate the water layer.

Experiment 6.20   Titrate a weak base with a strong acid in non-aqueous solution

Requirements: Vibret pH meter, glass and calomel electrodes, magnetic stirrer. Ephedrine hydrochloride [$C_6H_5 . CH(OH)CH(NH_2^+ . CH_3)CH_3Cl^-$], solid potassium phthalate, glacial acetic acid, 10% (w/v) solution of mercury (II) acetate in glacial acetic acid, 5% (w/v) solution of crystal violet in glacial acetic acid, 0.05 mol dm^{-3} solution of perchloric acid in glacial acetic acid. (Add slowly, with stirring, 8.5 cm^3 of 7% perchloric acid to 900 cm^3 of glacial acetic acid, followed by 30 cm^3 of acetic anhydride. Allow to stand 24 h.)

Procedure: Extreme care is needed in handling the chemicals; use a pipette bulb for handling solutions made up in glacial acetic acid:

1.   Titration using an indicator. Prepare a blank containing 50 cm^3 of glacial acetic acid, 10 cm^3 of the mercury(II)acetate solution and 3 drops of the crystal violet solution in a 250 cm^3 conical flask. Run in the standard perchloric acid solution from a 5 cm^3 burette until the indicator changes to a mid-green colour. Record the titre and retain the sample as a reference.

    Standardize the perchloric acid solution with potassium hydrogen phthalate (0.25 g dissolved in 50 cm^3 of glacial acetic acid) to which is added 10 cm^3 of mercury(II)acetate solution and 3 drops of the crystal violet solution. Titrate this solution with the perchloric acid solution to the same end-point and record the titre.

    Weigh out accurately about 0.015 g of ephedrine hydrochloride and dissolve 50 cm^3 of glacial acetic acid in a 250 cm^3 conical flask. Add 10 cm^3 mercury(II)acetate solution and 3 drops of the crystal violet solution. Titrate this solution with the perchloric acid solution to the same end-point and record the titre.

2.   Potentiometric titration. Weigh out accurately about 0.25 g of potassium hydrogen phthalate and dissolve in 50 cm^3 of glacial acetic acid in a 100 cm^3 beaker and add 10 cm^3 of the mercury(II)acetate solution. Dry the glass and calomel electrodes by *lightly* touching a tissue against them. Set the pH meter on the 2–8 mV scale and zero the meter. Run in the standard perchloric acid solution in 0.2 cm^3 portions, stir mechanically, and record the emf. of the cell after each addition. Continue adding perchloric acid until the end-point (as

determined with the indicator) has been passed. If it is necessary to change the emf. range on the meter, record the reading on both ranges for a given addition and make a correction for any observed difference.

Repeat the titration using about 0.015 g of ephedrine hydrochloride.

Treatment of experimental data and discussion

1.   For the potentiometric titrations plot the curves showing the variation of the emf. with the volume of acid added and also the first differential curve (Fig. 6.3). Hence locate the end-points and standardize the perchloric acid solution.

2.   Compare the results of the direct and potentiometric titrations and hence determine the purity of the sample of ephedrine hydrochloride.

3.   Explain the purpose of the mercury (II) acetate in the titrations.

4.   Why is it necessary to carry out a blank titration with the indicator?

5.   Discuss other instrumental methods which could be used to determine the end-point of this titration.

8   Thermometric titrations

A thermometric titration is one in which the temperature change occurring during the titration is measured. In the simplest method, the titrant is added discontinuously from a burette into a Dewar flask containing the solution to be titrated, a stirrer and a Beckmann thermometer. After each addition, the temperature rise is measured and the average temperature change per unit volume added is plotted against the total volume of titrant added. A sharp change in this average value occurs at the equivalence point. The results are plotted either as the derivative curve just described or as a thermogram in which the total temperature rise is plotted against the volume of titrant added (Fig. 6.15); the end-point being marked by a change in slope. This simple type of procedure, with an accuracy of about 1%, is suited to reactions which are complete and strongly exothermic.

In more recent titration calorimeters, the titrant is added continuously from a motor-driven syringe burette. The thermometer is replaced by a thermistor and the

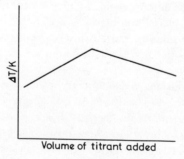

Fig. 6.15   Typical thermogram showing increase of the temperature as a function of the volume of titrant added

unbalanced potential from the bridge circuit, of which the thermistor is part, is continuously recorded. Thermometric titrations can now be achieved by this procedure as rapidly as by any other instrumental method.

Experiment 6.21   Determine the end-point in various titrations by the thermo-metric method

Theory: This experiment merely seeks to demonstrate the principle of the method using very simple apparatus; the sensitivity can be obviously improved in a well-equipped laboratory.

Requirements: Wide-necked Dewar flask, fitted with a cork carrying a stirrer, Beckmann thermometer (or thermometer calibrated to 0.05 °C) and hole for passing through the outlet stem of a 50 cm³ burette. 2.0 mol dm⁻³ solutions of sodium hydroxide and hydrochloric acid, 0.2 mol dm⁻³ solution of copper sulphate.

Procedure:
1.   Acid-base titration. Transfer 25 cm³ of the sodium hydroxide solution to the Dewar flask and add about 25 cm³ of water. Assemble the cork, thermometer (ensuring that the Beckmann thermometer is set for the range room temperature +6 °C, and that the bulb is completely covered) and the stirrer. Stir *gently* until the temperature remains constant; record this value. Run in 4 cm³ of the hydrochloric acid from the burette. Stir gently and after a standard time from the addition (1 min) read the thermometer again. Repeat this procedure of addition of 4 cm³ of acid, stirring, and reading the temperature after 1 min, until 40 cm³ of acid have been added.
      Carry out a standard titration of the acid and base using an indicator.
2.   Precipitation titration. Transfer 50 cm³ of the copper sulphate solution to the Dewar flask, stir and allow temperature equilibrium to be established. Using the same technique as in (1), run in 20 cm³ of 2.0 mol dm⁻³ sodium hydroxide in 1.5 cm³ portions.

Treatment of experimental data and discussion
1.   For each titration plot the graph of temperature against the titre, draw a smooth line through the points intersecting at the thermometric end-point. For the acid-base titration compare the end-point titre with that obtained using an indicator.
2.   For such a titration the water equivalent of the assembly is not required. What additional information can be obtained from your data if the water equivalent is known? What other precautions are necessary to increase the accuracy of the result?
3.   Suggest ways in which the accuracy of this very simple technique could be improved.
4.   Predict the usefulness of the thermometric technique for the titration of weak acids or weak bases.

7

Electrical measurements in chemistry

1 Transport numbers

Electricity is conducted through electrolytes, including solutions, and pure fused salts by the migration of both the anions ($-$) and cations ($+$) to their respective electrodes of opposite sign. The velocity of the ion is proportional to the potential gradient; the *mobility* ($m^2 s^{-1} V^{-1}$) of an ion is defined as the velocity ($m s^{-1}$) with which it moves under an applied field strength of $1 V m^{-1}$.

The conductivity of an electrolyte is the sum of the contributions made by the anions and cations and is related to the mobilities of the ions; thus the molar conductivity of a $1:1$ electrolyte is:

$$\Lambda = \frac{\kappa}{c} = \mathscr{F}(u_+ + u_-) \tag{7.1}$$

where $\kappa/\Omega^{-1} m^{-1}$ is the conductivity of a solution of concentration $c/mol\ m^{-3}$ and u_+ and u_- the mobilities of the cations and anions respectively. Λ increases with dilution; with weak electrolytes, the increase is due to an increase of dissociation, while, with strong electrolytes, the smaller change is attributed to changes in the interionic forces. Since the separate ions in an electrolyte generally have different mobilities their contributions to the total conductivity, $\mathscr{F}u_+$ or $\mathscr{F}u_-$, are not the same.

The *transport* or *transference number* (t) of an ion is defined as the fraction of the total current carried by the ion. Thus, since the total current carried is proportional to ($u_+ + u_-$) it follows that

$$t_+ = \frac{u_+}{u_+ + u_-} \quad \text{and} \quad t_- = \frac{u_-}{u_+ + u_-} \tag{7.2}$$
$$\text{(a)} \qquad\qquad\qquad \text{(b)}$$

whence $t_+ + t_- = 1$.

For a solution of concentration $c/mol\ m^{-3}$:

$$u = \frac{t\kappa}{\mathscr{F}c} \tag{7.3}$$

Owing to forces of interionic attraction, the transport numbers, and hence the ionic mobilities, vary with concentration. At high concentrations, the transport numbers of some ions change markedly in value and often in sign; this is generally due to complex ion formation (eg. $CdI_4^=$) or the formation of micelles (p. 272). The transport number also depends on the extent of solvation of the ion.

The two methods generally used for the determination of transport numbers are the Hittorf and the moving boundary methods. The ratio of the emf. of concentration cells with and without liquid transport (Expt. 7.8) provides another method.

Experiment 7.1   Determine the transport number of copper ions in copper sulphate solution by the Hittorf method

Theory: The 'Hittorf method' is illustrated diagrammatically in Fig. 7.1; after prolonged electrolysis, the electrolytes surrounding the anode and cathode are analysed to determine change in concentration. It is assumed that such changes in concentration occur only in the vicinity of the electrodes, and that the concentration of electrolyte in the intermediate or middle portion (separated by imaginary boundaries XX and YY) is unchanged. This means that electrolysis must be conducted under conditions such that the electrolyte in two, preferably three, compartments can be analysed. Excessive electrolysis must be avoided; the current is chosen sufficiently small to prevent heating and consequent mixing by convection, and sufficiently large to prevent diffusion affecting the small concentration changes. When 1 Faraday of electricity passes through the cell from left to right, t_+ equivalents of cations pass to the right and t_- or $(1 - t_+)$ equivalents of anions pass to the left.

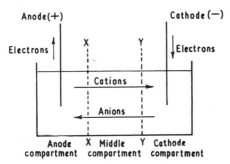

Fig. 7.1   Diagrammatic representation of Hittorf method

In the cell to be studied: $Cu|CuSO_4|Cu$; for the passage of 1 Faraday of electricity the following diagram shows the transfer of cations:

ANODE (+)	MIDDLE	CATHODE (−)
Cations		
1 equiv. gained from anode		1 equiv. lost by deposition
t_+ equiv. $\longrightarrow$		t_+ equiv. $\longrightarrow$
t_+ equiv. lost by transfer		t_+ equiv. gained by transfer
Net gain of $(1 - t_+) =$ t_- equiv. of Cu^{++}	No change	Net loss of $(1 - t_+) =$ t_- equiv. of Cu^{++}

Suppose that Q coulomb of electricity are passed during electrolysis and that the initial concentration of copper is w g in 1 kg of water. If, after electrolysis, the concentration of copper in the anode compartment is w' g in W g of solution, this concentration is w' g copper in $(W - w')$ g water. Thus the increase in concentration of copper/g kg^{-1} in the anode compartment is:

$$\left(w' - \frac{w(W - w')}{1000} \right) = A$$

This net gain is the result of copper ions passing into solution and copper ions migrating from the anode compartment. The weight of copper passing into solution for Q coulombs $= \dfrac{Q \times M_r(Cu)}{2\mathscr{F}}$ g $= B$ g. Thus, the amount actually migrating is $(B - A)$ g. The fraction of the current carried by this amount, ie. $t_+ = (B - A)/B$. B is determined from a silver or copper coulometer in series with the transport number cell. Similar reasoning applied to the cathode compartment also gives t_+.

Requirements: Transport number cell (Fig. 7.2a), electrical circuit (Fig. 7.3), 0–50 milliammeter, d.c., supply (100–200 V), silver coulometer (p. 307), copper sulphate, 0.05 mol dm^{-3} sodium thiosulphate solution, 15% silver nitrate solution, starch indicator.

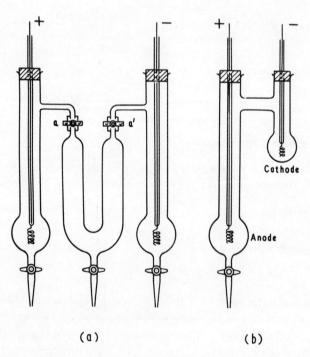

(a)   (b)

Fig. 7.2   Hittorf cells for transport number determination

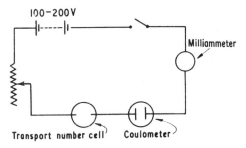

Fig. 7.3   Electrical circuit for transport number determination

Procedure: Prepare a standard solution of copper sulphate (approx. 0.05 mol dm^{-3}) and use this to standardize the sodium thiosulphate solution. Pipette 10 cm^3 of the copper solution into a weighed flask and weigh; add $1-2$ g solid potassium iodide and titrate the liberated iodine. Hence determine the weight of copper equivalent to 1 cm^3 of the sodium thiosulphate solution, and the weight of copper in 1 kg of water originally.

The electrodes in the cell are stout pieces of copper wire wound in the form of a spiral; these must be sheathed in glass tubing sealed at the end with sealing wax. Clean the electrodes with concentrated nitric acid and thoroughly rinse. Wash the cell, grease the taps and fill completely with the copper sulphate solution, replace the electrodes (clips a, a' open).

Set up the coulometer (p. 307) and record the combined weight of the platinum cathode and the sintered-glass crucible, after drying to constant weight. Connect the cell and the coulometer in series with a $0-50$ milliammeter (used only as an indicator), a variable resistance and the power supply (Fig. 7.3). Switch on the current and adjust to about 10 mA and leave for $2-3$ hours; the total quantity of electricity is about 100 C. Protect the apparatus from radiators, sunlight, etc., which might set up convection currents.

At the end of electrolysis, switch off the current, screw up clips a and a' and immediately drain off the solution in the anode, cathode and middle compartments into weighed flasks. Weigh each solution; add potassium iodide and titrate the liberated iodine in each with the standardized sodium thiosulphate solution. Finally, wash, dry and weigh the platinum cathode and sintered crucible in the coulometer.

(If the apparatus Fig. 7.2*b* is used, drain off about 10 cm^3 of solution from the anode and pipette out about 10 cm^3 from the cathode compartment and weigh. This apparatus, while being less accurate, has the advantage of a much lower resistance, and thus does not require such a high voltage supply.)

Treatment of experimental data and discussion

1.   From the titrations calculate the weight of copper contained in the known weight of solution, and, hence the weight of copper in a known weight of water. Using the theory, calculate the transport number of the copper ion from the change in concentration in both the anode and the cathode compartments. What is the transport number of the sulphate ion?

2. Did the concentration of copper remain constant in the middle compartment? Suggest reasons why this might vary and show how such a change would affect the final value of t_+.
3. Write down the reactions which occur at the anode and cathode in this cell.
4. Consider the cell Pt|CuSO$_4$|Pt, in which the electrodes are not attacked during electrolysis; write down the electrode reactions and show how this cell could be used to measure t_+.
5. What additional information is required to permit the calculation of the molar conductivity of each ion?
6. Outline the principles of other methods which may be used for the determination of transport numbers and list the relative advantages of each.

Experiment 7.2 Determine the transport number of the hydrogen ion in hydrochloric acid by the moving boundary method

Theory: The 'moving boundary method' is a direct method capable of high accuracy. Fig. 7.4 represents a section of tube containing suitably chosen electrolytes AX and BX with the common anion X. Let ab be the position of the boundary; the application of a potential difference across the ends of the tube results in the migration of the boundary to the new position cd. All A$^+$ ions will be removed from the volume $abcd$. Thus, the passage of a certain quantity of electricity results in the passage of an amount of A$^+$, across any plane, PQ, in the unchanged electrolyte solution, equal to that originally contained in the volume $abcd$.

Let V/m^3 be the volume swept through by the boundary for the passage of 1 Faraday of electricity through the cell. The number of moles of A passing PQ = Vc_A (where c_A/mol m^{-3} is the concentration of AX); thus, the transport number of A$^+$ = t_+ = Vc_A. If the boundary sweeps out a volume v/m^3, when a current of i A flows for t seconds then:

$$t_+ = \frac{vc_A\mathscr{F}}{it} \qquad (7.4)$$

For a tube of uniform cross-sectional area (A m^2), if the boundary moves a distance l/m, then:

$$t_+ = \frac{lA\mathscr{F}c_A}{it} \qquad (7.5)$$

In this theory it has been assumed that there are no disturbing effects due to interdiffusion or mixing of solutions, that the boundary is unaffected by the nature and concentration of the following ion B$^+$, and that there are no volume changes in the apparatus which offset the uniform movement of the boundary.

The choice of the following or indicator ion and its concentration is governed by the following conditions: (a) it must have a mobility less than that of A$^+$, (b) the conductance of BX must be less than that of AX, and (c) the denser solution must be beneath the less dense one to prevent mixing.

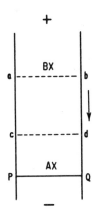

Fig. 7.4   Basis of moving boundary method

Since B^+ has the lower mobility, there will be a greater potential drop in BX than in AX. Thus, if some of the faster moving A^+ ions diffuse, or are carried back into the region BX, they will encounter a higher potential gradient, and will thus be sent forward to the boundary. On the other hand, any B^+ diffusing into AX will be overtaken by the boundary. The net effect is that a boundary of limited diffuseness is maintained. The position of the boundary is conveniently located and followed, either by having an indicator in solution, or by optical methods depending on the refractive index difference of the two solutions.

When a steady state is attained and the boundary is moving, the equation (7.4) derived for A^+ must also apply to B^+. Since the same amount of electricity is passing, it follows that $t_{A^+}/t_{B^+} = c_{A^+}/c_{B^+}$. Thus the concentration of the following solution (BX) is governed by the concentration of AX and the transport numbers of A^+ and B^+.

In this experiment lithium ions are used as the following ion, and methyl orange to reveal the boundary. The ratio $t_{H^+}/t_{Li^+} \approx 0.8/0.3$; thus the concentration of lithium chloride should be about 0.4 times that of the hydrochloric acid.

Requirements: Thermostat at 25.00 ± 0.02 °C, transport number cell (Fig. 7.5a), in which the graduated tube is a portion of a 1-cm³ microburette graduated in 0.01 cm³, cathode (C) is Ag/AgCl, anode (A) is platinum. Electrical circuit (Fig. 7.5b), multirange milliammeter, stop-watch. Standard 0.1 mol dm⁻³ hydrochloric acid solution (coloured with methyl orange), solid lithium carbonate and methyl orange.

Procedure: Clean the electrodes and anodize the cathode (p. 307). Add sufficient methyl orange to the hydrochloric acid so that the red colour is clearly visible in the graduated tube. Standardize this acid solution. To prepare the indicator, or following-ion solution, add solid lithium carbonate in small amounts to 25 cm³ of the coloured acid until the effervescence ceases and the solution just turns yellow. Make the solution up to 100 cm³ with distilled water.

Thoroughly clean the apparatus with chromic acid and rinse well. Dry and adequately grease tap B. Rinse the apparatus twice with the hydrochloric acid

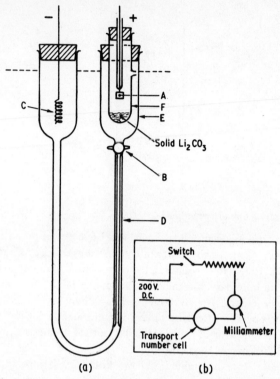

Fig. 7.5 Moving-boundary apparatus for the determination of transport numbers. (a) The cell. (b) The electrical circuit

solution, and completely fill with the coloured acid solution; leave B open. Insert the rubber bung carrying the cathode and make sure there is a tight fit with no trapped air bubbles, close B. Pour the surplus acid from E and wash thoroughly with distilled water and lithium chloride solution, and partially fill with the lithium chloride solution. Add a small amount of solid lithium carbonate to the inner tube F, and fill with the lithium chloride solution, replace the platinum anode in this tube. Slowly lower the compartment F into E and gradually fill E with lithium chloride solution by means of a drawn-out tube fitted with a teat. Immerse the whole apparatus in a glass-fronted thermostat to prevent boundary disturbances through temperature changes.

Connect up the circuit, open tap B and switch on the current (adjust the resistance to give a current of 3–5 mA). When the boundary (red below yellow) reaches the first graduation, start the stop-watch and record the current every 30 s until the boundary has moved at least half-way down the tube. Record the volume swept out by the boundary and the time taken for this.

Treatment of experimental data and discussion

1. Plot the graph of current/A against time/s and calculate the area under the graph. The area, expressed in A s = C, is the total quantity of electricity

passed during the movement of the boundary. Hence calculate t_+ and t_- (equation (7.4)).

2. Explain why a silver–silver chloride electrode is used, and why solid lithium carbonate is placed in the anode compartment.

3. Explain why as electrolysis proceeds the current decreases.

4. Comment on the most probable sources of error in this method and suggest ways in which they could be minimized.

2   Decomposition potential and over-potential

When two copper electrodes dipping in copper sulphate solution are connected to a battery, any emf., however small, causes electrolysis. If these electrodes are replaced by platinum electrodes, there is no appreciable electrolysis until the applied emf. has reached a certain value, the *decomposition potential*. Platinum electrodes in dilute acid solution behave in a similar fashion. The decomposition potential may be determined using the electrical circuit (Fig. 7.6). A variable emf. is supplied by the battery (B) and resistance or metre wire (R), to the platinum electrodes immersed in an electrolyte solution, in this case dilute acid. The applied potential to the electrodes is measured by the high-resistance voltmeter (V), and the current passing through the solution, by the milliammeter (A). When the circuit is complete, the current soon settles down to a steady value for a given applied potential; a typical graph of potential against current for increasing applied potential is shown in Fig. 7.7. The voltage at which the sharp rise of the curve begins is the decomposition potential, E_D; for sulphuric acid $E_D = 1.67$ V (p. 305), above this potential, the products of electrolysis (oxygen and hydrogen) are liberated. The current eventually reaches a limiting value independent of the applied potential; the limiting value is controlled by the rate of diffusion of the ions. At potentials less than E_D, no gas is evolved and a very small current (the

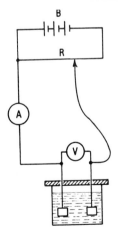

Fig. 7.6   Apparatus for the determination of potential–current curves

residual current) flows. This current is associated with the formation in solution of small amounts of oxygen and hydrogen which diffuse away from the electrodes. On the application of any potential less than E_D, a large current passes momentarily; this soon decays to the value shown in Fig. 7.7. The cell is now polarized, due to the accumulation of the electrolysis products on the electrode surfaces, and, a back emf. (opposing the applied potential) is set up.

If the potential (E) of an electrode is measured while ions flow across the metal/solution interface at a rate dependent on the current density, then without exception, E differs from the reversible electrode potential E_r. Thus

$$E - E_r = \eta \tag{7.6}$$

where η is the *over-potential* or *over-voltage* of the electrode process at the given current density. For cathodic processes, η is negative, and, for anodic processes, positive.

Three types of over-potential are recognized, each due to a different cause.

(*a*) *Concentration over-potential*, η_c, arises when the concentration of positive ions in the immediate vicinity of the cathode surface decreases due to deposition. The cathode potential dependent on this concentration is made less positive (conversely for the anode). If c_b and c_s represent the bulk and surface concentrations of ions of valency n, then the thermodynamic equations are still approximately valid. Thus

$$\eta_c = \frac{RT}{n\mathscr{F}} \ln \frac{c_s}{c_b}$$

η_c can be reduced to zero by adequate stirring of the solution; in the measurement of other kinds of over-potential, it is assumed that η_c has been reduced to negligible proportions.

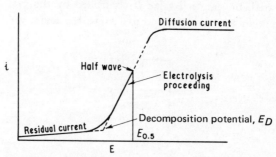

Fig. 7.7   Typical potential—current curves

(*b*) *Activation over-potential*, η_a, arises at electrodes where hydrogen, oxygen or nitrogen are evolved electrolytically. At a platinized platinum surface, hydrogen is liberated practically at the reversible hydrogen potential of the solution; with other electrodes, a more negative potential is required. This is the more common over-potential, which presents a great variety of problems.

(*c*) *Ohmic over-potential*, η_r, exists if an ohmic potential difference is

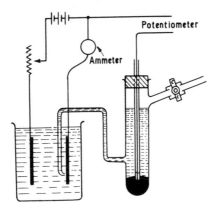

Fig. 7.8   Apparatus for the determination of the potential of an electrode during electrolysis

included in the measured potential. It occurs if the 'probe', used in the study of over-potentials (Fig. 7.8), leading from the working electrode to the reversible electrode, includes some current lines passing from the cathode to the anode. It is also present if the electrode surface is covered by an oxide film. $\eta_r = iR$, where i is the current and R the resistance between the electrode and a point in the probe of equal potential.

These three types of over-potential are easily distinguished by the manner of decay when the polarizing potential is switched off, η_c slowly, η_a exponentially, and, η_r instantaneously.

The apparatus (Fig. 7.8) using the capillary probe is preferred for accurate over-potential measurements, since the measured potential is not affected by the potential drop (iR) which must exist between two electrodes.

Since the over-potential may occur at both the anode $E(Oa)$ and cathode $E(Oc)$, the decomposition potential is given by:

$$E_D = E(\text{Cathode}) + E(Oc) - [E(\text{Anode}) + E(Oa)] \qquad (7.7)$$

where E(cathode) and E(anode) are defined by the Nernst equation.

If a 1.0 mol dm^{-3} solution of zinc bromide is electrolysed between platinum electrodes, the reversible emf. $[E(\text{cathode}) - E(\text{anode})]$, may be calculated as follows:

$$E(\text{cathode}) - E(\text{anode}) = E(Zn^{++}, Zn) - E(Br_2, Br^-)$$

$$= E^{\ominus}(Zn^{++}, Zn) - E^{\ominus}(Br_2Br^-) + \frac{RT}{2\mathscr{F}} \ln c_{Zn^{++}} c^2_{Br^-} + \frac{RT}{2\mathscr{F}} \ln \gamma_{Zn^{++}} \gamma^2_{Br^-}$$

$$\approx 1.80 \text{ V}$$

on substituting values for $E^{\ominus}(Zn^{++}, Zn)$ and $E^{\ominus}(Br_2, Br^-)$ and $\gamma_{\pm} = 0.23$.

This value is in agreement with the experimental value of 1.80 V.

In contrast, is the electrolysis of 1 mol dm^{-3} sulphuric acid between bright platinum electrodes. The ultimate process is the decomposition of water into gaseous hydrogen and oxygen; thus, the emf. required for decomposition should correspond to the free-energy change in the reaction $H_2 + \frac{1}{2}O_2 \rightarrow H_2O$ for which $\Delta G = -236.8$ kJ. Since $\Delta G = -nE\mathscr{F}$, $E_D = 1.23$ V. The measured decomposition potential of acids (or bases) with bright platinum electrodes is about 1.7 V. This can be regarded as the theoretical decomposition potential (1.23 V) + anode over-voltage (0.44) + cathode over-voltage (0.04) = 1.71 V.

Experiment 7.3   Plot the current-potential curves with a series of electrodes in different electrolytes

Requirements. Two platinized platinum, silver, copper, nickel and zinc electrodes, 1 cm^2 of stiff foil mounted on stiff wire and sealed in glass tubing; 0.05 mol dm^{-3} solutions of sulphuric acid, 0.1 mol dm^{-3} sodium sulphate, copper sulphate, zinc sulphate, zinc bromide, 2 mol dm^{-3} hydrochloric acid. Apparatus (Fig. 7.6), milli-ammeter reading to 0.2 A, 0–5 V voltmeter of high resistance, slide wire potentio-meter with tapping key or sliding resistance.

Procedure: Set up the apparatus, but, before connecting the electrodes, take a series of readings of the current for various recorded applied potentials at 0.1 V intervals from 0–3 V. This calibration enables the subsequent ammeter readings to be corrected for voltmeter current, and the voltmeter readings for iR drop in the solution.

Thoroughly clean and prepare the electrodes and mount rigidly about 2 cm apart in the solution under test. Record the current passing at applied voltages in the range 0–2 V (or higher if there is no sudden increase in current) in increments of 0.2 V. Allow at least 2 min for the current to attain a steady value before in-creasing the potential. Observe the potential at which visible electrolysis first occurs.

Set up the following cells, in all cases using a platinized platinum anode, and record the potential–current curves:
1. sulphuric acid; cathode: platinized platinum;
2. sodium sulphate, or zinc sulphate or copper sulphate; cathode: platinized platinum;
3. hydrochloric acid, of the following concentrations, 2.0, 0.5, 0.1, 0.025 mol dm^{-3}; cathode: platinized platinum;
4. sulphuric acid; cathode: silver, copper or nickel;
5. zinc bromide; cathode: platinized platinum.

Treatment of experimental data and discussion
For each of the cells, plot the corrected current against the potential and determine E_D for each.
Cell 1. The cathode can be assumed to be reversible, thus the hydrogen discharge potential is given by $(RT/\mathscr{F}) \ln a_{H^+} = -0.8$ V $(a_{H^+} \approx 0.04)$. Hence, using equation (7.7), calculate the discharge potential of oxygen $[E(\text{Anode}) + E(Oa)]$ at the anode. $E(\text{Anode})$ for the reversible oxygen electrode is given by $E^{\ominus}(O_2, OH^-) + (RT/\mathscr{F}) \ln a_{H^+}$, $E^{\ominus}(O_2, OH^-) = 1.23$ V; hence, calculate the over-potential of oxygen at the anode.

Cell 2. All have the same anode process as Cell 1, thus $[E(\text{Anode}) + E(\text{Oa})]$ is that already calculated. From the observed E_D values for the different solutions, calculate the cathode potential of the cell. This method avoids the use of the more complicated probe (Fig. 7.8).

Cell 3. Determine the E_D values for the four different concentrations of hydrochloric acid. Explain why there is the trend with increasing dilution of acid.

Cell 4. All have the same platinized platinum anode and hence same $[E(\text{Anode}) + E(\text{Oa})]$ as Cell 1. From the E_D values, calculate the hydrogen over-potential for each metal.

Cell 5. Compare the observed decomposition potential of this cell with that calculated for a reversible cell (mean ionic activity coefficients, p. 316).

3 Polarography

The polarographic method of chemical analysis is based on the unique characteristics of the current—voltage curves obtained when a continuously increasing negative potential is applied to a mercury dropping cathode immersed in a solution containing reducible ions, using a mercury pool as anode.

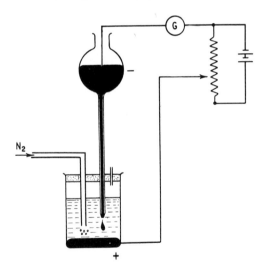

Fig. 7.9 Schematic apparatus for polarography

The dropping electrode (Fig. 7.9) consists of a capillary glass tube (0.03−0.05 mm diameter) supplied with mercury from a reservoir. The height of the reservoir is adjusted so that the mercury issues in drops (0.5 mm diameter) at a rate of one every 3 s. The electrolysis current seldom exceeds 50 μA and it is therefore necessary to use a sensitive current measuring device; a long-period galvanometer is commonly used. Oxygen must be swept out of the solution, either by passing through an inert gas (nitrogen) or by adding sodium sulphite to the solution.

Figure 7.7 shows a typical polarogram obtained with an air-free solution (eg. zinc sulphate) in 0.1 mol dm^{-3} potassium chloride solution. The current determining cathode reaction is the reduction of the metal ion to form a very dilute metal amalgam on the surface of the mercury drops, eg.

$$Zn^{++} + 2\epsilon + Hg \rightleftharpoons Zn \, (Hg)$$

The limiting current is directly proportional to the concentration of metal ions; this may therefore be used to measure small concentrations of metal ions in solution. Owing to the periodic change in area during the formation and breaking away of the drop, the current oscillates between a minimum value (near to zero) and a maximum value at the moment a drop falls. The current measuring device must therefore have a period several times greater than the drop time.

In a solution containing several reducible substances, each will produce its own wave, and provided their reduction potentials are sufficiently separated, it is possible to detect and determine each concentration from a polarogram.

The decomposition potential varies with the concentration of the substance and therefore cannot be precisely defined. It is standard practice to report the reduction potential as the potential at the mid-point of the wave when the current is half its limiting value. This half-wave potential ($E_{0.5}$) is related to the standard electrode potential of the electrode reaction and is independent of the concentration.

In principle, the apparatus required for polarographic measurements is simple, consisting of (i) a means of applying a variable and known voltage from 0 to 2 or 3 V to the cell, (ii) a means of measuring the resulting current which is usually less than 100 μA and (iii) a cell with a dropping mercury cathode. The applied voltage should be known to ±0.001 V while the sensitivity of the current measuring device should be at least 0.01 μA.

Manual and recording polarographs are commercially available.

The dropping mercury electrode and polarographic cells

The dropping electrode consists of a glass capillary tube supplied with mercury from a reservoir under a head of 30—60 cm. The recommended characteristics of such an electrode are: drop time 2—5 s, rate of flow of mercury 1—3 mg s^{-1} under an applied pressure of 30—60 cm. The radius (r) of the capillary is more critical than its length (l) in determining the drop time (t):

$$t = \frac{5.5 \times 10^{-7} l}{r^3 h}$$

where h is the applied pressure (cm of mercury).

Electrical connection to the mercury in the reservoir is made by platinum wire sealed into glass tubing, partly filled with mercury and held in place by the stopper of the reservoir. The simple arrangements shown in Fig. 7.10 are very satisfactory; the anode in (a) is a mercury pool, and in (b) it is a standard reference electrode, eg. calomel. Both cells are equipped with a tube to displace dissolved oxygen, the inert gas must not be bubbled during a measurement.

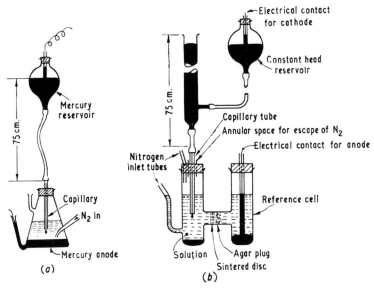

Fig. 7.10   Polarographic cells. *(a)* Simple arrangement devised by Heyrovsky. *(b)* Cell employing permanent external anode

Care must be taken that the capillaries do not become contaminated and clogged with dirt. The dropping electrode should never be allowed to stand in a solution when mercury is not flowing. After use the capillary should be withdrawn from the cell and washed thoroughly with distilled water while the mercury is still issuing. The reservoir is then lowered until the mercury just stops dropping from the jet; it is then allowed to stand in distilled water.

Experiment 7.4   Study the polarographic waves produced by dissolved oxygen

Theory: In all aqueous solutions there is always a certain amount of dissolved oxygen, unless steps have been taken to eliminate it. This oxygen is reducible at the dropping mercury cathode giving rise to polarographic waves.

The concentration of dissolved oxygen varies according to the concentration of dissolved electrolyte, but, with solutions generally used, the diffusion current is comparable with those produced with a 10^{-3} mol dm^{-3} solution of a metal. Dissolved oxygen must therefore be eliminated in analyses of other reducible species in solution. This is best achieved by degassing the solution for 10–15 min with a steady stream of hydrogen or oxygen-free nitrogen. Alkaline solutions (but not acid) can be deoxygenated by adding sodium sulphite (0.1 g in 10 cm^3 of solution) and allowing to stand for 15 min.

Requirements: Polarograph and dropping mercury assembly, 0.1 mol dm^{-3} potassium chloride solution, cylinder of hydrogen or nitrogen.

Procedure: Set up the polarograph and electrode assembly and standardize according to the instructions. Add some 0.1 mol dm^{-3} potassium chloride solution to the cell; insert the capillary and adjust the height of the mercury reservoir until the drop time is 3—6 s per drop. Plot the polarogram (ie. $i - E$ curve) of this solution from 0 to -1.9 V using the correct sensitivity setting to give the maximum displacement on the chart, or maximum reading on the microammeter.

Pass a slow stream of hydrogen or nitrogen through the solution for 15 min (with the capillary tube inserted). The gas passed into the cell must be previously saturated with water at the same temperature as the cell. If this is omitted, then there will be a change in concentration of the metal ion solution; this is serious in quantitative work. Stop the flow of gas and plot the polarogram of this solution on the previous graph for comparison.

Treatment of experimental data and discussion
1. Compare the polarograms before and after degassing the potassium chloride solution.
2. From the first polarogram state the number of stages by which gaseous oxygen is reduced and determine the half-wave potentials. Write down the reactions which give rise to the polarographic waves in both acidic and alkaline solution.

Experiment 7.5   Determine the half-wave potential of the cadmium ion in 0.1 mol dm^{-3} potassium chloride solution

Theory: In elucidating the nature of a reduction process occurring at the dropping mercury cathode, it is necessary to know the number of electrons involved. The shape of the wave depends on this quantity and therefore provides a means for its determination.

On theoretical grounds, Heyrovsky and Ilkovic derived the equation:

$$E = E_{0.5} - \frac{RT}{n\mathcal{F}} \ln \frac{i}{i_d - i} \tag{7.8}$$

where i_d is the limiting diffusion current, and i the diffusion current at some point on the wave when the applied potential is E, and n is the number of electrons involved. The half-wave potential, $E_{0.5}$, characteristic of the reducible ions in solution, but independent of their concentration, is equal to the applied potential, when $i = i_d/2$.

Requirements: Polarograph and electrode assembly, 0.1 mol dm^{-3} cadmium sulphate solution, 1.0 mol dm^{-3} potassium chloride solution, 0.2% (w/v) gelatine solution; cylinder of hydrogen or nitrogen.

Procedure: Set up the polarograph and electrode assembly and standardize according to the instructions. Pour some 0.1 mol dm^{-3} potassium chloride solution into the cell, insert the capillary and adjust the height of the mercury reservoir until the drop time is 3—6 s per drop.

Dilute the cadmium sulphate solution to give a 2.5×10^{-2} mol dm^{-3} solution. To 10 cm^3 of this solution, in a 100-cm^3 volumetric flask, add 10 cm^3 of 1.0 mol dm^{-3} potassium chloride solution, 10 cm^3 of gelatine solution and make up to the mark. Transfer a portion of this to the cell and degas with hydrogen or nitrogen for 10–15 min. Turn off the gas and plot the polarogram, with the sensitivity set at a low value, over the range 0 to −1.5 V. Repeat with an increased sensitivity (this may now be calculated), so that the limiting diffusion current gives almost full deflection on the recording paper. If the polarogram now consists of a series of waves of varying height (caused by the excursions of the pen which occur with the formation and fall of the drop), adjust the damping until a smooth curve is obtained. Repeat this procedure several times on the same solution.

Finally, plot a polarogram on a degassed solution prepared as above, but, with the omission of the gelatine.

Treatment of experimental data and discussion

1. Compare the polarograms measured in the presence and absence of gelatine. What is the purpose of gelatine?
2. From the polarograms, measured in the presence of gelatine, determine the current, $i/\mu A$ (from the chart reading and known sensitivity) at various values of E and tabulate values of i_d, i, and $\log(i/i_d - i)$. Plot the graph of E against $\log i/(i_d-i)$ and determine n and $E_{0.5}$ (equation (7.8)). Quote the value of $E_{0.5}$ relative to the saturated calomel electrode.
3. Why is it necessary to have a high concentration of inert electrolyte (KCl) present in the solution for polarographic analysis?
4. Explain why it is possible to plot many polarograms on the same sample of solution.
5. Explain the advantages of a dropping mercury electrode over a stationary electrode.

Experiment 7.6   Construct a wave-height–concentration curve for the cadmium ion

Theory: Using the dropping mercury cathode and a non-polarizable electrode in a solution of a simple depositable ion, it is possible to obtain complete concentration polarization and a limiting current flow. The limiting current is composed of (*a*) a residual current, i_r, (*b*) a migration current, i_m, (*c*) a diffusion current, i_d, and (*d*) a possible adsorption current, i_a.

In the presence of an excess of base or supporting electrolyte, of concentration at least 100 times that of the simple ion, and a maximum suppressor, i_m and i_a are eliminated. The i_r factor is independent of the concentration and can be allowed for. The i_d factor, arising from the diffusion of the simple ions, alone is responsible for the wave, the height of which serves as a basis of quantitative analysis. Ilkovic derived the equation:

$$i_d = kn \mathscr{F} cD^{1/2} m^{2/3} t^{1/6} \tag{7.9}$$

relating the diffusion current to (*n*) the number of electrons involved in the reduction of one mole of the electroactive substance, (*c*) the concentration, (*D*) the diffusion coefficient of the reducible substance, (*m*) the weight of mercury, flowing through the capillary per second and (*t/s*) the time necessary for the formation of one drop of mercury; *k* is a constant. With all other factors remaining constant, i_d is proportional to the concentration, ie. $i_d = Kc$.

Requirements: Polarograph and electrode assembly, 0.1 mol dm^{-3} cadmium sulphate solution, 1.0 mol dm^{-3} potassium chloride solution, 0.2% (w/v) gelatine solution; cylinder of hydrogen or nitrogen.

Procedure: Using the procedure of Expt. 7.5, plot the polarograms, each at suitable sensitivity, for solutions of the following final concentrations of cadmium sulphate: 10^{-2}, 2×10^{-3}, 10^{-3}, 5×10^{-4}, 2.5×10^{-4}, 10^{-4}, 5×10^{-5}, 2.5×10^{-5} and 10^{-5} mol dm^{-3}, always including in each 100 cm^3 of solution 10 cm^3 of 1 mol dm^{-3} potassium chloride and 10 cm^3 of gelatine solutions.

Treatment of experimental data and discussion

1. From the chart reading and the sensitivity, calculate i_d for each concentration. Plot graphs of i_d against the concentration and $\log i_d$ against $\log c$. Why is the latter a better plot?
2. Is the half-wave potential for Cd^{++} constant for solutions of CdSO$_4$ of different concentration?
3. Discuss the errors involved in the analysis of an unknown solution by this method.
4. Is there a change in concentration of the Cd^{++} in solution when a polarogram has been plotted?

Experiment 7.7   Plot the polarographic waves for a mixture of simple ions

Theory: In the case of simple ions, where no complex formation occurs, $E_{0.5}$ for the reduction process is independent of the metal ion concentration, and is practically unchanged by the alteration of the concentration of the supporting electrolyte, or by the presence of other reducible ions. Thus, a mixture of reducible ions can be analysed provided that the $E_{0.5}$ values are at least 0.2 V apart.

Requirements: Polarograph and electrode assembly, 0.01 mol dm^{-3} solutions of cadmium sulphate, zinc sulphate and manganese sulphate, 1.0 mol dm^{-3} potassium chloride solution and 0.2% (w/v) gelatine solution, cylinder of hydrogen or nitrogen.

Procedure: Using the procedure of Expt. 7.5, plot the polarograms of solutions containing 10 cm^3 potassium chloride, 10 cm^3 of gelatine and (a) 10 cm^3 cadmium sulphate, (b) 10 cm^3 zinc sulphate, (c) 10 cm^3 manganese sulphate, and (d) 10 cm^3 of each metal solution, made up to 100 cm^3 in each case.

Carefully pipette a suitable quantity of solution (d), eg. 10 cm^3 into the cell and plot the polarogram; to this solution, add exactly 0.1 cm^3 of one of the metal solutions, degas, and again plot the polarogram. Repeat for other small additions.

Treatment of experimental data and discussion

1. Compare the half-wave potential and the limiting diffusion current for each metal ion when measured separately and in the mixture.
2. Demonstrate how, by the addition of small amounts of a standard metal solution, the concentration originally present can be calculated. Is a volume correction necessary?

3. How might ions with nearly the same half-wave potentials be identified?
4. Why is it necessary to have a high concentration of inert electrolyte (KCl) in the solution for polarographic analysis?

4 Electromotive force of cells

All cells consist of a series of conducting phases in contact, ie. electrodes and one or more liquid electrolytes. Phase boundary potentials exist at any boundary between phases of different composition; thus, the emf. of a cell is the algebraic sum of all the phase boundary potentials, including any metal–metal contact potentials which may be present.

In the representation of cells, phase boundaries are shown by vertical lines; these are dotted for boundaries between two electrolyte solutions. The emf. is given a positive sign, by convention, if the electrode at the right is positive. Thus the Daniell cell is written

$$\ominus \quad Cu'|Zn|ZnSO_4 \vdots CuSO_4|Cu \quad \oplus \quad E \text{ (298 K)} = 1.1 \text{ V}$$

in addition, there must be a copper–zinc boundary (Cu'/Zn) which is not usually indicated, the Cu' representing the wire connection between the electrodes. The emf. is the potential difference between two pieces of metal with identical composition, ie. the ends of the chain of conducting phases (Cu–Cu').

When the external circuit of the Daniell cell is completed, positive current flows from left to right inside the cell, and right to left outside; the electron flow outside is left to right, ie. Zn to Cu via Cu'. For the spontaneous operation of the cell the electrode reactions are:

$$\text{Zn electrode:} \quad Zn \longrightarrow Zn^{++} + 2\epsilon \text{ (oxidation)}$$

$$\text{Cu electrode:} \quad Cu^{++} + 2\epsilon \longrightarrow Cu \text{ (reduction)}$$

The overall reaction occurring in the cell is:

$$Zn + Cu^{++} \longrightarrow Zn^{++} + Cu$$

from which the energy of the cell is obtained.

The decrease in the Gibbs' free energy $(-\Delta G)$ is related to the emf., E, for a *reversible cell* by the equation

$$-\Delta G = nE\mathscr{F} \text{ joules} \tag{7.10}$$

where n is the number of Faradays of electricity passing through the cell for the number of moles, indicated in the chemical equation, to react. Thus for the Daniell cell, $\Delta G = -212.3$ kJ.

This free-energy decrease is the same whether the chemical reaction occurs in a cell, or, by the displacement of copper from copper sulphate by zinc.

Thus ΔG can be calculated from the emf. of a reversible cell discharging infinitely slowly. The open-circuit emf. of the cell must therefore be measured with a potentiometer (whereby an opposing emf. is applied so that no current flows) or with a digital voltmeter.

Substituting equation (7.10) in the Gibbs–Helmholtz equation:

$$\Delta G - \Delta H = T \left(\frac{\partial(\Delta G)}{\partial T} \right)_P \tag{7.11}$$

gives

$$n\dot{E}\mathscr{F} + \Delta H = n\mathscr{F}T \left(\frac{\partial E}{\partial T} \right)_P \tag{7.12}$$

and

$$\Delta S = - \left(\frac{\partial(\Delta G)}{\partial T} \right)_P = n\mathscr{F} \left(\frac{\partial E}{\partial T} \right)_P \tag{7.13}$$

Thus, the measurement of the emf. and the temperature coefficient of the emf. enables ΔG, ΔH and ΔS to be calculated.

All reversible electrode equilibria involve two opposing reactions, an oxidation and a reduction reaction, and can be represented generally as:

$$\text{Oxidized state} + n\epsilon \rightleftharpoons \text{Reduced State}$$

The electrode potential $E(O, R)$ for such a system is given by the Nernst equation:

$$E(O, R) = E^{\ominus}(O, R) + \frac{RT}{n\mathscr{F}} \ln \frac{a \text{ (oxidized)}}{a \text{ (reduced)}} \tag{7.14}$$

where $E^{\ominus}(O, R)$ is the standard electrode potential.

Thus for the hydrogen electrode:

$$\text{H}^+ + \epsilon \rightleftharpoons \tfrac{1}{2}\text{H}_2(\text{g})$$

$$E(\text{H}^+, \text{H}_2) = E^{\ominus}(\text{H}^+, \text{H}_2) + \frac{RT}{\mathscr{F}} \ln \frac{a_{\text{H}^+}}{p_{\text{H}_2}^{1/2}} \tag{7.15}$$

and for a metal electrode:

$$\text{M}^{n+} + n\epsilon \rightleftharpoons \text{M}$$

$$E(\text{M}^{n+}, \text{M}) = E^{\ominus}(\text{M}^{n+}, \text{M}) + \frac{RT}{\mathscr{F}} \ln a_{\text{M}^+} \tag{7.16}$$

since the metal is in its standard state and $a_M = 1$.

Single electrode potentials cannot be measured: the accepted convention is that the standard hydrogen electrode is the arbitrary zero; thus all other electrode potentials are related to this. Thus, if an electrode, when coupled to the standard hydrogen electrode, is the positive electrode of the cell of emf. $+E$, the electrode potential is $+E$. The algebraic sum of the separate electrode potentials of the electrodes in the cell gives the emf. E:

$$E = E_2 - E_1$$

where the electrode potentials E_1 and E_2 may be either positive or negative with respect to the solution.

When two reversible electrode systems are connected without free mixing of solutions, electron transfer takes place through the external circuit when the electrodes are connected, oxidation occurring at the left-hand electrode and reduction at the right-hand electrode (cf. Daniell cell).

For the general cell reaction which occurs for n Faradays:

$$aA + bB \rightleftharpoons lL + mM$$

$$\Delta G = \Delta G^{\ominus} + RT \ln \frac{a_L^l a_M^m}{a_A^a a_B^b} \tag{7.17}$$

whence, the emf. E is given by:

$$E = E^{\ominus} - \frac{RT}{n\mathscr{F}} \ln \frac{a_L^l a_M^m}{a_A^a a_B^b} \tag{7.18}$$

where a_L, a_M, a_A, and a_B are the activities of the various species present in the cell. This same equation can be obtained from the algebraic sum of the two separate electrode potentials (equation (7.14)).

Since

$$\Delta G^{\ominus} = -RT \ln K_{Therm} \tag{7.19}$$

it follows from equation (7.10), that

$$E^{\ominus} = \frac{RT}{n\mathscr{F}} \ln K_{Therm} \tag{7.20}$$

Thus emf. measurements of simple galvanic cells provides data for the determination of activities of ions in solution, standard electrode potentials and equilibrium constants.

In contrast to simple cells, no overall chemical reaction occurs in *concentration cells*; the process resulting in the production of an emf. is merely the transfer of some substance from one concentration to another. If two electrodes can be found, each reversible to one of the ions of an electrolyte, then a concentration cell without transport can be constructed:

$$\ominus A \mid \underset{m_1}{AX \text{ aq.}} \mid \underset{\text{solid}}{BX} \mid B\!-\!B \mid \underset{\text{solid}}{BX} \mid \underset{m_2}{AX \text{ aq.}} \mid A \oplus \tag{7.21}$$

where molality $m_2 > m_1$. The net result for the passage of one Faraday through the cell is the transfer of one mole of AX from m_2 to m_1. The emf. E of such a cell, when AX is a $1:1$ electrolyte, is:

$$E = \frac{2RT}{\mathscr{F}} \ln \frac{a_2}{a_1} = \frac{2RT}{\mathscr{F}} \ln \frac{m_2\gamma_2}{m_1\gamma_1} \tag{7.22}$$

where m, a and γ (with the appropriate subscripts) are the mean ionic molality, activity and activity coefficient respectively, defined as $m = m_\pm = (m_+ \, m_-)^{1/2}$, $a = a_\pm = (a_+ \, a_-)^{1/2}$, etc., for $1:1$ electrolytes.

This concentration cell is simply a combination of two separate cells connected back to back, ie.

$$\begin{array}{ll} \text{B} \mid \text{BX} \mid \text{AX aq.} \mid \text{A} & \text{of emf.} \, E_1 \\ \quad \text{solid} \quad m_1 & \end{array}$$

$$\begin{array}{ll} \text{B} \mid \text{BX} \mid \text{AX aq.} \mid \text{A} & \text{of emf.} \, E_2 \\ \quad \text{solid} \quad m_2 & \end{array}$$

whence $E = E_1 - E_2$. E_1 and E_2 can obviously be measured separately.

In equation (7.22), E, m_1 and m_2 can be measured; it is thus possible to calculate the ratio of the mean ionic activity coefficients (γ_2/γ_1) at the two concentrations. If m_1 could be made so dilute that $\gamma_1 = 1$, then γ_2 at any molality m_2 could be determined from a single emf. measurement. This is not possible experimentally, and, instead, the emf. values of a series of cells are measured in which m_2 is kept constant and m_1 varied; an extrapolation to zero ionic strength using the Debye–Hückel theory permits the calculation of the activity coefficient.

It is not always possible to construct such a concentration cell, as electrodes reversible to both ions are not available. If only one is available, a concentration cell with transport can be constructed:

$$\begin{array}{cc} \text{A} \mid \text{AX aq.} \mid \text{AX aq.} \mid \text{A} \\ \quad m_1 \qquad m_2 \end{array} \tag{7.23}$$

the emf. of which is given by:

$$E = 2t_- \frac{RT}{\mathscr{F}} \ln \frac{a_2}{a_1} = 2t_- \frac{RT}{\mathscr{F}} \ln \frac{m_2\gamma_2}{m_1\gamma_1} \tag{7.24}$$

in the derivation of which, it is assumed that the transport number of the anion (t_-) does not vary between m_1 and m_2.

Comparison of the emf. of cells with and without transport, from equations (7.22) and (7.24) gives:

$$\frac{E_{\text{trans.}}}{E} = t_- \tag{7.25}$$

Construction of cells and measurement of emf.

When the cell consists of two electrodes in a single electrolyte, it is most convenient to mount them in the same vessel of electrolyte. When two electrolytes or the same

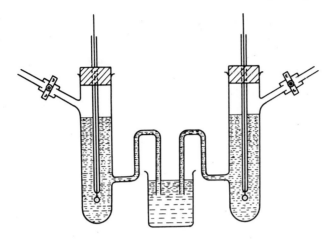

Fig. 7.11   Simple method of constructing a cell

electrolyte at two different concentrations are used in the cell, a liquid junction will inevitably be present. The electrodes are then most conveniently mounted in boiling tubes which carry side arms (Fig. 7.11) dipping into a central beaker. This beaker may contain either a saturated solution of KCl, KNO_3 or NH_4NO_3, forming a salt bridge between the two solutions, thereby minimizing the liquid junction potentials, or, the more dense of the two solutions of the same electrolyte, as in concentration cells with transport. Care must be taken that the clips on the side arms are tight, or siphoning will occur.

In the measurement of the emf. of a cell, the following details must be rigidly followed.

1.   The measurements must be made on duplicate sets of electrodes and duplicate aliquots of solution. The cause of any discrepancy in the results should be located and remedied, eg. clean and prepare electrodes again, more thorough washing of glassware on changing solutions.

2.   The emf. must not be measured immediately; this applies particularly to cells using gas electrodes, as some time is necessary for equilibrium to be established. A series of measurements should be made until three concordant results are obtained.

3.   If a commercial potentiometer is available, it must be used on the most sensitive range appropriate to the emf. under measurement. Further, the circuit must only be closed momentarily, to avoid polarizing the cell under test or the standard cell.

4.   Commercial digital voltmeters, of high resistance may be used for measurement of emf. values. Such instruments have the additional advantage that they indicate the establishment of equilibrium conditions.

5.   The temperature of the cell must be known; for preference it should be thermostatted.

Experiment 7.8   Measure the emf. of some simple and concentration cells using the hydrogen and silver—silver chloride electrodes

Theory: The following cells are to be studied:

1 PtH_2 | HCl 0.01 mol dm^{-3} | AgCl(s) | Ag E_1

2 PtH_2 | HCl 0.10 mol dm^{-3} | AgCl(s) | Ag E_2

3 PtH_2 $\left|\begin{array}{c} HCl \\ 0.01\ mol\ dm^{-3} \end{array}\right|$ AgCl(s) | Ag—Ag | AgCl(s) $\left|\begin{array}{c} HCl \\ 0.10\ mol\ dm^{-3} \end{array}\right|$ H_2Pt E_3

4 PtH_2 | HCl 0.01 mol dm^{-3} ⫶ HCl 0.10 mol dm^{-3} | H_2Pt E_4

5 Ag | AgCl(s) | HCl 0.10 mol dm^{-3} ⫶ HCl 0.01 mol dm^{-3} | AgCl(s) | Ag E_5

6 Ag | AgCl(s) $\left|\begin{array}{c} HCl \\ 0.10\ mol\ dm^{-3} \end{array}\right.$ ⫶ Satd. KCl ⫶ $\left.\begin{array}{c} HCl \\ 0.01\ mol\ dm^{-3} \end{array}\right|$ AgCl(s) | Ag E_6

Cells 1 and 2 are typical galvanic cells from which free energy changes can be determined. Cell 3 is a concentration cell without transport and the emf. is given by equation (7.22). Cells 4 and 5 are concentration cells with transport, having electrodes reversible to cation and anion respectively (equation (7.24)). In cell 6 the solutions are separated by a saturated solution of KCl which, more or less, eliminates the liquid junction potential, so the emf. is equal to the difference of two Ag, AgCl, Cl$^-$ electrodes. Thus:

$$E_6 = E(0.01) - E(0.1) = \left(E^{\ominus} - \frac{RT}{\mathscr{F}} \ln a_1 \right) - \left(E^{\ominus} - \frac{RT}{\mathscr{F}} \ln a_2 \right)$$

$$= \frac{RT}{\mathscr{F}} \ln \frac{a_2}{a_1} \tag{7.26}$$

Requirements: Potentiometer and galvanometer or digital voltmeter, two hydrogen electrodes (p. 123), two silver—silver chloride electrodes (p. 307), cell (Fig. 7.11), standard 0.1 mol dm^{-3} and 0.01 mol dm^{-3} hydrochloric acid, saturated potassium chloride solution, supply of pure washed hydrogen.

Procedure: Set up the potentiometer circuit and standardize with a Weston cell, or standardize the digital voltmeter. First check the two hydrogen electrodes against one another by dipping both in the same beaker of acid (0.1 mol dm^{-3}). If the emf. is greater than 0.5 mV, replatinize the electrodes. Similarly check the two silver—silver chloride electrodes against one another, using the apparatus shown in Fig. 7.11.

Make sure that the hydrogen gas is saturated by bubbling it through a bottle of the solution before it enters this solution in the cell; keep the pressure constant at atmospheric pressure. Set up four half-cells, one hydrogen and one silver—silver chloride containing 0.1 mol dm^{-3} hydrochloric acid dipping into a beaker of 0.1 mol dm^{-3} acid, and the other pair similarly using 0.01 mol dm^{-3} hydrochloric acid. Measure the emf. of each of the six cells, allowing sufficient time for the emf.

to become steady. Duplicate readings on different samples of the acid should be the same. Record the polarity of each electrode and also the temperature.

Treatment of experimental data and discussion

1.   Write down the electrode reactions and hence the overall cell reaction for cells 1 and 2. Deduce the equation for the emf. of these cells in terms of the mean ionic activity of hydrochloric acid and the pressure of hydrogen. Determine the free energy of formation of HCl (aq) at the two concentrations, comment on the values and compare these with $\Delta G_f^{\ominus}(\text{HCl, aq})$ obtained from tabulated data.

2.   Calculate the ratio of the mean ionic activity coefficients of HCl at concentrations of 0.1 and 0.01 mol dm^{-3}, from E_3.

3.   From the values of E_3, E_4 and E_5 calculate the transport numbers of H$^+$ and Cl$^-$; these values refer to a concentration which is approximately the geometric mean of the two concentrations used.

4.   From theoretical considerations prove that: $E_3 = E_1 - E_2$, $E_3 = E_4 + E_5$, and $E_3 = 2E_6$. Use the experimental data to confirm these identities.

Experiment 7.9   Determine the electrode potentials of some metal electrodes

Theory: The aim of this experiment is to measure the emf. of various cells, determine the electrode potential of metal electrodes and consider the application of the Nernst equation (7.16) to the metal electrode potentials.

Requirements: Potentiometer and galvanometer or digital voltmeter, electrodes of various metals, eg., Cd, Cu, Pb, Ag, Zn, saturated calomel electrode (dip type with ceramic disc), standard 0.1 mol dm^{-3} solutions of CdCl$_2$, CuSO$_4$, Pb(NO$_3$)$_2$, AgNO$_3$, ZnSO$_4$, salt bridge, either 2 mol dm^{-3} NH$_4$NO$_3$ solution in H-tube (p. 307) or 2 mol dm^{-3} NH$_4$NO$_3$ solidified with 4% agar.

Procedure: Before use, clean the electrodes by dipping in 1:1 nitric acid, wash thoroughly and store in distilled water until required. Set up and standardize the potentiometer or digital voltmeter.

Pour some of the 0.1 mol dm^{-3} CdCl$_2$ solution into a small beaker and dip in the cadmium and calomel electrodes. Measure the emf. between these electrodes at regular intervals until it attains a steady value, record this emf., the polarity of the electrodes and the temperature. Repeat this determination using standard CdCl$_2$ solutions of 0.01 and 0.001 mol dm^{-3}.

Set up similar cells for the other metal electrodes and their appropriate solutions at the three different concentrations. For the lead and silver electrodes it will be necessary to interpose a NH$_4$NO$_3$ salt bridge between the calomel electrode and the nitrate solution.

Finally measure the emf. and record the polarity of the electrodes in the following cells in which an ammonium nitrate bridge separates the 0.1 mol dm^{-3} metal solutions: (*a*) Zn/Zn^{++} with Cu/Cu^{++}, (*b*) Ag/Ag$^+$ with Cu/Cu^{++} and (*c*) Ag/Ag$^+$ with Zn/Zn^{++}.

Treatment of experimental data and discussion

1. From the measured emf. of the simple cells, the known polarity of the electrodes and E(calomel) calculate the electrode potential for each metal in the three solutions of different concentration. Using the Nernst equation (7.16) and the activity coefficient (p. 316) estimate the standard electrode potential for each metal from the data on each of the three cells.

    In one well constructed table present the following using one line for each cell (ie. each concentration of electrolyte): the emf. of the cell, the electrode potential, $E(M^+, M)$, the standard electrode potential, $E^{\ominus}(M^+, M)$, the spontaneous reaction and ΔG for this reaction.

2. Comment on the applicability of the Nernst equation to the experimental data.

3. For one of the electrodes, plot a graph of $E(M^+, M)$ against $\log c$, determine the slope of the line and compare it with that predicted by equation (7.16).

4. For the three cells, a, b, and c, formulate the cells according to the Stockholm convention and write down the spontaneous reaction for each. Compare the emf. of each cell with the algebraic sum of the electrode potential determined previously, comment on your observations. The emf. values of these three cells are related; deduce this relation and test it with the experimental data.

5. Explain the function of a salt bridge. What properties must it possess to fulfil its function? Explain why KCl acts as a salt bridge in some of the cells.

6. Explain the significance of the standard electrode potential for a reversible electrode.

Experiment 7.10   Determine the mean ionic activity coefficient of HCl from emf. studies of a galvanic cell

Theory: The emf. of the cell:

$$\ominus\ Pt, H_2\ (1\ atmos.)\ \left|\ \underset{m}{HCl\ (aq.)}\ \right|\ AgCl, Ag\ \oplus$$

is given by:

$$E = E^{\ominus} - 2\,\frac{RT}{\mathscr{F}}\,\ln a_{\pm}$$

$$= E^{\ominus} - 2\,\frac{RT}{\mathscr{F}}\,\ln m_{\pm}\gamma_{\pm} \tag{7.27}$$

where $a_{\pm}$, $m_{\pm}$ and $\gamma_{\pm}$ are the mean ionic activity, molality and activity coefficient respectively of HCl. Measurements of E for several values of m can be used to determine $E^{\ominus}$ and hence $\gamma_{\pm}$.

Requirements: Potentiometer and galvanometer, hydrogen electrode (p. 123), silver–silver chloride electrode (p. 307), source of pure, oxygen-free hydrogen, cell assembly (Fig. 7.12), thermostat at $25.00 \pm 0.01\ ^{\circ}C$, standard approx. $1.0\ mol\ dm^{-3}$ hydrochloric acid.

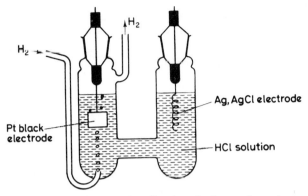

Fig. 7.12   Simple cell assembly for galvanic cells using a hydrogen electrode

Procedure: Very careful work is necessary to obtain the required precision; particular care should be taken in accurately diluting the solutions, in temperature control and in the measurement of emf. Accurately dilute by volume the standard 1.0 mol dm^{-3} hydrochloric acid to give about 250 cm^3 of solutions of about 0.15, 0.11, 0.07, 0.03, 0.01 and 0.005 mol dm^{-3}.

Standardize the potentiometer with the standard Weston cell and measure the emf. of the cell with each of the solutions 0.15 to 0.005 mol dm^{-3} as follows. Wash the electrodes with distilled water and then several times with small quantities of the relevant HCl solution. Fill the cell with this solution and place in a thermostat. Insert the electrodes and bubble hydrogen (purified in the usual way) over the platinum electrode at a rate not exceeding about 2 bubbles per second. Measure the emf. of the cell when it has attained a steady value, recording several readings taken every 2 min to demonstrate constancy. Record the barometric pressure.

When the experiment is finished return the electrodes to the cell and leave filled with distilled water.

Treatment of experimental data and discussion

1. Correct all the concentrations c/mol dm^{-3} to molality m/mol kg^{-1} by the use of the relation:

$$m = c - 0.0030\, c$$

This relation is sufficiently accurate over the concentration range studied.

2. The Debye–Hückel relationship for $\gamma_\pm$ for a 1 : 1 electrolyte is:

$$-\log \gamma_\pm = \frac{A\sqrt{I}}{1 + B \mathring{a}\sqrt{I}} - CT \qquad (7.28)$$

where $A\,(= 0.509$ mol$^{-1/2}$ dm$^{-3/2}$ at 25 °C), B and C are constants, and $\mathring{a}$ the effective diameter of the hydrated ions; in this case $B\mathring{a}$ has a value of 1.

Substituting (7.28) into equation (7.27) and rearranging (for a 1 : 1 electrolyte $m_\pm = m = I$) gives:

$$E + \frac{2 \times 2.303\, RT}{\mathscr{F}} \left[\log m - \frac{A\sqrt{m}}{1 + \sqrt{m}} \right] = E^{\ominus} - \frac{2 \times 2.303\, RT}{\mathscr{F}} Cm \qquad (7.29)$$

For each concentration evaluate the left-hand side of equation (7.29) and plot this against m. (In the interests of clarity it is advisable to construct a well designed table containing all the various functions.) Determine the intercept; this is $E^{\ominus}$. Use this value of $E^{\ominus}$ in equation (7.27) to calculate $\gamma_{\pm}$ at each concentration.

3.   Plot the graph of $- \log \gamma_{\pm}$ against $\sqrt{m}$; also include the line corresponding to the Debye–Hückel limiting law (ie. $- \log \gamma_{\pm} = A\sqrt{m}$). Comment on the resulting graphs.

4.   Write down the electrode and cell reactions and derive equation (7.27) from first principles.

5.   Explain the basis of the extrapolation.

6.   What is the standard electrode potential of the silver–silver chloride electrode?

Experiment 7.11   Determine the mean ionic activity coefficient of hydrochloric acid at different concentrations using a concentration cell without transport

Theory: The cell to be used is:

$$\ominus \quad PtH_2 \mid HCl \mid AgCl(s) \mid Ag–Ag \mid AgCl(s) \mid HCl \mid H_2Pt \quad \oplus$$
$$\qquad\quad m_1 \qquad\qquad\qquad\qquad\qquad\qquad m_2$$

where $m_2 > m_1$. The emf. of this cell is measured keeping m_2 constant and varying m_1, the molality of the more dilute solution. It is then possible by extrapolation to determine the activity coefficient at any acid concentration.

Requirements: Potentiometer and galvanometer, two hydrogen electrodes (p. 123), two silver–silver chloride electrodes (p. 307), source of pure oxygen-free hydrogen, two cell assemblies (Fig. 7.12), thermostat at 25.00 ± 0.01 °C, standard approx. 0.1 mol dm^{-3} hydrochloric acid.

Procedure: Very careful work is necessary to obtain the required precision, particular care should be taken in accurately diluting the solutions, in temperature control and in the measurement of the emf. Accurately dilute by volume the standard 0.1 mol dm^{-3} hydrochloric acid in two-fold dilution in the range 0.1 to 0.0001 mol dm^{-3}.

Set up the potentiometer circuit and standardize with a Weston cell. Check the two hydrogen electrodes against one another using the same acid solution (0.1 mol dm^{-3}). If the emf. exceeds 0.5 mV, replatinize the electrodes. Similarly check the silver–silver chloride electrodes against one another.

Set up the cell as shown with one hydrogen and one silver–silver chloride electrode in each of the two assemblies containing the acid aolution, and connect the silver–silver chloride electrodes externally. The hydrogen passing to the two electrodes, which are connected to the potentiometer, must come from the same supply; rate of bubbling about 2 bubbles per second.

Wash the electrodes and assemblies several times with the appropriate HCl solution. Fill the cells with the appropriate solution and place in the thermostat. Measure the emf. of the concentration cell when it has attained a constant value, recording several values taken every 2 min to demonstrate constancy.

Measure the emf. of the cell with 0.1 mol dm^{-3} hydrochloric acid in both assemblies and check that it is zero. Now determine the emf. of cells, always carrying out duplicate readings with a fresh sample of solution, in which m_2 is constant (0.1 mol dm^{-3}) and m_1 varied. The concentrations most suitable for use (m_1) are two-fold dilutions of the standard 0.1 mol dm^{-3} acid down to approximately 0.0001 mol dm^{-3}. Record the temperature.

Treatment of experimental data and discussion

1. Correct all the concentrations c/mol dm^{-3} to molality m/mol kg^{-1} by the use of the relation:
$$m = c - 0.0030\, c$$

2. For the left-hand cell, in which γ_1 is the mean ionic activity coefficient of HCl, the Debye–Hückel relation (equation (7.28)) becomes:

$$-\log \gamma_1 = \frac{A\sqrt{m_1}}{1 + \sqrt{m_1}} - Cm_1 \qquad (7.30)$$

Combining equation (7.30) with the identity:

$$-\log \gamma_1 = -\log \gamma_2 + \log \gamma_2/\gamma_1$$

it follows that:

$$\log \frac{\gamma_2}{\gamma_1} - \frac{A\sqrt{m_1}}{1 + \sqrt{m_1}} = \log \gamma_2 - Cm_1 \qquad (7.31)$$

$\log \gamma_2/\gamma_1$ can be calculated from measured values of E at known m_1 and m_2 using equation (7.22) which on rearrangement becomes:

$$\log \frac{\gamma_2}{\gamma_1} = \frac{\mathscr{F}E}{2 \times 2.303\, RT} - \log \frac{m_2}{m_1} \qquad (7.22a)$$

Evaluate the left-hand side of equation (7.31), using the values of $\log \gamma_2/\gamma_1$ obtained from the experimental data, and plot this against m_1. Determine the intercept ($= \log \gamma_2$), and using this value in equation (7.22a), determine γ_1 at all the different m_1 values.

3. Plot the graph of $-\log \gamma_1$ against $\sqrt{m_1}$; also include the line corresponding to the Debye–Hückel limiting law (ie. $-\log \gamma_\pm = A\sqrt{m}$). Comment on the resulting graphs.

4. Explain the advantages of using a concentration cell, instead of a galvanic cell (Expt. 7.10) for the determination of $\gamma_\pm$.

5. How would a change of hydrogen gas pressure affect the emf.?

6. Explain how this cell could be used to measure the variation of γ, for HCl, with the concentration of an inert electrolyte, eg. $NaNO_3$.

Experiment 7.12   Determine the activity coefficient of silver ions using a concentration cell with a salt bridge

Theory: The cell to be used is:

$$\ominus \; Ag \mid AgNO_3 \; aq. \mid Salt \; bridge \mid AgNO_3 \; aq. \mid Ag \oplus$$
$$\qquad\qquad m_1 \qquad\qquad NH_4NO_3 \qquad\quad m_2$$

in which the two solutions are separated by a salt bridge which eliminates the liquid junction potential. The emf. is equal to the difference of the two single electrode potentials·

$$E = \frac{RT}{\mathscr{F}} \ln \frac{(a_{Ag^+})_2}{(a_{Ag^+})_1} = \frac{RT}{\mathscr{F}} \ln \frac{m_{+2}}{m_{+1}} + \frac{RT}{\mathscr{F}} \ln \frac{\gamma_{+2}}{\gamma_{+1}} \qquad (7.22a)$$

E is measured keeping m_2 constant and varying m_1, the molality of the more dilute solution. The activity coefficients are determined by an extrapolation method.

Requirements: Potentiometer and galvanometer; two silver electrodes, calomel electrode (p. 126), standard 0.1 m silver nitrate solution, cell assembly (Fig. 7.11), saturated ammonium nitrate solution, solid potassium nitrate.

Procedure: Set up the potentiometer circuit and standardize with a Weston cell. Clean and electroplate the silver electrodes (p. 306) and check against one another when both are dipping in 0.1 m silver nitrate solution. If the emf. is not zero, replate them.

Measure the emf. of each silver electrode in 0.1 m silver nitrate solution, against the calomel electrode, using a salt bridge:

$$Hg|Hg_2Cl_2|Satd.KCl \mid NH_4NO_3 \mid AgNO_3 \; 0.1 \; m|Ag$$

These values should check to 0.1 mV.

In this experiment, the emf. measurements must be made with high accuracy; the accuracy can be still further increased if oxygen is removed from all solution by bubbling hydrogen or nitrogen through them. Set up the concentration cell and measure the emf. when $m_1 = m_2 = 0.1$ m. Prepare 2–3 fold dilutions (wt/wt) of the silver nitrate solution down to 0.0001 m, and measure the emf. (of duplicate samples of each solution) of the cells so formed (m_2 is always 0.1 m). Record the temperature.

Repeat the experiment using a similar concentration cell in which the silver nitrate concentration is constant in both compartments (0.01 m) but in which the concentration of potassium nitrate is varied from 0–1.0 m. Record the polarity of the electrodes and the temperature.

Treatment of experimental data and discussion

1.   From the measured values of E (for each experiment) at the known m_1 and m_2 calculate $\log \gamma_2/\gamma_1$ (equation (7.22a)). Evaluate the quantity

$$\left\{ \; \log \frac{\gamma_2}{\gamma_1} - \frac{A\sqrt{I_1}}{1 + \sqrt{I_1}} \; \right\}$$

and plot this against I_1 (equation (7.31)), in the first experiment $I_1 = m_1$, but in the second I_1 is due to both $AgNO_3$ and to KNO_3. Determine the intercepts (log γ_2) and hence calculate γ_1 in all the solutions of different ionic strengths.

2.   For the two sets of data, plot graphs of $-\log \gamma_1$ against $\sqrt{I_1}$; also include the line corresponding to the Debye–Hückel limiting law (ie. $-\log \gamma_+ = A\sqrt{I}$). Comment on the results for the two experiments and their agreement with the theoretical graph.

3.   Comment on the significance of the mean ionic activity coefficient of an ion.

4.   Calculate $E^{\ominus}(Ag^+, Ag)$ from the emf. measurements against the calomel electrode, using the appropriate value of γ_+.

Experiment 7.13   Determine thermodynamic functions from emf. measurements of galvanic cells

Theory: Measurements of the emf. and the temperature coefficient of emf. of galvanic cells under reversible conditions provides a means of obtaining numerical values of ΔG, ΔH and ΔS for chemical reactions (equations (7.10), (7.12) and (7.13)). Two cells which can be conveniently studied are:

1.   The Clark cell: Zn, $ZnSO_4$ (s) | $ZnSO_4$ aq. | Hg_2SO_4 (s), Hg;

2.   Ag, AgCl (s) | KCl aq. | Hg_2Cl_2 (s), Hg.

Requirements: Potentiometer and galvanometer or digital voltmeter, Clark cell (Fig. 7.13a) or H-shaped cell (Fig. 7.13b), mercury, mercury(I)chloride, silver–silver chloride electrode (p. 306), 0.1 mol dm^{-3} potassium chloride, and thermostat adjustable over range 5–40 °C.

Procedure: Set up the cell as shown in Fig. 7.13 and place in the beaker containing water cooled with ice to about 5 °C, keep the water level below the corks in the H-tubes. Stir continuously and keep the temperature as constant as possible by the addition of ice; allow 15–20 min for the attainment of thermal equilibrium. Standardize the potentiometer or digital voltmeter and connect to the cell. When the temperature has become steady measure the emf. of the cell, check that it remains constant for about 5 min.

Raise the temperature of the bath about 5 °C, remove the source of heat, stir the bath continuously and when thermal equilibrium has been established measure the emf. Repeat the measurement of emf. every 5 °C to a maximum value of 40 °C.

Cool the bath at intervals of about 5 °C by removing hot water and replacing it with cold water, measure the emf. at each temperature when equilibrium has been established (this takes longer on decreasing the temperature).

Treatment of experimental data and discussion

1.   From the tabulated emf. values at different temperatures, plot the graph of E/V against $T/°C$, distinguishing the values of E for increasing and decreasing temperature. From the graph determine E and $(\partial E/\partial T)_p$ and hence calculate ΔG, ΔH and ΔS for the cell reaction, at 298 K.

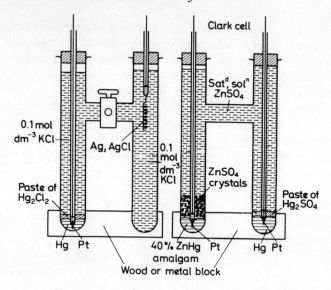

Fig. 7.13   Typical galvanic cells for measurement of thermodynamic functions

2.   For the cell write down the reaction occurring at each electrode and the over-all reaction. From tabulated thermodynamic data calculate $\Delta G^{\ominus}$, $\Delta H^{\ominus}$ and $\Delta S^{\ominus}$ and comment on any differences from the measured values.

3.   Why is the sign of ΔG negative for the cell reaction? Comment on the value of ΔS with reference to the changes occurring in the cell.

4.   Compare the accuracy of thermal and electrochemical methods for the determination of thermodynamic functions.

5.   For cell 2, explain how the emf. of the cell varies with the concentration of the potassium chloride solution. Is it possible to calculate the value of standard thermodynamic functions?

5   Capacitance and dielectric phenomena

The capacitance of a condenser depends on the geometry of the condenser and the nature of the medium between the plates. The ratio of the capacitance of the condenser containing the dielectric between the plates (C) to the capacitance of the same condenser with a vacuum between the plates (C_0) is the relative permittivity, $\epsilon = C/C_0$.

The increase in capacitance when a dielectric is between the plates is due to induced polarization of the molecules of the dielectric in the electric field between the plates, ie. the molecules have a temporary dipole moment with the axis of the dipole parallel to the lines of force of the field. A dipole is a pair of equal opposite charges separated by short distance; the dipole moment is the product of one of the charges and the distance between the charges. All molecules are, to some extent, polarizable in an electric field, but besides this induced (distortion) polarization, many molecules have a permanent dipole moment, μ, due to the asymmetric distribution of electrons and atomic nuclei. In the absence of an electric field these permanent dipoles are arranged at random and there is no polarization of the dielectric as a whole. In an electric field the permanent dipoles become oriented with their axes parallel to, but opposite to the direction of, the field and superimpose an orientation polarization, P_O, on the induced polarization, P_d, caused by temporary dipoles set up by the temporary displacement of electrons and nuclei.

For a non-polar molecule in the gas phase the induced molar polarization is related to the polarizability, α, and the relative permittivity by the Clausius–Mossotti equation:

$$\frac{\epsilon - 1}{\epsilon + 2} \frac{M_r}{\rho} = \frac{4\pi N_A \alpha}{3} = P_d \tag{7.32}$$

where M_r and ρ are the relative molecular mass and density of the dielectric respectively and N_A is the Avogadro constant.

Debye has shown that, for a polar molecule, the total molar polarization, P_T, is the sum of the induced and orientation polarizations, ie.:

$$P_T = P_d + P_O = \frac{\epsilon - 1}{\epsilon + 2} \frac{M_r}{\rho} = \frac{4\pi N_A}{3} \left\{ \alpha + \frac{\mu^2}{3kT} \right\} \tag{7.33}$$

This only strictly applies to gases but may be used for other very dilute dipolar media.

The α term in equation (7.33) is due mainly to the electronic polarizability with a small contribution from the distortion of the atomic framework.

For a solid, for which the dipoles can no longer rotate

$$P_T \text{ (solid)} = \tfrac{4}{3} \pi N_A (\alpha_{electronic} + \alpha_{atomic}) \tag{7.34}$$

The relative permittivity of the solid falls to a value near the optical value. From Maxwell's theory $\epsilon = n^2$, if both measurements are made at the same frequency.

Relative permittivity is usually measured in the 10^6 Hz region whereas refractive index n_D is at the sodium D line, 5×10^{14} Hz. The relationship is then only true for non-polar substances or, as mentioned above, polar substances in the solid state.

For gases, equation (7.33) provides a most useful method of determining the molecular dipole moment. Measuring the relative permittivity, the pressure and temperature of a gas (both required to define M_r/ρ), over a range of temperatures gives a range of values of total polarization. From a graph of P_T against T^{-1}, α may be calculated from the intercept and μ from the slope. A rough value may be obtained from measurements at two temperatures. If measurements can be made only at one temperature, the constant term $\tfrac{4}{3}\pi N_A \alpha$ may be obtained from the molar refraction

$$\left(\frac{n_D^2 - 1}{n_D^2 + 2} \right) \frac{M_r}{\rho},$$

which neglects α_{atomic}, or from measurements of the solid, equation (7.34).

The dipole moment of non-volatile compounds may be determined from the relative permittivity of dilute solutions in a non-polar solvent. The specific polarization of a solution of a polar solute B in a non-polar solvent A is given by

$$p_{AB} = p_A w_A + p_B w_B = \left(\frac{\epsilon_{AB} - 1}{\epsilon_{AB} + 2} \right) v_{AB} \qquad (7.35)$$

where w_A, w_B are the weight fractions of A and B, and v_{AB} is the specific volume of the solution.

From the approximations of Halverstadt and Kumler, the specific polarization of B at infinite dilution $(p_B)_0$ is:

$$(p_B)_0 = \frac{3\alpha v_A}{(\epsilon_A + 2)^2} + \frac{\epsilon_A - 1}{\epsilon_A + 2} (v_A + \beta) \qquad (7.36)$$

where α is the slope of the line ϵ_{AB} versus w_B and β is the slope of the graph of specific volume against weight fraction.

$$(P_B)_0 = (p_B)_0 M_r(B) \qquad (7.37)$$

The specific refraction is given by equation (7.38), where ν is the slope of the graph of the square of the refractive index of the solutions, measured at the sodium-D line, against weight fraction.

$$r_B = \frac{3\nu v_A}{(n_A + 2)^2} + \left(\frac{n_A^2 - 1}{n_A^2 + 2} \right) (v_A + \beta) \qquad (7.38)$$

and
$$P_d = r_B M_r(B) \qquad (7.39)$$

Then

$$P_0 = (P_B)_0 - P_d = \frac{4\pi N_A \mu^2}{9kT} \qquad (7.40)$$

Hence μ may be determined.

If B is a liquid then

$$P_d = \frac{n_B^2 - 1}{n_B^2 + 2} \frac{M_r(B)}{\rho_B} \qquad (7.41)$$

Using SI units,

$$\mu = (P_0 9 k T \epsilon_0 N_A^{-1})^{1/2} \qquad (7.42)$$

where ϵ_0 is permittivity of a vacuum.

Experiment 7.14   Determine the dipole moment of an organic liquid

Requirements: Wayne Kerr Universal Bridge, B.221B Wayne Kerr permittivity cell, refractometer, pyknometer or Westphal balance, pure benzene and one of the following: chlorobenzene, 2-chloro-2-methylpropane, nitrobenzene, 1-chloro-4-nitrobenzene, 1-chloro-2-nitrobenzene, acetone, benzophenone.

Procedure: The capacitance and hence the relative permittivity can be accurately measured with a transformer ratio-arm bridge. Details for the use of the instrument are given in the manufacturer's operational manual.

Connect the permittivity cell using short leads and join the screening of these two leads together. In the absence of the glass tube, measure the capacitance of the cell, this is the capacitance of air, C_{air}. Use low sensitivity initially and adjust the capacitance and, if necessary, the conductance controls to give a balance. Increase the sensitivity and obtain the final balance. Insert the glass tube, note its exact position, and again measure the capacitance, $C_{air+glass}$; the difference between these capacitance readings is the capacitance of glass, C_{glass}. Fill the glass tube to the mark with pure benzene and replace the tube in the cell in exactly the same position as previously, measure the capacitance. The capacitance of benzene is obtained by subtracting C_{glass}, and the relative permittivity of benzene from the ratio $\epsilon \approx C_{benzene}/C_{air}$, assuming that the relative permittivity of air is unity.

Prepare six solutions of say chlorobenzene in benzene (by weight) of weight fraction of chlorobenzene in the range 0–0.3. Use 20 cm^3 of benzene in each solution. Measure the capacitance of the cell containing pure chlorobenzene and each solution in turn. Make sure that the glass tube is adequately washed out between each determination and that it is always located in the correct position in the cell. Record the temperature of each solution in the cell.

For the other liquids suggested for measurement arrange the weight fractions of the solutions such that the relative permittivity of the strongest solution is only 0.2 greater than that of benzene.

Determine the density of the two pure liquids and of each of the solutions at the temperature of the determination (Expt. 2.1 or 2.2).

Finally determine the refractive index of chlorobenzene at the same temperature (Expt. 5.1).

Treatment of experimental data and discussion

1. Subtract C_{glass} from the observed capacitance of each solution and hence calculate $\epsilon_{obs.}$ ($\approx C_{soln.}/C_{air}$) at t °C. Using the standard value of the relative permittivity of benzene as reference correct these values of $\epsilon_{obs.}$:

$$\frac{\epsilon_{AB}}{\epsilon_{obs.}} = \frac{\epsilon_{benzene,\ stand}}{\epsilon_{benzene,\ obs.}} = \frac{2.284 - 0.002(t - 20)}{\epsilon_{benzene,\ obs.}}$$

2. Tabulate the values of relative permittivity, weight fraction, specific volume, refractive index and the square of refractive index.
3. Plot a graph of ϵ_{AB} against w_B and determine α from the slope. Compare the accepted value of the relative permittivity of benzene with the value of ϵ_{AB} at $w_B = 0$.
4. Plot a graph of n_{AB}^2 against w_B, slope equals ν.
5. Plot a graph of v_{AB} against w_B, slope equals β.
6. Compare the value of P_d obtained using equation (7.41), with that from equation (7.39). Account for any difference.
7. Calculate the total polarization at infinite dilution and using equation (7.40)

calculate the dipole moment. Convert your answer into SI units. Compare your value with that for the dipole moment in the gas phase and account for the difference.

8.   Why is it necessary to determine the molar polarization at infinite dilution?
9.   What error is involved in comparing the capacitance of the cell containing the solution to the capacitance in air instead of in vacuum?
10. Outline the principles of other methods which may be used for the determination of dipole moments and list the relative advantages of each.
11. If this experiment is carried out as a class exercise, each group should measure a different liquid. Discuss the dipole moment values obtained, mentioning the significance of the difference of the moment of the C-Cl and CO bonds in aliphatic and aromatic compounds. Demonstrate that to a certain extent the dipole moments of groups of atoms can be added as vectors. State the conditions under which this may be tested. Predict the dipole moment of 1-chloro-3-nitrobenzene.

8
Chemical kinetics

1 Chemical reactions

The simple theory of reaction velocity states that the rate of a chemical reaction is proportional to the active masses (concentrations) of the reacting species; thus if for the reaction:

$$A + B + \cdots = M + N \ldots$$

the rate of reaction may be written as:

$$-\frac{1}{a}\frac{dc_A}{dt} = -\frac{1}{b}\frac{dc_B}{dt} = \frac{1}{m}\frac{dc_M}{dt} = \frac{1}{n}\frac{dc_N}{dt} = k\, c_A^a\, c_B^b \tag{8.1}$$

then the quantity $(a + b + \ldots)$ is known as the total order of the reaction, whilst a, b, etc., are the orders of the reaction with respect to the reactants A, B, ... Since many complex reactions proceed in stages via intermediates (eg. ions, molecules or free radicals), and, since the measured rate is always that of the slowest step, a knowledge of the overall order, or better, the separate orders with respect to the individual reactants, often enables the steps to be elucidated. In such circumstances, the rate of reaction cannot be expressed as a simple equation.

In an empirical study of the kinetics of a reaction it is necessary to determine if side or consecutive reactions are present; if it is homogeneous or heterogeneous; if it is catalysed; and if (in ionic reactions), there are any salt effects.

The rate constant, k, of reactions increases with temperature, the change usually conforming to the Arrhenius equation:

$$k = Ae^{-E/RT}$$

or
$$2.303 \log k = -E/RT + \text{Const.} \tag{8.2}$$

where A, a constant, and E, the energy of activation, are practically independent of temperature. The energy of activation is most simply interpreted as the minimum energy which the molecules must possess before they can undergo reaction. This energy is needed, either, to rupture a chemical bond, eg. in free radical gas reactions, or to allow rearrangements when the molecules collide.

A determination of the rate constant of the reaction, at a series of temperatures, enables E to be calculated.

Since the rate of a given reaction depends upon the concentration of the reactants, the speed of the process falls off as the reaction proceeds, for the reactants are being continuously consumed. The reaction is becoming slower and slower but theoretically never ceasing. It is, therefore, not possible to define the general rate of a reaction, and so in practice the velocity is considered at a particular instant.

The rate may be defined in any convenient way, usually, the rate of change of concentration (c) of one of the reactants or products is chosen. The experimental observations will then show how that concentration varies with time, and, the rate at any instant will be given by the tangent of the $c-t$ curve.

2   First-order reactions

In a first-order reaction of the kind:

$$A \longrightarrow \text{Products}$$

the rate is directly proportional to the concentration of the reacting substance, c, ie.

$$-\frac{dc}{dt} = kc \tag{8.3a}$$

or

$$\ln c = -kt + B \tag{8.3b}$$

where B is the integration constant. If a is the initial concentration, and x is the decrease after time t, then $c = (a - x)$ and equation (8.3a) becomes:

$$\frac{dx}{dt} = k(a - x) \tag{8.4}$$

The quantity dx/dt, which is a measure of the rate of decomposition is called the *reaction velocity*, and, the constant k, the *rate constant*.

Experimental methods for determining the rate constant for first-order reactions

(*a*) *Graphical method:* Rearranging equation (8.4) and integrating ($t = 0$ when $x = 0$; to $t = t$, $x = x$) gives:

$$\ln \frac{a}{(a - x)} = 2.303 \log \frac{a}{(a - x)} = kt \tag{8.5}$$

For first-order reactions, k has the dimensions of time^{-1}.

Instead, however, of calculating k for various time intervals, a graphical method which covers the whole reaction period is to be preferred. Rearrangement of equation (8.5) gives

$$2.303 \log a - 2.303 \log (a - x) = kt \tag{8.6}$$

Since, for a given reaction, the first term on the left-hand side of the equation is a constant, it follows that, for a first-order reaction, the graph of log $(a - x)$ against t is linear and of slope $-k/2.303$, from which the rate constant can be calculated.

(*b*) *Guggenheim's method* (*Phil. Mag.* (1926), **2**, 538): An alternative method of using the results from a kinetic study, particularly when the initial and/or the infinity readings are not known with the same accuracy as the rest of the readings, is that due to Guggenheim.

A series of 'readings', r (eg. titres, angles of rotation, levels in a dilatometer tube, etc.), are taken at regular time intervals, thus $r_1, r_2, r_3 \ldots r_n$ at times $t_1, t_2, t_3 \ldots t_n$. A second set of readings r' are taken at exactly the same time interval (T) after the corresponding reading, r, ie. $r'_1, r'_2, r'_3 \ldots r'_n$ at times $t_1 + T$, $t_2 + T$, $t_3 + T \ldots t_n + T$. The value of T should, for best results, have a value of at least twice the time of half reaction.

For the two sets of readings, the kinetic equation (8.3*b*) may be used, in which c is proportional to either $(r_\infty - r_n)$ or $(r_\infty - r'_n)$, where r_∞ is the common infinity reading. Upon substitution the following equations are obtained:

$$-kt_n = \ln (r_\infty - r_n) + D \qquad (8.7a)$$

and

$$-k(t_n + T) = \ln (r_\infty - r'_n) + D \qquad (8.7b)$$

where D is a constant, incorporating the original integration constant and the proportionality constant.

Taking antilogs, and, subtracting to eliminate r_∞ gives:

$$e^{-kt_n}\left\{\frac{1 - e^{-kT}}{e^D}\right\} = (r'_n - r_n) \qquad (8.8)$$

or

$$2.303 \log (r'_n - r_n) = -kt_n + 2.303 \log\left\{\frac{1 - e^{-kT}}{e^D}\right\} \qquad (8.9)$$

The final term in this equation is a constant, so the value of the rate constant, k, can be obtained by plotting log $(r'_n - r_n)$ against t and determining the slope $(-k/2.303)$.

(*c*) *Time of half reaction:* Another important conclusion can be obtained from a consideration of the time taken to complete a definite fraction of the reaction (usually half). If τ is the time required for the concentration to be reduced to half the initial value, then $x = a/2$. On substitution in equation (8.5), the equation:

$$\tau = \ln 2/k = (2.303 \log 2)/k \qquad (8.10)$$

is obtained showing that τ is independent of the initial concentration of the reactant.

3   Second-order reactions

When the velocity of a reaction depends on two concentration terms, the process is of the second order. In the bimolecular reaction:

$$A + B \longrightarrow \text{Products}$$

let a and b represent the initial concentrations of A and B respectively, and x the decrease of each after time t, then:

$$\frac{dx}{dt} = k(a - x)(b - x) \tag{8.11a}$$

which on integration, between the limits $0, t$ and $0, x$ gives:

$$\frac{2.303}{(a - b)} \log \frac{b(a - x)}{a(b - x)} = kt \tag{8.12a}$$

where k, the second-order rate constant, has the dimensions of time^{-1} concentration^{-1} (eg. $dm^3\,mol^{-1}\,s^{-1}$, $m^3\,mol^{-1}\,s^{-1}$).

When the initial concentrations a and b are the same, the corresponding equations are

$$\frac{dx}{dt} = k(a - x)^2 \tag{8.11b}$$

and

$$\frac{x}{a(a - x)} = kt \tag{8.12b}$$

(a) *Graphical determination of rate constant:* In addition to the direct method of determining whether equation (8.12a) or (8.12b) is applicable (according to circumstances), a graphical method may be employed to demonstrate that a reaction is of the second order. For a process involving two reactants at different concentrations, equation (8.12a) may be arranged to take the form:

$$\frac{1}{(a - b)} \log \frac{b}{a} + \frac{1}{(a - b)} \log \frac{(a - x)}{(b - x)} = \frac{kt}{2.303} \tag{8.13}$$

Since for the given reaction a and b are constants, it follows that the graph of $\log [(a - x)/(b - x)]$ against t is linear, and of slope $k(a - b)/2.303$; the rate constant may thus be evaluated.

If the initial concentrations are identical, or, if two molecules of one reactant take part in the reaction, then equation (8.12b) may be rewritten:

$$\frac{1}{(a - x)} - \frac{1}{a} = kt \tag{8.14}$$

whence it can be seen that the graph of $(a - x)^{-1}$ against t is linear and of slope k.

(b) *Time of half reaction:* For second-order reactions it can be shown (by substitution in equation (8.12b)) that τ is inversely proportional to the initial concentration,

ie.
$$\tau = 1/ka \tag{8.15}$$

It is thus possible to distinguish in a simple manner between first and second-order reactions, particularly when only one substance is involved.

4   Reactions of higher orders

Expressions for the rate of reactions of higher order may be obtained in a similar manner. Experimentally, higher orders have been reported for ionic reactions, but it seems that other disturbing factors are prominent and influence the overall kinetics.

5   Methods available for the determination of rate constants

The progress of a chemical reaction may be followed by any convenient method. Physical methods are to be preferred as they may be used without disturbing the concentrations of the reacting materials. Any physical property which undergoes a change during the course of a reaction may be utilized; thus, in gas reactions involving a change in the number of molecules, the variation of pressure (at constant volume) is used. Changes in volume, optical rotation, electrical conductance and extinction coefficient are amongst the simpler physical methods described in this book. Other properties that are available in particular cases include thermal conductivity (gas mixtures), mass, refractive index, pH, viscosity (particularly in studies involving polymerization) and mass spectrometry.

In methods using chemical analysis, samples are removed and titrated. If necessary, the reaction taking place in the sample may be stopped, either by cooling or the addition of a suitable reagent. A variation of this method involves the addition of an indicator to the reaction mixture.

(a) Chemical methods

Experiment 8.1   Study the kinetics of the hydrolysis of methyl acetate, catalysed by hydrochloric acid

Theory: Methyl acetate is readily hydrolysed to give methanol and acetic acid:

$$CH_3COOCH_3 + H_2O + H^+ = CH_3COOH + CH_3OH + H^+$$

The reaction is catalysed by hydrogen ions; it does not proceed with any measurable velocity in pure water. Although two molecules are involved in the reaction, the water is present in such excess that only the methyl acetate appears to change in concentration. In addition, the large excess of water prevents any appreciable back reaction. As each molecule of methyl acetate is hydrolysed, one molecule of acetic acid is produced, and so the increase in acidity is a direct measure of the amount of ester that has reacted. The amount of hydrochloric acid remains unchanged throughout the experiment. If T_0, T_t and T_∞ are the titres/cm^3 (of both hydrochloric acid and acetic acid) at times 0, t and 48 hours respectively, then the concentration of methyl acetate at time t (ie. $(a - x)$) is proportional to $(T_\infty - T_t)$, and the initial concentration (a) is proportional to $(T_\infty - T_0)$. Substituting these values in equation (8.6) gives:

$$\log (T_\infty - T_0) - \log (T_\infty - T_t) = \frac{kt}{2.303}$$

Requirements: Thermostats at varying temperatures, 250-cm^3 and 150-cm^3 conical flasks, stop-clock, methyl acetate, 0.1 mol dm^{-3} sodium hydroxide solution and 0.5 mol dm^{-3} hydrochloric acid.

Procedure: The concentration of methyl acetate at any time is determined by titration of aliquots (removed from the mixture) with 0.1 mol dm^{-3} sodium hydroxide solution. Careful pipetting and titrating are an essential feature of the experiment.

Transfer 100 cm^3 of the hydrochloric acid into a dry stoppered 250-cm^3 flask clamped in a thermostat bath at 25 °C. Also place a stoppered tube containing about 20 cm^3 of methyl acetate in the same bath. When temperature equilibrium has been established, pipette exactly 5 cm^3 of the methyl acetate into the flask of acid and note the time. Shake thoroughly, and, immediately withdraw a 5-cm^3 sample of the mixture and run into about 25 cm^3 of ice-cold CO$_2$ free water (to arrest the reaction). Note the time, to the nearest 15 s, at which the pipette has been half discharged into the water of the titration flask. Titrate the acid in the sample as soon as possible with the standard sodium hydroxide solution. Repeat the titration of further 5 cm^3 samples after 10, 20, 30, 40, 60, 90, 120 and 180 min. Reserve at least 25 cm^3 of the mixture in a stoppered flask at 25 °C for 48 hours to give a final titration after the reaction has gone to completion.

Repeat the experiment at different temperatures, say, 30, 35 and 40 °C. The reaction will be faster, and, it is therefore necessary to take readings at more frequent intervals.

Treatment of experimental data and discussion
1.   Tabulate T_t/cm^3, t/s, and log $(T_\infty - T_t)$. For each temperature plot a graph of log $(T_\infty - T_t)$ against t, from the slope of the best straight line through the points determine the rate constant of the reaction.
2.   From the values of the rate constant at different temperatures, calculate the activation energy of the reaction; plot a graph of log k against T^{-1}, hence E can be determined from the slope (equation (8.2)).
3.   Calculate the pre-exponential constant A (equation (8.2)).
4.   Discuss the mechanism of the acid hydrolysis of methyl acetate.
5.   Write the rate equation for the forward reaction and discuss the factors which influence the rate of the reaction.
6.   Do you consider that the 'quenching' of the reaction by the suggested method is adequate?

Experiment 8.2   Determine the rate constant of the hydrolysis of ethyl acetate by sodium hydroxide

Theory: This experiment illustrates a bimolecular reaction for which a second-order rate constant can be calculated:

$$CH_3COOC_2H_5 + Na^+ + OH^- = CH_3COO^- + Na^+ + C_2H_5OH$$

The reacting species are ethyl acetate and the hydroxyl ion.

As the reaction proceeds, each hydroxyl ion, removed in the formation of ethanol, removes one molecule of ethyl acetate, x being the number of mol dm^{-3} of either OH^- or ethyl acetate so removed. The value of x at any time, t, is determined by subtracting the concentration of sodium hydroxide at time t from the initial value, a. This concentration of sodium hydroxide is determined by the back titration with alkali, after the addition of excess hydrochloric acid.

The number of moles of ethyl acetate, b, originally present is determined by subtracting the final concentration of sodium hydroxide from the initial value, a. The final (infinity) titration thus gives the excess sodium hydroxide over the ethyl acetate present initially, ie. $(a - b)$.

Requirements: Thermostat, 150-cm^3 and 1000-cm^3 stoppered conical flasks, stopclock, 0.02 mol dm^{-3} sodium hydroxide (free from carbonate), 0.02 mol dm^{-3} hydrochloric acid and ethyl acetate.

Procedure: The technique is similar to that used for the acid hydrolysis of methyl acetate (Expt. 5.1), but the reaction is more rapid.

Prepare 500 cm^3 of an approximately 0.01 mol dm^{-3} solution of ethyl acetate using CO_2-free distilled water. Transfer 250 cm^3 of this solution and 250 cm^3 of 0.02 mol dm^{-3} sodium hydroxide solution to separate dry flasks in a thermostat at 25 °C. After temperature equilibrium has been established, start the reaction by rapidly pouring the ethyl acetate solution into the sodium hydroxide solution. Shake thoroughly and note the time of mixing.

Withdraw a 50-cm^3 sample after 2 min and discharge as rapidly as possible into a known excess of 0.02 mol dm^{-3} hydrochloric acid. Titrate the excess acid with the standard sodium hydroxide solution using phenolphthalein as indicator. Take further samples and treat in the same manner after 5, 10, 20, 40, 80, 160, etc., min. Reserve the last 100 cm^3 in a stoppered flask to give a final titration after the reaction has gone to completion (24 hours).

Repeat the experiment, using half the quantity of ethyl acetate solution, but maintain the volume constant by the addition of 125 cm^3 of distilled water.

Treatment of experimental data and discussion
1.   Tabulate t/min, titre/cm^3, x/cm^3, $(a - x)$/mol dm^{-3}, $(b - x)$/mol dm^{-3} and $\log (a - x)/(b - x)$. Plot graphs of $\log (a - x)/(b - x)$ against t for the two experiments, and hence calculate k from the slopes (equation (8.13)). Comment on the values of the two calculated rate constants. Is k the true rate constant of the reaction?
2.   Is the reaction first order with respect to each reactant?
3.   What variables besides the concentration of the OH^- ion and ester might affect the rate of reaction?
4.   Suggest a mechanism for the reaction studied.

Experiment 8.3   Determine the rate constant and the energy of activation of the reaction between hydrogen peroxide and hydrogen iodide

Theory: The reaction:

$$H_2O_2 + 2HI = I_2 + 2H_2O$$

in which iodine is liberated, may be followed chemically by incorporating an indicator in the reaction mixture.

   If the conditions of the experiment are suitably adjusted, it is possible to study the rate of the reaction in the presence of a constant excess of hydrogen iodide; the reaction with respect to hydrogen peroxide will then be first order. This constant excess of hydrogen iodide is achieved experimentally by continually adding small amounts of sodium thiosulphate to remove the iodine, and, hence regenerate the iodide ions. The course of the reaction is readily followed by timing the appearance of iodine (indicated by starch) after the addition of a small known volume of sodium thiosulphate solution. The amount of iodine liberated by the reaction at a series of times corresponds to the volume of thiosulphate added. The total amount of iodine liberated at infinite time can be determined from a standardization of the hydrogen peroxide used. Thus, it is possible to determine the concentration of hydrogen peroxide at any time, since 1 mol of iodine is liberated for every mol of hydrogen peroxide destroyed.

Requirements: Thermostat, 500-cm^3 conical flask, 2 vol hydrogen peroxide solution, potassium iodide solution (4 g dm^{-3}), 100 cm^3 of sulphuric acid (1 vol concentrated acid to 2 vol of water), standard 0.1 mol dm^{-3} sodium thiosulphate solution, freshly prepared starch solution.

Procedure: To 10 cm^3 of the hydrogen peroxide solution, add about 2 g solid potassium iodide and 10 cm^3 of sulphuric acid. Titrate the liberated iodine with standard sodium thiosulphate solution, and hence standardize the peroxide solution.

   To 250 cm^3 of the potassium iodide solution in a 500-cm^3 flask add 15 cm^3 of the diluted acid and place in a thermostat at about 15 °C. At the same time, have 10 cm^3 of the peroxide and 10 cm^3 of the starch solution (in separate boiling tubes) at the same temperature. Arrange the 50-cm^3 burette, containing the sodium thiosulphate solution, above the flask in the thermostat so that it will deliver directly into the solution. At time '0', add the starch and hydrogen peroxide solution and shake vigorously; if a blue colour appears, add 1 cm^3 of the thiosulphate solution from the burette. Note the time at which the blue colour reappears, and immediately add 1 cm^3 (only) of thiosulphate solution to discharge the colour. Shake well, and note the time of reappearance of the colour. Continue the addition of 1 cm^3 aliquots of sodium thiosulphate until the blue colour takes five to six times the initial time to reappear. It is essential that the reaction mixture be shaken continuously.

   Repeat the experiment at various temperatures in the range 0—30 °C.

   To determine the order of the reaction with respect to hydrogen iodide, repeat the experiment (at 15 °C), now using half and double the amounts of sulphuric acid and potassium iodide in the same total volume of reaction mixture.

Treatment of experimental data and discussion

1. Determine the concentration of hydrogen peroxide c/mol dm^{-3} at each time. Plot a graph of log c against t, at each temperature and hence calculate the rate constants (equation (8.3b)) for the reaction of H_2O_2 with HI at constant iodide ion concentration.

2. From the values of the rate constant at different temperatures, calculate the activation energy of the reaction (equation (8.2)).

3. From the values of the rate constant at different iodide concentrations determine the order of the reaction with respect to hydrogen iodide. The overall reaction is second order.

4. Discuss the mechanism of the reaction.

5. Is the volume of sodium thiosulphate added sufficient to cause a significant increase in the volume of the solution and hence change the I$^-$ concentration.

(b) Polarimetric method

Experiment 8.4 Determine the rate constant of the mutarotation of glucose in the presence of acid or alkali

Theory: A freshly prepared solution of α-glucose does not retain its initial optical activity, the rotation falling steadily to a constant value. This 'mutarotation' is due to the change of α-glucose into an equilibrium mixture of α- and β-glucose. The change has proved to be strictly first order, under all conditions of catalysis by acids and bases. Since the reaction is a balanced one, the first order constant measured is equal to the sum of the separate constants, for the forward and reverse reactions.

For a reversible first-order reaction it can be shown that:

$$2.303 \log x_e/(x_e - x) = k_2 t \qquad (8.16)$$

where x_e is the equilibrium concentration of the reactant and $k_2 = k_1 + k_{-1}$, the sum of the forward and reverse rate constants.

If α_0 represents the initial angle, α_∞ the final angle when equilibrium has been attained, and α_t the rotation at time t, then x_e is proportional to $(\alpha_0 - \alpha_\infty)$ and $(x_e - x)$, ie. the extent to which the equilibrium has proceeded is proportional to $(\alpha_t - \alpha_\infty)$. Substituting these values in equation (8.16), and rearranging gives:

$$\log (\alpha_0 - \alpha_\infty) - \log (\alpha_t - \alpha_\infty) = \frac{k_2 t}{2.303} \qquad (8.17)$$

The total rate constant for the mutarotation of glucose may be expressed:

$$k = k_{H^+} c_{H^+} + k_{OH^-} c_{OH^-} + k_{H_2O} \qquad (8.18)$$

where k_{H^+} and k_{OH^-} are the catalytic constants for H^+ and OH^- and k_{H_2O} is the contribution to the rate constant due to catalysis by water.

Requirements: Polarimeter, sodium lamp, thermostat at 25 $^\circ$C and circulating pump, water-jacketed (20 cm) polarimeter tube, pure dextrose, 0.2, 0.1 and

0.05 mol dm^{-3} solutions of hydrochloric acid, approx. 0.05 mol dm^{-3} ammonia solution and solid ammonium chloride.

Procedure: Before commencing the experiment, confirm that the adjustment and reading of the polarimeter can be made rapidly and accurately.

Determine the zero reading of the polarimeter with water in the tube. Keep the temperature of the tube constant to within 0.1 °C during the experiment, by circulating water from a thermostat.

(a) Catalysis by H$^+$

Allow 90 cm^3 of distilled water in a small flask, and about 20 cm^3 of 0.2 mol dm^{-3} hydrochloric acid in a separate tube, to attain temperature equilibrium in a thermostat at 25 °C. Dissolve 10 g of dextrose in the water as rapidly as possible, and add exactly 10 cm^3 of the acid. Shake thoroughly, transfer a sample to the clean, dry, jacketed polarimeter tube, and commence taking readings of the angle of rotation at minute intervals, as soon as possible. The time of this operation should not exceed 3 min. Continue taking readings at minute intervals for about 30 min. Preserve a sample of the mixed glucose and acid in a stoppered flask in the thermostat for a 72 hour (equilibrium) reading of the angle of rotation.

Repeat the experiment, using 0.1 and 0.05 mol dm^{-3} hydrochloric acid, taking minute readings for longer times, as the reaction is slower.

(b) Catalysis by OH$^-$

Standardize the ammonia solution. Weigh out 1.1 g of ammonium chloride and dissolve in 100 cm^3 of the ammonia solution. This gives a buffer solution in which the concentration of hydroxyl ions is given by:

$$c_{OH^-} = \frac{K_{NH_3} c_{NH_3}}{c_{NH_4Cl}}$$

the dissociation constant of ammonia, $K_{NH_3} = 2.0 \times 10^{-5}$ mol dm^{-3} at 25 °C in a solution of ionic strength 0.2 mol dm^{-3}.

Dissolve 10 g of dextrose in this buffer solution, previously brought to temperature equilibrium at 25 °C, and follow the rate of mutarotation as previously described.

Treatment of experimental data and discussion

1. Plot a graph of log $(\alpha_t - \alpha_\infty)$ against t/min for each of the different concentrations of acid or alkali (due allowances must be made for the sign of the angle of rotation). Determine a value for k_2 from each graph (equation (8.17)).
2. Plot a graph of total rate constant (obtained under acid conditions) against the concentration of hydrogen ions c_{H^+} and hence determine k_{H^+} and k_{H_2O} (equation (8.18)). Is it valid to neglect the catalysis by OH$^-$ in this case?
3. Is it possible to obtain k_{OH^-} from the values of total rate constant obtained under alkali conditions?
4. How else may the rate of interconversion be followed?

Experiment 8.5   Determine the effect of acid concentration and temperature on the rate of inversion of sucrose

Theory: The inversion of dextrorotatory sucrose in the presence of acid gives rise to a laevorotatory mixture of glucose and fructose. This may be conveniently followed in the polarimeter as described in Expt. 8.4.

$$\text{The rate of reaction} = k' [\text{sucrose}] [H_2O] [H_3O^+]$$

In dilute acid solutions the concentrations of the hydrogen ion and the water remain effectively constant during the reaction, and hence the reaction shows first order kinetics, ie.

$$\text{Rate of reaction} = k [\text{sucrose}]$$

where

$$k = k' [H_2O] [H_3O^+] = k_{H^+} [H_3O^+]$$

Requirements: Polarimeter, sodium lamp, thermostat adjustable over the temperature range 25–45 °C and circulating pump, water-jacketed (20 cm) polarimeter tube, pure sucrose, 2 mol dm^{-3} hydrochloric acid.

Procedure: Dissolve 20 g of sucrose in 100 cm^3 of distilled water. Transfer 25 cm^3

Temperature/ $t\,°C$	Final acid concentration/ mol dm^{-3}	Intervals between readings/ min	Final Reading/ h
25	0.2*	15, 30, 30, 45, 45, 45, 45, 45	72
	0.9	5, 10, 15, 15, 20, 20, 30, 45	24
25	0.3	15, 30, 30, 45, 45, 45, 45, 45	48
	0.7	5, 10, 15, 20, 20, 30, 45, 45	24
25	0.4*	5, 10, 15, 20, 30, 45, 45, 60	48
	0.5	5, 10, 15, 20, 20, 30, 45, 45	48
30	0.2*	15, 30, 30, 45, 45, 45, 45, 45	48
	0.4*	5, 10, 15, 20, 20, 30, 45, 45	24
35	0.05	15, 30, 30, 45, 45, 45, 45, 45	72
	0.9 (fast)	5, 5, 5, 5, 5, 5, 10, 10	4
35	0.2*	5, 10, 15, 20, 20, 30, 45, 45	48
	0.7 (fast)	5, 5, 5, 5, 10, 10, 10, 10	4
35	0.4*	5, 5, 10, 10, 15, 15, 20, 20	24
	0.5	5, 5, 10, 10, 15, 15, 20, 30	24
45	0.2*	5, 5, 10, 10, 15, 15, 20, 30	24
	0.4* (fast)	2, 3, 5, 5, 5, 5, 5, 5	2

* It is essential that the acid concentration is exactly the same each time for 0.2 mol dm^{-3}, and each time for 0.4 mol dm^{-3}, as these results are to be used to determine the energy of activation.

of this to a 50-cm^3 volumetric flask and place in a thermostat at the required temperature (t °C) together with samples of water and the hydrochloric acid solution in separate tubes. When temperature equilibrium has been established transfer the calculated quantity of 2 mol dm^{-3} hydrochloric acid to the sucrose solution and make up to 50 cm^3 with the warmed water. Rapidly fill the polarimeter tube, also thermostatted at t °C, determine the angle of rotation and at the same time start a stop-watch. Take further readings at the times suggested. After 30 min start the second reaction using a second polarimeter tube (the pairs of experiments, and the time intervals, have been selected so that this can be easily accomplished). The whole experiment is best completed as a class experiment, each small group of students studying two acid concentrations at one temperature.

Treatment of experimental data and discussion
1.   Tabulate the angles of rotation α_t at each time t/min, $(\alpha_t - \alpha_\infty)$ and log $(\alpha_t - \alpha_\infty)$, for each run. Plot log $(\alpha_t - \alpha_\infty)$ against t for each run and hence determine the rate constants for the reactions catalysed by the different concentrations of acid at the different temperatures.
2.   Plot graphs of the rate constant against acid concentration, for the catalysed reactions at 25 and 35 °C and from the slope determine k_{H^+}.
3.   Determine the energy of activation and the value of the pre-exponential term A (equation (8.2)), from the rate constants (both in 0.2 and 0.4 mol dm^{-3} acid) at the different temperatures.

(c) Methods involving volume changes

During the course of many chemical reactions, volume changes occur; these may be used to study the rate of the reaction. Volume changes may occur in gas reactions (carried out at constant pressure) and also during reactions in solution.

Experiment 8.6   Determine the rate constants of the decomposition of benzene diazonium chloride at different temperatures

Theory: Benzene diazonium chloride on hydrolysis with water decomposes according to the equation:

$$C_6H_5 . N_2Cl + H_2O = C_6H_5OH + HCl + N_2$$

The rate of decomposition is readily followed by measuring the rate of evolution of nitrogen.

Requirements: Small flask with side arm connected to Hempel gas burette (Fig. 8.1), mechanical shaker, thermostat, aniline, 10 mol dm^{-3} hydrochloric acid, sodium nitrite.

Procedure: Prepare the solution of benzene diazonium chloride by dissolving 0.035 mol (3.3 g per 3.25 cm^3) of aniline in 10.7 cm^3 of 10 mol dm^{-3} hydrochloric acid, cool well in ice-water. To this ice-cold solution, add gradually from a dropping funnel a cold solution of 0.035 mol (2.45 g) of sodium nitrite dissolved in about

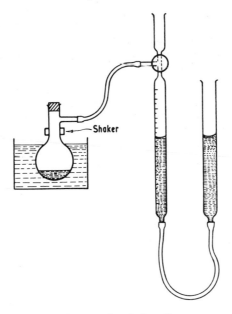

Fig. 8.1   Apparatus for studying the rate of evolution of gas

35 cm³ of water. Finally, make up the solution to 500 cm³ in a standard flask. Keep in the cold until required.

Fill the Hempel burette completely with water by raising the reservoir; turn off the tap. Accurately pipette 25 cm³ of the stock solution of benzene diazonium chloride into the flask and connect the side arm to the Hempel gas burette, but do not yet start collecting the gas. Close the neck of the flask with a rubber stopper (or better, use a flask with ground-glass joint). Attach the flask to a suitable shaker, so adjusted that the flask may be shaken whilst still immersed in the thermostat at 35 °C. After the reaction has proceeded for about 5 min, start collecting the gas, by turning the tap at the top of the burette to the correct position. This is zero time. Observe the volume of gas collected at 10 min intervals, after adjusting the pressure to atmospheric each time, by the compensating tube. Note the barometer reading and the temperature at which the gas is collected.

The end-point of the reaction may be hastened by immersing the flask containing the reaction mixture in a large beaker of hot water, until on cooling again to 35 °C no further increase in volume is recorded.

Repeat at 45, 55 and 65 °C.

Treatment of experimental data and discussion

1.   The initial concentration of benzene diazonium chloride in solution is directly proportional to the total volume of nitrogen finally collected (V_∞). The concentration at any time, t, is given by $V_\infty - V_t$ (where V_t is the volume of gas collected at time t). Plot a graph of log $(V_\infty - V_t)$ against t/min (for each temperature), and hence calculate the rate constant of the reaction (equation (8.5)) at each temperature.

2.   Determine the energy of activation of the reaction (equation (8.2)).

3.   Why can the order of reaction with respect to benzene diazonium chloride be determined by this procedure only if the solution is dilute?

4.   Suggest a means of determining the order of reaction with respect to water.

Experiment 8.7   Investigate the catalytic effect of sodium hydroxide on the depolymerization of diacetone alcohol

Theory: The reaction to be studied is the decomposition of diacetone alcohol according to the equation:

$$(CH_3)_2C(OH)CH_2COCH_3 \rightleftharpoons 2CH_3COCH_3$$

The forward reaction is catalysed by hydroxyl ions, and goes practically to completion in dilute solution. There is an increase in volume, and so a measure of the rate of increase in volume is a measure of the rate of reaction.

Requirements: Glass-fronted thermostat at 25 °C controlled at 25.00 ± 0.01 °C, dilatometer (Fig. 8.2), diacetone alcohol and 0.25, 0.10, 0.05 and 0.025 mol dm^{-3} solutions of sodium hydroxide.

Procedure: The dilatometer consists of a bulb (A) of about 30 cm^3 capacity, a capillary stem (B) about 25 cm in length and a funnel (C). Calibrations on the stem (B), or, a scale mounted directly behind the stem may be used to determine the rate of movement of the meniscus.

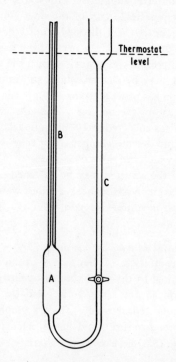

Fig. 8.2   Dilatometer

Thoroughly clean the dilatometer, wash with water and acetone, and dry by sucking a stream of filtered air through. Clean the tap and barrel with carbon tetra-chloride, and grease (not in excess) so that there is no leakage. Clamp the dilato-meter in the thermostat (with as much stem immersed as possible) to attain temperature equilibrium. Pipette into separate boiling tubes 10 cm^3 of diacetone alcohol and 40 cm^3 of 0.25 mol dm^{-3} sodium hydroxide and allow these to attain temperature equilibrium in the thermostat. Transfer, by pipette, 2.0 cm^3 of the diacetone alcohol to the sodium hydroxide solution, mix and pour into the funnel of the dilatometer. By sucking or blowing at B, adjust the height of the liquid until the meniscus is near the bottom of the capillary tube; close the tap. The liquid must be completely free from air bubbles. Take readings of the meniscus level every minute for 30 min. In this case, T is 15 min. Repeat the experiment as follows: 0.10 mol dm^{-3} NaOH, readings every 2 min for 60 min ($T = 30$ min); 0.05 mol dm^{-3} NaOH, readings every 2 min for 30 min, leave for 20 min, then readings every 2 min for a further 30 min ($T = 50$ min); 0.025 mol dm^{-3} NaOH, readings every 4 min for 60 min, leave for 30 min, then readings every 4 min for a further 60 min ($T = 90$ min).

Treatment of experimental data and discussion

1. Using the 'Guggenheim' method, determine the rate constant of the reaction at each concentration of sodium hydroxide.
2. Plot k against c_{OH^-}, and from this determine the order with respect to hydroxyl ion and the specific rate constant.
3. Why is it necessary to control the temperature to ± 0.01 °C?
4. Given that the dilatometer bulb has a volume of about 30 cm^3, and the diameter of the capillary is 1 mm, calculate the expected volume change and compare it with the observed change.
5. Discuss the mechanism of the reaction.
6. Why is the 'Guggenheim' method of determining rate constants the most appropriate to use in this case? What determines T?
7. Does the volume change lead to any significant change in the hydroxyl ion concentration?

(d) Conductometric methods

Experiment 8.8   Study the kinetics of the solvolysis of tertiary aliphatic halides (eg. t-butyl chloride, t-butyl iodide or t-amyl iodide)

Theory: These reactions are conveniently followed by measurements of electrical conductance. The solvolysis of tertiary aliphatic halides in aqueous solvents does not require the presence of hydroxyl ions, being kinetically of zero order with respect to them, and of first order with respect to the halide. Thus the net change with t-amyl iodide in water is:

$$C_5H_{11}I + H_2O = C_5H_{11}OH + H^+ + I^-$$

On this basis, conductometric measurements of the rate of formation of hydrogen iodide indicate the course of the simple first-order ionization process.

Requirements: Thermostat at 20 °C, direct-reading conductance bridge, dip-type conductance cell with platinum black electrodes, large boiling tubes (to take the dip cell), aqueous ethanol (80% ethanol by volume), and t-aliphatic halide.

Procedure: Pipette 50 cm^3 of the aqueous ethanol into a large, corked boiling tube in a thermostat at 20 °C. A second tube, also in the thermostat, contains the dip-type cell. When temperature equilibrium is established, pipette 0.3 cm^3 of the tertiary halide into the alcohol, stir well, insert the electrodes and commence taking conductance readings of the solution at regular intervals (half-minute intervals for the first 5 min and then minute intervals) for about 30 min. Return the electrodes to the empty tube, cork the tube containing the reaction mixture, and place in water at 60 °C for 5–10 min. Replace the tube in the thermostat bath, allow temperature equilibrium to be re-established, and determine the final conductance. Repeat this operation to ensure that the reaction has gone to completion.

To establish that the reaction is of the first order, repeat the experiment using different initial concentrations of the t-amyl iodide (0.2 and 0.4 cm^3) and demonstrate that the time of half reaction is independent of the initial concentration.

Repeat the experiment at different temperatures to determine the energy of activation of the reaction. Suitable conditions are as follows:

halide	$t/°C$	$t/°C$	$t/°C$
t-amyl iodide	10	20	30
t-butyl iodide	10	20	30
t-butyl chloride	25	35	45
conductance readings every ½ min for first 5 min and then every min for	1 h	30 min	30 min

Treatment of experimental data and discussion

1. Tabulate G_0, G_t and G_∞ the conductance at times 0, t, and infinity respectively, $(G_\infty - G_0)$ and $(G_\infty - G_t)$. Show that

$$\log \frac{G_\infty - G_0}{G_\infty - G_t} = \frac{kt}{2.303}$$

Plot a graph of $\log (G_\infty - G_0)/(G_\infty - G_t)$ against t/min and determine k.
2. Show that the reaction is first order with respect to the halide.
3. From the values of k at the different temperatures calculate the activation energy and the pre-exponential constant (equation (8.2)).
4. Will traces of electrolyte impurities affect the value of k?
5. Does the rate of reaction depend on the solvent used?
6. Discuss the mechanism of this reaction.
7. Indicate why it is not necessary to determine the cell constant of the dip-type cell for this experiment.

Experiment 8.9 Determine the second-order rate constant for the hydrolysis of ethyl acetate by sodium hydroxide

Theory: In this reaction, the initial concentrations of the two reactants are made the same (ie. 0.01 mol dm^{-3}). Under these conditions, equation $(8.12b)$ is applicable for determining the rate constant. The conductance initially will be that of 0.01 mol dm^{-3} sodium hydroxide, and finally, that of 0.01 mol dm^{-3} sodium acetate.

Requirements: Thermostat at 25 °C, direct-reading conductance bridge, dip-type conductance cell with platinum black electrodes (p. 306), large boiling tubes (to take the dip cell), 0.05 mol dm^{-3} sodium hydroxide solution, 0.2 mol dm^{-3} solution of ethyl acetate, 0.01 mol dm^{-3} solution of sodium acetate (prepared by neutralizing 20 cm^3 of 0.05 mol dm^{-3} sodium hydroxide solution with acetic acid and making up to 100 cm^3). All the solutions should be prepared using CO_2-free distilled water.

Procedure: Measure the conductances of 0.01 mol dm^{-3} sodium hydroxide and 0.01 mol dm^{-3} sodium acetate solutions at 25 °C. These represent the initial (G_0) and final (G_∞) conductances of the mixture, respectively.

Suspend two boiling tubes in the thermostat, one containing 20 cm^3 of sodium hydroxide solution and 50 cm^3 of distilled water, and the other, 5 cm^3 of the ethyl acetate solution and 25 cm^3 of water; allow temperature equilibrium to be established. Mix the two solutions, shake thoroughly, place the electrodes in the mixture and take readings of the conductance of the mixture at minute intervals during the course of the reaction.

Treatment of experimental data and discussion
1. The total possible amount of reaction, a, is proportional to $(G_0 - G_\infty)$ and the amount of reaction after time $t(x)$ is proportional to $(G_0 - G_t)$. Substitution of these values in equation $(8.12b)$ gives:

$$\frac{1}{0.01} \times \frac{G_0 - G_t}{G_t - G_\infty} = kt$$

Plot a graph of $G_0 - G_t/G_t - G_\infty$ against t/min and from the slope deduce k.
2. Compare this experiment with Expt. 8.2; list their relative merits.
3. Discuss the mechanism of the reaction.
4. Is it necessary to make the initial concentrations of the two reactants equal?
5. Will the conductance of the water affect the accuracy of your result?

(e) Method involving change in molar extinction coefficient

Experiment 8.10 Study the kinetics of the iodination of cyclohexanone

Theory: The reaction between an aqueous solution of cyclohexanone and iodine (I_2 and I_3^-) in the presence of acid may be represented as follows.

I_3^- + [cyclohexanone structure] $\longrightarrow$ [2-iodocyclohexanone structure] + H^+ + $2I^-$

In a solution of I_2 in KI the equilibrium

$$I_2 + I^- \rightleftharpoons I_3^- \ (K = 710 \ \text{dm}^3 \ \text{mol}^{-1} \ \text{at } 25 \ °C)$$

is very rapidly established. Both I_2 and I_3^- are effective iodinating agents for the ketone, therefore their total concentration must be known. Both species absorb visible radiation and at two wavelengths 565 and 468 nm their extinction coefficients are equal (isosbestic points). Thus if I_2 and I_3^- are present the absorbance at these wavelengths is directly proportional to the sum of their concentrations.

The course of this reaction may be followed by making absorbance measurements at the wavelength of one of the isosbestic points. In the 468 nm region the absorbance of I_3^- is changing very rapidly with wavelength, therefore 565 nm is the chosen wavelength for this experiment.

Requirements: Spectrophotometer fitted with a thermostatted cell compartment, 1 cm glass cells, 100 cm³ volumetric flask, 100 cm³ conical flasks, cyclohexanone, 0.5 mol dm³ hydrochloric acid, 0.05 mol dm³ iodine in 10% potassium iodide solution (thus $[I_3^-] \approx 0.05 \ \text{mol dm}^{-3}$). Thermostat (15, 25, and 35 °C).

Procedure: Set the temperature of the cell compartment at 25 °C. Prepare accurately 100 cm³ of a 0.46 mol dm⁻³ solution of cyclohexanone in water. Prepare the solution to be used in the reference cell from 1 part cyclohexanone solution, 1 part water and 3 parts acid solution. From solutions thermostatted at 25 °C prepare the reaction mixture in a conical flask, accurately adding each solution by pipette in the following order: 5 cm³ cyclohexanone solution, 10 cm³ water, 5 cm³ hydrochloric acid and 3 cm³ iodine in iodide solution. When all the iodine solution has been added, thoroughly mix the solution, note the time and immediately fill the spectrophotometer cell with the solution. Record the absorbance of the solution at 565 nm and continue measuring the absorbance at 2 min intervals until the absorbance reading is zero. (There may be some scatter in the initial readings.) Repeat the experiment at 15 °C (taking readings at 5 min intervals) and at 35 °C (taking readings at 1 min intervals).

Keeping the total volume constant, investigate the effect of change in the concentration of (*a*) iodine, (*b*) cyclohexanone and (*c*) acid, on the rate of the reaction. Do not use more than 10 cm³ of either the acid or the cyclohexanone solutions.

Treatment of experimental data and discussion
1.   From the results at each temperature plot a graph of absorbance against t/min. Determine the order of reaction with respect to iodine, cyclohexanone

and acid. Calculate the extinction coefficient at 565 nm (at each temperature). Calculate the value of k; what are the units of k? From this data suggest
a possible mechanism for the reaction.

2. From the rate constants at the three temperatures calculate the activation
    energy of the reaction and the pre-exponential term (equation (8.2)). Why is
    it better to determine E graphically and A by substitution in the equation?
3. Gives reasons for the curvature in the graphs for the first few min.
4. Why is it necessary to have a ten-fold excess of iodide?
5. How would you test for the occurrence of a reaction involving further iodina-
    tion of the iodocyclohexanone?
6. With the experimental conditions in this study is there a measurable concen-
    tration of I_2 present at any stage in the reaction?

6 Determination of the order of reaction in solution

The most convenient method for obtaining the order of a reaction (particularly an
ionic reaction in solution) is to carry out the reaction with the reactants at two
different concentrations (one usually half the other) and to determine the ratio of
the times necessary for the completion of a definite fraction of the reaction.

For a reaction of the nth order:

$$\frac{dx}{dt} = k(a - x)^n \tag{8.19}$$

which on integration (limits 0, t and 0, x) gives:

$$kt = \frac{1}{(n-1)} \left\{ \frac{1}{(a-x)^{n-1}} - \frac{1}{a^{n-1}} \right\} \tag{8.20}$$

For initial concentrations a_1 and a_2, the corresponding times of half reaction τ' and
τ'' are given by:

$$\tau' = \frac{1}{(n-1)k} \left\{ \frac{2^{n-1} - 1}{a_1^{n-1}} \right\} \tag{8.21a}$$

and

$$\tau'' = \frac{1}{(n-1)k} \left\{ \frac{2^{n-1} - 1}{a_2^{n-1}} \right\} \tag{8.21b}$$

whence

$$\tau'/\tau'' = (a_2/a_1)^{n-1} \tag{8.22}$$

Thus, by determining the times of half reaction for two different initial concentrations, the order of the reaction may be determined.

The rates of reactions between ions are markedly affected by the presence of
neutral salts. Generally, reactions between ions of the same sign are accelerated, and
reactions between ions of opposite sign are retarded, by the addition of neutral
salts. The effect is explained by the Brønsted equation in terms of an alteration in

the thermodynamic activity coefficient of the reacting ions by the electrical forces of the added ions. Thus, in the ionic reaction

$$A + B \rightleftharpoons X \rightleftharpoons \text{Products}$$

where X is the transition complex, it can be shown that:

$$k = k_0 \frac{\gamma_A \gamma_B}{\gamma_X} \tag{8.23}$$

where k is the rate constant, k_0 the ideal rate constant at $I = 0$, γ_A, γ_B and γ_X the activity coefficients of the species A, B and X respectively. Taking logarithms of equation (8.23), and using the Debye–Hückel relationship for activity coefficients, it follows that

$$\ln k = \ln k_0 - A(z_A^2 + z_B^2 - z_X^2)\sqrt{I}$$

where z_A, z_B and z_X are the algebraic values of the charges on A, B and X, $z_X = z_A + z_B$ hence

$$\ln k = \ln k_0 + 2A z_A z_B \sqrt{I} \tag{8.24}$$

This is only strictly true if quantitative allowance is made for the incidence of ion association.

The ions taking part in an ionic reaction will also exert a salt effect on their own reaction. For this reason, the measured order of a reaction, found by experiments at two dilutions, often differ from an integer. Thus, in a reaction between ions of the same sign, the more concentrated solution will have a rate greater than that expected and the measured order will be slightly higher than the true integer.

Experiment 8.11   Determine the order of the reaction between bromate and bromide ions in acid solution

The overall reaction to be studied may be represented:

$$BrO_3^- + 5Br^- + 6H^+ = 3H_2O + 3Br_2$$

Requirements: Thermostat at 25 °C, large conical flasks, 0.2 mol dm^{-3} solutions of potassium bromide and potassium bromate, 0.1 mol dm^{-3} sulphuric acid, 0.5 mol dm^{-3} sodium sulphate solution and standard 0.01 mol dm^{-3} sodium thiosulphate solution.

Procedure: Prepare the following mixtures from solutions which are at 25 °C. The quantities may be measured with a cylinder with the exception of the potassium bromate solution. Add the sulphuric acid last and take the time of this addition as zero time.

Mixture number	KBr/cm^3	H$_2$SO$_4$/cm^3	KBrO$_3$/cm^3	Na$_2$SO$_4$/cm^3	water/cm^3
I	125	150	25	0	200
II	125	150	25	0	450
III	125	150	25	200	0

Thermostat the mixtures, in stoppered flasks (to prevent the escape of bromine vapour), at 25 °C. Withdraw 50-cm³ samples every 5 min at the start, and later at longer intervals, and rapidly titrate with 0.01 mol dm⁻³ sodium thiosulphate solution to the disappearance of the yellow colour. The titration reaction is:

$$Na_2S_2O_3 + 4Br_2 + 5H_2O = Na_2SO_4 + H_2SO_4 + 8HBr$$

compare

$$2Na_2S_2O_3 + I_2 = 2NaI + Na_2S_4O_6$$

Treatment of experimental data and discussion

1. Multiply the titration figures for mixture II by $^3/_2$ and plot concentration of bromine (as sodium thiosulphate titres) against t/min for the three mixtures. Calculate the order of the reaction, equation (8.22).
2. Compare the results for mixtures II and III and discuss the effect of the added salt on the rate of the reaction.
3. Indicate other methods of following this reaction.

Experiment 8.12   Study the autocatalytic reaction between potassium permanganate and oxalic acid

Theory: The reaction between potassium permanganate and oxalic acid:

$$2KMnO_4 + 5H_2C_2O_4 + 3H_2SO_4 = K_2SO_4 + 2MnSO_4 + 10CO_2 + 8H_2O$$

is catalysed by the manganese ion formed.

Requirements: Thermostat at 25 °C, large conical flasks; 0.1 mol dm⁻³ oxalic acid, 0.02 mol dm⁻³ potassium permanganate, 1 mol dm⁻³ sulphuric acid, 0.2 mol dm⁻³ manganese sulphate solutions, standard 0.01 mol dm⁻³ sodium thiosulphate solution, 10% potassium iodide solution.

Procedure: Prepare the following mixtures from solutions thermostatted at 25 °C. Add the solutions in the order given. Start your stop-clock when half the permanganate solution has been added.

Mixture number	$H_2C_2O_4$/cm³	H_2SO_4/cm³	$MnSO_4$/cm³	water/cm³	$KMnO_4$/cm³	Total volume/cm³
I	200	10	0	40	50	300
II	200	10	0	65	25	300
III	200	10	5	35	50	300
IV	200	10	5	335	50	600

Withdraw 25 cm³ samples from time to time and run into excess potassium iodide solution. Titrate the liberated iodine (after 5 min) with standard sodium thiosulphate (0.01 mol dm⁻³) solution. To obtain an initial titre mix 40 cm³ of oxalic acid, 2 cm³ sulphuric acid, 5 cm³ manganese sulphate and 13 cm³ water. Place in the thermostat to equilibrate, then take 25 cm³ sample and run into excess KI solution; leave for 5 min and then titrate the liberated iodine with standard thiosulphate solution.

Treatment of experimental data and discussion

1.   Plot a graph of titre of sodium thiosulphate (equivalent to the permanganate in the mixture), against time, for each reaction mixture. Discuss the shape of the curves. Using the curves for III and IV, determine the order of the reaction (equation (8.22)).
2.   Why are the solutions left for 5 min before titrating with the thiosulphate?
3.   Discuss possible mechanisms for this reaction.

Experiment 8.13   Investigate the reaction between acetone and iodine

Theory: In aqueous solution, acetone and iodine react according to the equation:

$$CH_3COCH_3 + I_2 = CH_3COCH_2I + HI$$

The reaction is catalysed by both acids and bases. The aim of the experiment is to determine the order in each reactant and suggest a possible mechanism for the acid catalysed reaction.

Requirements: Thermostat at 25 °C, 300 cm^3 conical flasks, acetone 1 mol dm^{-3} sulphuric acid, 0.05 mol dm^{-3} iodine in 10% potassium iodide solution, 1 mol dm^{-3} sodium acetate solution and standard 0.01 mol dm^{-3} sodium thiosulphate solution.

Procedure: Prepare the mixture at 25 °C, from solutions thermostatted at that temperature, in the following order: 20 cm^3 acetone, 150 cm^3 water, 10 cm^3 sulphuric acid and 20 cm^3 iodine in KI. Remove 20 cm^3 samples every 5 min and run into about 10 cm^3 of sodium acetate solution. Titrate the residual iodine with the standard sodium thiosulphate solution.

Keeping the total volume constant, investigate the effect of changes in the concentration of (*a*) iodine, (*b*) acetone and (*c*) acid on the rate of the reaction.

Treatment of experimental data and discussion

1.   Plot a graph of titre of sodium thiosulphate (equivalent to the residual iodine) against time, for each mixture. Determine the order of the reaction with respect to iodine, acetone and acid.
2.   From the data you obtain, derive a possible mechanism for the reaction. Suggest possible ways in which the mechanism may be tested.
3.   What is the purpose of the sodium acetate solution?

Experiment 8.14   Study the reaction between potassium peroxodisulphate and potassium iodide in solution

Theory: The reaction in solution proceeds with the liberation of iodine according to the equation:

$$2I^- + S_2O_8^{--} \longrightarrow I_2 + 2SO_4^{--}$$

Requirements: Thermostat at 25 °C, 0.06 mol dm^{-3} potassium peroxodisulphate solution, 0.12 mol dm^{-3} potassium iodide solution, 0.9 mol dm^{-3} potassium chloride solution, 0.9 mol dm^{-3} sodium chloride solution. 0.3 mol dm^{-3} magnesium chloride solution, 2.0 mol dm^{-3} acetic acid and 0.01 mol dm^{-3} sodium thiosulphate solution. These solutions should be within 1% of the recommended concentrations.

Procedure:
(a) To determine the rate constant, the order of the reaction and the effect of a neutral salt, use mixtures 1, 2 and 3.

(b) To study the influence of ionic strength on the reaction use mixtures 1, 3, 4, 5 and 6.

(c) To study the effect of the cation of a neutral salt use mixtures 1, 6 and 7.

(d) To study the effect of ion-pairing use mixtures 3, 4, 6, 8, 9 and 10.

Reaction mixtures, volumes of solutions/cm^3

Mixture	1	2	3	4	5	6	7	8	9	10
$K_2S_2O_8$	50	50	50	50	50	50	50	50	50	50
KI	50	50	50	50	50	50	50	50	50	50
KCl	0	0	20	50	100	150	0	0	0	0
NaCl	0	0	0	0	0	0	150	0	0	0
$MgCl_2$	0	0	0	0	0	0	0	20	50	150
H_2O	200	505	180	150	100	50	50	180	150	150
CH_3COOH	5	5	5	5	5	5	5	5	5	5

The procedure for studying each reaction mixture is as follows. Equilibrate the iodide plus any other reagents and water to 25 °C for at least 20 min in one vessel, add this solution to the similarly equilibrated peroxodisulphate; note the time. Immediately pour the solutions back into the vessel which originally contained the iodide (in order to mix the reagents thoroughly). Pipette out 25 cm^3 samples from the reaction mixture at regular intervals, quench the reaction by discharging the contents of the pipette on to crushed ice and water mixture. Quickly titrate with 0.01 mol dm^{-3} sodium thiosulphate solution using starch indicator. Suitable time intervals are 5–10 min. Make sure that the flasks are fitted with corks to prevent the loss of iodine. Keep a sample of each mixture to take an infinity reading (24 hours).

Treatment of experimental data and discussion

(a)
1. Multiply the titration figures for the dilute mixture (2) by 2 to make the results comparable. Plot a graph of concentration of potassium peroxodisulphate (measured by the titration difference $T_\infty - T_t$) against t, for each of the mixtures and draw smooth curves through the points. Calculate the order of the reaction, equation (8.22).
2. Determine the slope of the tangent ($-dc/dt$) to these curves at different values of c; plot $-dc/dt$ against c, and hence determine k.
3. Compare the results for mixtures 2 and 3, and hence deduce the effect of the neutral salt.

(b)
1. Calculate the ionic strength of each mixture assuming complete ionization of all salts.

2.   For the different mixtures, plot log $(a - x)$ against t. a and x may be expressed in cm^3 of sodium thiosulphate solution; a is obtained from the infinity reading. Determine k (equation (8.6)).

3.   Test the validity of equation (8.24).

4.   From your observations discuss a possible mechanism for the reaction.

(*c*)

1.   Determine the rate constant from the titration figures for each mixture. Indicate whether or not the cation present plays any specific ionic role in catalysing the reaction.

(*d*)

Runs (8), (9) and (10) may be compared with (3), (4) and (6). The ionic strengths are nominally the same but Mg^{++} replaces K^+. If ion-pairing takes place it would be expected that the $Mg^{++}S_2O_8^{--}$ pair would have the largest formation equilibrium constant of all the possible ion pairs in the system. Such ion-pairing might assist or inhibit the reaction. Investigate this point.

7   Heterogeneous gas reactions

The velocities of homogeneous reactions in the gaseous phase and in solution are generally governed by the rate of collisions of the reacting species. The rate of heterogeneous gas reactions may be governed by one or more of the following factors:

(i)   rate of adsorption of reactant molecules on the surface;

(ii)   rate at which reaction of adsorbed molecules occurs;

(iii)   rate of desorption of product molecules, leaving vacant sites on the surface for fresh molecules of reactant.

Experiment 8.15   Study the thermal decomposition of ammonia on a tungsten surface

Theory: The heterogeneous decomposition of ammonia on a tungsten surface is conveniently studied by electrically heating a tungsten filament in an atmosphere of the gas, contained in a reaction vessel connected by tubes of negligible volume, to a capillary manometer. The rate of change of pressure, as registered by the manometer, gives the rate of decomposition of the gas, according to the equation:

$$2NH_3 = N_2 + 3H_2$$

Thus, the manometer reading gives directly the amount of ammonia decomposed.

Requirements: Apparatus and electrical circuit, Fig. 8.5, solid sodium hydroxide, 0.880 ammonia solution.

Procedure: Place about 5 g of solid sodium hydroxide in the 250-cm^3 round-bottomed flask; insert the rubber bung, and fill the funnel with 0.880 ammonia

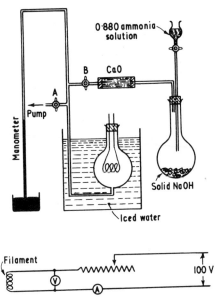

Fig. 8.3   Apparatus for studying the decomposition of ammonia on a tungsten surface

solution. Close taps A and B and start evacuating; now carefully open first A and then B, and evacuate the system as far as possible. Close A and check that the apparatus is airtight by watching the manometer for a few minutes. Run the ammonia solution dropwise on to the sodium hydroxide, thereby generating ammonia gas, until a pressure of about 400 mmHg is recorded; close B and pump out the ammonia by opening A for a few minutes. Close A and admit ammonia, until the required initial pressure is attained, by carefully opening B. Surround the reaction vessel with icc/water mixture and allow temperature equilibrium to be established before taking the initial pressure on the manometer. Both taps A and B must now remain closed during the experiment. This evacuation and flushing opera-tion must be carried out between successive experiments.

The temperature of the filament (about 30 cm of 0.01 cm diameter tungsten wire) must be kept constant throughout the experiment. Owing to the production of hydrogen during the experiment, the thermal conductivity of the gas in the reac-tion chamber will rise, resulting in a decrease in the temperature of the filament. Counteract this tendency by continually decreasing the rheostat resistance, thereby increasing the current flowing. The resistance of the filament depends upon its temperature and this is kept constant by maintaining the ratio of volts/amps con-stant (use slide rule for this). Shunt the voltmeter so that one division represents 1 V, and, the ammeter, so that one division is 0.02 A.

With the ammonia in the reaction vessel, at the measured initial pressure, turn the current on and adjust the resistance until the filament just glows. Note the ratio V/I, and keep this value constant throughout the experiment. After about 60 s, switch the current off, stir the iced water, tap and read the manometer. Turn the current on again (constant V/I ratio), switch off after 60 s and read the manometer.

Repeat this procedure at convenient time intervals. After sufficient readings have been taken, heat the wire for 2–3 min to decompose any residual ammonia. Using this technique, it is assumed that the filament cools down and heats up instantaneously as the current is switched off or on respectively.

Repeat the experiment with different initial pressures of ammonia at constant temperature (ie. at constant V/I ratio). The effect of temperature may be studied by repeating with different values of this ratio.

Treatment of experimental data and discussion

1. Assuming that all the ammonia has been decomposed, the initial pressure of ammonia is given by the difference between the initial and final manometer readings. Plot graphs showing the amount of ammonia decomposed after different times, and hence, deduce the initial rate of the reaction.

2. Determine the order of the reaction at the different initial pressures, this can be most conveniently found by determining the value of the ratio:

$$\frac{\text{time of half reaction}}{\text{time of quarter reaction}}.$$

Compare your answer with the theoretical value of this ratio for a zero and for a first order reaction.

3. Discuss the shape of the graphs obtained in (1); indicate causes for non-linearity, with particular reference to the tungsten surface.

4. What is the effect of replacing the tungsten wire by a platinum wire?

8   Photochemical reactions

Many reactions, which normally do not proceed to any great extent, occur readily when irradiated with ultraviolet radiation. The radiation is absorbed by the reactant molecules which are then raised to a higher electronic state (primary reaction). The excited molecule may undergo one or more of several secondary processes; thus it may:

(a) re-emit light of the same or different wavelength;

(b) return to its normal state, the energy being dissipated as heat;

(c) decompose into free radicals, or

(d) react with another molecule in a chemical reaction.

The primary quantum yield, defined as:

$$\phi = \frac{\text{Number of molecules activated}}{\text{Number of quanta of energy absorbed}}$$

is always unity, but the overall quantum yield, defined as:

$$\Phi = \frac{\text{Number of molecules finally decomposed}}{\text{Number of quanta of energy absorbed}}$$

can vary widely, from less than unity to about 10^6 (for chain reactions), depending on the nature of the secondary reactions.

In detailed photochemical studies, the amount of energy absorbed by the reacting material must be known; this requires the use of thermopiles or bolometers. The photolysis of uranyl oxalate in solution can, however, be studied in a simple manner without resort to absolute measurements of energy.

Experiment 8.16 Study the photolysis of uranyl oxalate

Theory: Oxalic acid is transparent except in the shorter regions of the ultraviolet. Uranyl ions absorb radiation of wavelength shorter than 440 nm, but when they absorb this radiation no reaction occurs, the energy being simply dissipated as heat. The absorption of radiation by the uranyl ions is, however, increased by the addition of the transparent oxalic acid. As a result the oxalic acid in the mixed solution is rapidly decomposed into carbon monoxide and carbon dioxide This transfer of energy from an absorbing molecule to another molecule which then undergoes reaction is known as 'Photocatalysis'.

The reaction:

$$UO_2^{++} + H_2C_2O_4 + hv = UO_2^{++} + H_2O + CO + CO_2$$

may be followed by titrating the residual oxalic acid, at any time, with standard potassium permanganate solution.

Requirements: Mercury vapour lamp and choke, six Pyrex (or quartz) 6×1 in tubes, shielding box with support for tubes, 0.01 mol dm^{-3} oxalic acid solution, 0.02 mol dm^{-3} uranyl sulphate solution, standard 0.004 mol dm^{-3} potassium permanganate solution.

Procedure: Arrange the supporting rack for the tubes so that there are two circles of holes (10 and 20 cm) from the lamp, placed centrally. The tubes must hang vertically around the lamp. Paint the inside of the box and the supporting rack with optical black. Use special goggles when the lid is removed while the mercury lamp is on.

Fill the six tubes with 25 cm^3 of the oxalic acid and 25 cm^3 of the uranyl sulphate solutions, mix well and place them at the 10 cm distance from the lamp. Replace the lid and switch on the lamp for 30 min. Remove the tubes, and titrate each solution with the standard solution of potassium permanganate and check that there is little variation of the titres for the different positions around the lamp.

Repeat the experiment with the same solutions, but now remove the tubes one at a time after 5, 10, 20, 30, 40 and 50 min. Carry out further experiments, again using the same six tubes to determine the influence of (*a*) volume, (*b*) concentration of oxalic acid, (*c*) concentration of uranyl sulphate, and (*d*) distance from the lamp. Only change one variable at a time.

Treatment of experimental data and discussion

1. From the potassium permanganate titrations, calculate the amount of oxalic acid decomposed at different times. Plot graphs of the amount of oxalic acid decomposed against *t*. Determine the order of the reaction.

2. Represent the influence of the other variables in suitable form, ie. either tabular or graphical and explain your results.
3. Explain why the absorption of radiation by the uranyl ions is increased by the addition of oxalic acid.

9

Equilibrium constants

Many chemical reactions (or physical processes) do not go to completion, but proceed to a position of equilibrium where the reaction (or process) apparently ceases. Under a standard set of conditions of temperature, pressure and concentration the equilibrium position is always the same; ie. there is a fixed relationship between the active masses of reactants and products. This is a dynamic state of equilibrium, in which the rate of the forward process is equal to the rate of the reverse process. Thus for the reaction:

$$aA + bB \rightleftharpoons lL + mM$$

the van't Hoff isotherm relating the free energy change in a reaction to the active masses of the reactants and products (designated [ ]) can be written:

$$\Delta G = \Delta G^{\ominus} + RT \ln \frac{[L]^l [M]^m}{[A]^a [B]^b} \qquad (9.1)$$

but for a system in equilibrium, $\Delta G = 0$, hence:

$$\Delta G^{\ominus} = -RT \ln \left\{ \frac{[L]^l [M]^m}{[A]^a [B]^b} \right\}_e = -RT \ln K \qquad (9.2)$$

where subscript 'e' indicates the active masses of the species present at equilibrium.

Since G is a state function, the ratio of active masses of the products and reactants is constant and independent of their individual values. Hence the equilibrium constant is defined as:

$$K = \left\{ \frac{[L]^l [M]^m}{[A]^a [B]^b} \right\}_e = e^{-\Delta G^{\ominus}/RT} \qquad (9.3)$$

In actual practice the active mass is replaced by an activity, concentration or pressure term, giving K_{therm}, K_c and K_p respectively. K_{therm}, the thermodynamic equilibrium constant, is a true constant for a reaction at a given temperature. On the other hand the activity term, in equation (9.4), is a quantity

$$K_{therm} = K \times \frac{\gamma_L^l \gamma_M^m}{\gamma_A^a \gamma_B^b} \qquad (9.4)$$

(K can be K_c, or K_p and γ the activity coefficient) whose value for gas reactions depends on the gases involved and the pressure; and, for reactions in solution, on the nature of the solvent and solute and the ionic strength. Thus K_p and K_c are not true constants but vary with P or I, approaching the appropriate value of K_{therm} as $P \to 0$ or $I \to 0$.

The magnitude of K determines the extent to which the reaction or process can proceed under given conditions. A large value for $K(\Delta G^{\ominus} \ll 0)$ indicates that the forward process is favoured. A small value for $K(\Delta G^{\ominus} > 0)$ indicates that the process does not proceed to any appreciable extent under the given conditions (eg. the dissociation constant of a weak acid). Thus the only condition under which chemical equilibrium is possible is $(dG)_{T, P} = 0$.

K for a given reaction varies with the temperature according to the van't Hoff isochore (equation (9.5)).

$$\left(\frac{\partial \ln K}{\partial T} \right) = \frac{\Delta H^{\ominus}}{RT^2} \tag{9.5}$$

Thus from a knowledge of the variation of K with temperature, $\Delta H^{\ominus}$ for the process may be calculated.

Many equilibrium constants have specific names, among those discussed in this book are Partition or Distribution Coefficients (Expt. 9.4 to 9.9); Acid Dissociation or Ionization Constants (Expt. 9.11, 9.16, 9.18, 9.19, 9.20); Hydrolysis Constants (Expt. 9.13, 9.17); Ionic Product of water (Expt. 9.14); Stability, Formation or Association constants (Expt. 9.10, 9.12, 9.15, 9.24) of complex, which for the system:

$$q M^+ + r A^- \rightleftharpoons M_q A_r^{\pm}$$

is defined as:

$$K_s = \frac{a_{M_q A_r^{\pm}}}{a_{M^+}^q \, a_{A^-}^r} \tag{9.6}$$

and the Dissociation or Instability constant of a complex ion ($K_i = 1/K_s$).

There are many different experimental methods, both chemical and physical, by which the value of an equilibrium constant may be determined; the merits and applicability of each depend on the particular system under study.

1   Chemical methods

In this method the reactants are mixed, equilibrium established and the resulting mixture analysed. The determination of solubility products (Expt. 4.9, 4.10) is an example of an analytical procedure.

Experiment 9.1   Determine the equilibrium constant for the keto-enol tautomerism of ethylacetoacetate

Theory: Ethylacetoacetate either pure or in solution exists as an equilibrium mixture of the two tautomeric forms:

$$CH_3-\overset{\overset{O}{\|}}{C}-CH_2-\overset{\overset{O}{\|}}{C}-O\cdot CH_2CH_3 \rightleftharpoons CH_3-\overset{\overset{OH}{|}}{C}=CH-\overset{\overset{O}{\|}}{C}-O\cdot CH_2CH_3$$
(keto)   (enol)

Equilibrium is rapidly established and there is no need for a catalyst. The enol form reacts rapidly with bromine while the keto form does not. The concentration of the enol form in the equilibrium mixture cannot be obtained by direct titration with bromine since the hydrogen bromide liberated by this reaction catalyses the keto-enol transformation and, in consequence, the end-point drifts. The indirect method is more accurate. To the ethylacetoacetate solution is added first excess bromine and then β-naphthol immediately. The β-naphthol removes any bromine remaining after the reaction with the enol form originally present. Thus, all the enol form is brominated and the excess bromine destroyed. The concentration of bromoester, and hence the original concentration of the enol form, is determined by adding potassium iodide and titration of the liberated iodine with standard sodium thiosulphate.

Requirements: 4 × 250 cm^3 stoppered bottles, thermostat at 25 °C, 0.4 mol dm^{-3} solution of ethylacetoacetate in methanol, 10% β-naphthol in methanol, 0.1 mol dm^{-3} bromine in methanol (freshly prepared), 0.1 mol dm^{-3} aqueous potassium iodide solution, standard 0.1 mol dm^{-3} sodium thiosulphate solution.

Procedure: Determine the exact concentration of ethylacetoacetate in the stock solution: transfer 5 cm^3 of this solution to a 500 cm^3 conical flask and add 100 cm^3 bromine solution (measured in a cylinder). Allow the solution to stand for some minutes so that the keto form is completely transformed to the enol and the latter brominated. Add 10 cm^3 β-naphthol (measured in a cylinder) and 50 cm^3 of potassium iodide; allow to stand for 15 min. Titrate the liberated iodine with the standard sodium thiosulphate solution to a colourless end-point which persists for at least 3 min. Starch cannot be used as an indicator. Make a duplicate determination.

Accurately dilute the stock ethylacetoacetate solution with methanol to give concentrations of about 0.2 and 0.1 mol dm^{-3}. Suspend stoppered bottles each containing about 200 cm^3 of the stock or diluted solutions in a thermostat at 25 °C and leave for about 1 hour, shake periodically. Transfer a 50-cm^3 aliquot of one of the solutions to a large conical flask and add bromine solution (50 cm^3 for 0.4 and 0.2 mol dm^{-3}, and 25 cm^3 for 0.1 mol dm^{-3} ester solution), shake well and *immediately* add 10 cm^3 β-naphthol solution. The addition of the bromine and β-naphthol solutions must be completed within 1 min; use measuring cylinders for the additions. Now add 50 cm^3 of potassium iodide solution; allow to stand for 15 min and titrate with sodium thiosulphate solution to a colourless end-point. Make duplicate determinations on each solution.

Treatment of experimental data and discussion

1.   From the titration figures calculate the concentration of the enol form

present at equilibrium and hence, by difference from the total ester concentration, the concentration of the keto form. Calculate $K = c_{enol}/c_{keto}$. (2 mole sodium thiosulphate $\equiv$ 1 mole I_2 $\equiv$ 1 mole ester.)

2. Does K vary with the ester concentration? Explain any variation.
3. Why cannot starch be used as an indicator?
4. Calculate the value of $\Delta G^{\ominus}$ for this equilibrium. What other experimental information is required to calculate $\Delta H^{\ominus}$ and $\Delta S^{\ominus}$?

Experiment 9.2   Determine the equilibrium constant of an esterification reaction

Theory: An acid and an alcohol react to form ester and water, whilst the ester is hydrolysed to give the original acid and alcohol. These opposing reactions proceed until a state of equilibrium is reached. The equilibrium constant obtained in this method, is in error since the activity coefficients are neglected.

The equilibrium to be studied:

$$C_2H_5OH + CH_3COOH \rightleftharpoons CH_3COOC_2H_5 + H_2O$$

is reached only very slowly; it is necessary to add hydrochloric acid as a catalyst, which, while hastening the reaction, takes no other part in it.

Requirements: Glass-stoppered bottles, 1, 2 and 5 cm^3 pipettes, standard 0.5 mol dm^{-3} sodium hydroxide solution, 3 mol dm^{-3} hydrochloric acid solution, ethyl acetate, glacial acetic acid and absolute ethanol.

Procedure: Prepare the following mixtures in the bottles:

No.	Volume of HCl /cm^3	Volume of ethyl acetate /cm^3	Volume of water /cm^3	Volume of absolute ethanol /cm^3	Volume of acetic acid /cm^3
1.	5	0	5	0	0
2.	5	5	0	0	0
3.	5	4	1	0	0
4.	5	2	3	0	0
5.	5	4	0	1	0
6.	5	4	0	0	1

and stopper immediately. Allow to stand for a week in a warm place, with occasional shaking. It is not essential to thermostat the mixtures. When equilibrium is established, titrate the contents of each bottle with the standard sodium hydroxide solution.

For the calculation, the weight of each reactant in each mixture must be known. This is achieved most easily by pipetting known volumes of reagents directly into a weighed stoppered bottle and weighing. Determine the weight of 5 cm^3 of 3 mol dm^{-3} hydrochloric acid, 5, 4 and 2 cm^3 of ethyl acetate, 1 cm^3 of ethanol, 1 cm^3 of glacial acetic acid and 1, 3 and 5 cm^3 of water. In the pipetting, take care that the draining procedure is carefully repeated each time.

Treatment of experimental data and discussion

1. The original weight of ester in each mixture and absolute ethanol (in 5) and glacial acetic acid (in 6) is directly known from the weights. From the titration of the hydrochloric acid with alkali (in 1), calculate the weight of hydrogen chloride present, and hence, by difference, the weight of water in 5 cm^3 of the acid. Add this value to the weights of water. Calculate the amount of acetic acid in each equilibrium mixture, first subtracting the volume of sodium hydroxide used in (1), since (1) gives only hydrochloric acid, and the remainder hydrochloric and acetic acids. In (6) subtract the acetic acid originally present to give the amount produced during the reaction.

   For each mole of acetic acid produced, 1 mole of ethanol is produced and 1 mole each of water and ester disappear. Calculate the total number of moles of each reagent in the original mixture and hence the final number of each at equilibrium. Assuming the final volume of each mixture to be 10 cm^3, calculate the concentration of each reagent and hence K_c for each mixture:

$$K_c = \frac{c_{ester} \times c_{water}}{c_{acid} \times c_{alcohol}}$$

2. Comment on any variations of K_c for the different mixtures.
3. Comment on the assumption that the total volume of each mixture is 10 cm^3.
4. Explain why thermostatting is not essential in the study of this equilibrium.
5. What is the value of K_x, the equilibrium constant in which the concentrations are expressed as mole fractions?

Experiment 9.3 Determine the equilibrium constant of the system
$$2Ag^+ + CaSO_4(s) \rightleftharpoons Ag_2SO_4(s) + Ca^{++}$$

Theory: The equilibrium systems are prepared by mixing solutions of silver and calcium nitrates and then partially precipitating the silver and calcium as sulphates. At equilibrium, the silver is estimated by titration with standard ammonium thiocyanate solution, and the total silver and calcium, by addition of an excess of standard ammonium oxalate solution and back-titrating with potassium permanganate solution.

Requirements: Stoppered bottles, thermostat at 25 °C, standard solutions of silver nitrate (0.35 mol dm^{-3}), ammonium thiocyanate (0.1 mol dm^{-3}), ammonium oxalate (0.1 mol dm^{-3}), potassium permanganate (0.02 mol dm^{-3}), and solutions of calcium nitrate (0.25 mol dm^{-3}), potassium sulphate (0.25 mol dm^{-3}) and a saturated solution of ammonium iron (III) sulphate as indicator.

Procedure: Prepare the following mixtures in the stoppered bottles:

	1	2	3	4
Vol. of silver nitrate solution/cm^3	25	25	25	25
Vol. of calcium nitrate solution/cm^3	25	25	25	25
Vol. of potassium sulphate solution/cm^3	20	25	30	35

After the addition of the potassium sulphate solution, carefully seed each solution with a trace of mixed solid silver and calcium sulphates.

Stopper the bottles and shake vigorously for 15 min, place in the thermostat at 25 °C and shake periodically for 2—3 hours. Filter each solution in turn through a No. 3 sintered glass crucible. Remove a 25-cm^3 aliquot, and titrate with the standard ammonium thiocyanate solution, using ammonium iron (III) sulphate as indicator. This provides a measure of the total silver ions in solution. Remove another 25-cm^3 aliquot and treat as follows, for the total silver and calcium ions in solution: dilute with about 25 cm^3 of distilled water and heat to 70 or 80 °C. Slowly and with constant shaking, add 25 cm^3 of the standard ammonium oxalate solution. Allow the solution to cool and filter through a No. 3 sintered glass crucible, wash the precipitate with small quantities of cold distilled water and add to the filtrate. Titrate the filtrate and washings with the standard potassium permanganate solution.

Treatment of experimental data and discussion
1. From the titration figures, calculate the individual concentrations of silver and calcium in solution. Hence evaluate the equilibrium constant, $K_c = c_{Ca^{++}}/c_{Ag^+}^2$.
2. Explain how K_{therm} may be obtained from the experimental data.
3. Calculate $\Delta G^{\ominus}$ for this equilibrium and compare it with the value obtained from tabulated values of $\Delta G_f^{\ominus}$.

2   Partition or distribution methods

When a small quantity of a solute is distributed between two immiscible phases (an organic solvent —o, and water or an aqueous solution —w) with which it does not react, then at equilibrium the chemical potential of the solute is the same in both phases, ie.

$$\mu_o = \mu_w \quad \text{or} \quad RT \ln a_o = RT \ln a_w$$

whence

$$a_o/a_w = \text{Const.} \tag{9.7a}$$

In dilute solution:

$$c_o/c_w = D \tag{9.7b}$$

where D is the distribution or partition coefficient between the two solvents. Equations (9.7a) and (9.7b) are only valid when the solute does not react with either solvent and undergoes no association or dissociation. If in one or both solvents the solute consists, to some extent, of molecules of different relative molecular mass or composition, then the distribution law is applicable only to the molecular species common to both solvents.

In the distribution of a weak organic acid between an organic solvent and water, the acid in the organic solvent may exist partly as the dimer, while in water it is partly dissociated. The species common to both solvents is the undissociated monomer. If α_o and α_w represent the fraction of solute associated or dissociated in each solvent then:

$$D = \frac{(1 - \alpha_o)c_o}{(1 - \alpha_w)c_w} \qquad (9.8)$$

If the solute, B, has a normal molecular mass in water but associates in the organic solvent:

$$nB \rightleftharpoons B_n$$

the association constant is given by:

$$K = \frac{\alpha_o c_o}{n(1 - \alpha_o)^n c_o^n} \qquad (9.9)$$

Combining equations (9.8) and (9.9) ($\alpha_w = 1$) gives:

$$D = \frac{(\alpha_o c_o/nK)^{1/n}}{c_w} \qquad (9.10)$$

when the molecules in the organic phase are almost completely associated, $\alpha_o \sim 1$,

$$c_o^{1/n}/c_w = \text{Const.} \qquad (9.11a)$$

or

$$\log c_w = \frac{1}{n} \log c_o + \text{Const.} \qquad (9.11b)$$

In a similar way if the solute has a normal molecular mass in the organic solvent but dissociates in the aqueous phase then:

$$c_o/c_w^2 = \text{Const.} \qquad (9.12)$$

For the weak acid which exists in the monomeric form in the organic solvent but which in water dissociates:

$$HA + H_2O \rightleftharpoons H_3O^+ + A^-$$

$$D = [HA]_o/[HA]_w = c_o/[HA]_w$$

but the measured concentration of the acid in the aqueous phase $c_w = [HA]_w + [A^-]$. Whence it follows that:

$$[A^-] = c_w - [HA]_w = c_w - c_o/D$$

The dissociation constant for the weak acid is given by:

$$K_c = \frac{[H_3O^+][A^-]}{[HA]_w} = \frac{(c_w - c_o/D)^2}{c_o/D} \qquad (9.13)$$

Experiment 9.4 Study the distribution coefficients of two dyes between an organic solvent and water

Requirements: Stoppered separating funnels, spectrophotometer, and 1 cm glass cells, solid bromocresol green (BCG), solid phenol red (PR), butan-1-ol, 0.05 mol dm^{-3} sodium carbonate solution.

Procedure: Shake together equal volumes of butan-1-ol and 0.05 mol dm^{-3} sodium carbonate solution in a large separating funnel to produce saturation of each solvent in the other. Separate the two phases and use as stock solutions (about 1 dm^3 of each is required).

(a) Preparation of calibration curves for the dyes: two calibration curves are required for each dye, one in each phase. Dissolve about 2 mg of each dye in 100 cm^3 of each solvent (20 μg cm^{-3}) and from these prepare by dilution standard solutions of known concentration in the range 0.5 to 15 μg cm^{-3}. Plot the absorption spectrum for an intermediate concentration of each dye in each solvent (BCG from 450 to 700 nm and PR from 450 to 600 nm). Select the wavelengths of maximum absorption for each dye solution and determine the absorbances of the remaining solutions in the series. Hence construct a calibration curve for each dye in each solvent (4 in all) and show that the Beer–Lambert law (equation (5.10)) is valid.

(b) Determination of the Partition Coefficients: prepare stock solutions of PR; 3–4 mg PR in 100 cm^3 sodium carbonate solution (saturated with butan-1-ol), and BCG; 4–5 mg BCG in 100 cm^3 butan-1-ol (saturated with carbonate).

(i)   Transfer 20 cm^3 of the stock PR solution to a separating funnel containing 20 cm^3 of butan-1-ol. Equilibrate for 3 min and carefully separate the two layers. Measure the absorbance of each solution at the appropriate wavelength of maximum absorbance (dilute the carbonate solution 10 times).

(ii)   Determine the partition coefficient as for PR using 20 cm^3 of stock solution and 20 cm^3 of carbonate for partition.

Treatment of experimental data and discussion

1.   From the appropriate calibration curves calculate the concentration and amount of dye in each phase. Check that the total amount of each dye is the same as that added originally. Determine the partition coefficient of each dye.

2.   Comment on the value of the distribution coefficients for the two dyes and discuss the possibility of separating a mixture of these dyes by repeated solvent separation. (See also Expt. 11.7.)

3.   Explain why it is essential to saturate each solvent with the other.

Experiment 9.5   Study the distribution of succinic acid between water and ether

Requirements: 4×250 cm^3 glass stoppered flasks, 0.55 mol dm^{-3} succinic acid, standard 0.2 mol dm^{-3} sodium hydroxide, diethyl ether.

Procedure: Using a graduated cylinder prepare the following mixtures in the stoppered flasks:

Flask	Vol. succinic acid/cm^3	Vol. water/cm^3	Vol. ether/cm^3
1	65	15	40
2	70	10	40
3	75	5	40
4	80	0	40

Stopper the flasks and shake thoroughly for 30–40 min in a thermostat at 25 °C to establish equilibrium. Allow the flasks to stand for a further 10 min for the phases to separate.

With a dry pipette withdraw 20 cm^3 of the ether (upper) phase, add about 20 cm^3 water and titrate with the standard sodium hydroxide solution (using phenolphthalein as indicator) while shaking vigorously. Withdraw a 10 cm^3 portion of the aqueous phase, keep a finger over the pipette until the tip has entered the aqueous phase. Titrate this solution with the sodium hydroxide solution.

Treatment of experimental data and discussion
1. Tabulate the concentration of succinic acid in each phase and calculate the distribution coefficient in each flask (equation (9.7b)).
2. Do the results give any information about the association of the acid in ether?
3. Account for any deviations from a constant value for D.
4. Explain why it is not necessary to make up the original mixtures accurately.
5. How would you expect the value of D to vary with temperature?

Experiment 9.6   Study the association of benzoic acid from its distribution between benzene and water

Requirements: 6 × 250 cm^3 stoppered flasks, saturated solution of benzoic acid in benzene (70 g in 1 dm^3 at 18 °C), standard 0.05 mol dm^{-3} sodium hydroxide.

Procedure: Using burettes prepare the following mixtures in the stoppered flasks:

	1	2	3	4	5	6
Vol. of saturated benzoic acid/cm^3	40	35	30	25	20	15
Vol. of benzene/cm^3	10	15	20	25	30	35
Vol. of water/cm^3	50	50	50	50	50	50

Stopper the flasks and shake thoroughly for 30–40 min to establish equilibrium in a thermostat at 25 °C (or at room temperature – record this). Allow the bottles to stand for a further 10 min for the phases to separate.

Titrate 20 cm^3 of each aqueous phase, with a finger over the top of a 20-cm^3 pipette (to prevent any of the benzene layer entering the pipette), immerse the end of the pipette in the lower layer. If any benzene layer enters the pipette gently blow it out. Wipe off all liquid from the outside of the pipette before transferring it to the titration vessel. Titrate with the standard alkali (phenolphthalein indicator).

Using a dry pipette, withdraw 5 cm^3 of each benzene layer, add 20 cm^3 of water and titrate with the standard sodium hydroxide solution; shake vigorously during the titration. Carry out duplicate titrations.

Treatment of experimental data and discussion

1. From the titration figures calculate the concentration of benzoic acid in the aqueous layer (c_w) and the benzene layer (c_o) and using equation (9.7*b*) establish that c_o/c_w is not constant.
2. Plot log c_w against log c_o (equation (9.11*b*)) and from the slope determine the association number, *n*. Hence determine the partition coefficient defined as $c_o^{1/n}/c_w$.
3. To what extent would the accuracy be improved if the dissociation of benzoic acid in water had been taken into account?
4. Explain why it is important that none of the benzene layer contaminates the sample of the aqueous layer taken for titration. During the titration of the benzene layer with standard alkali the flask must be vigorously shaken; explain why.
5. Suggest reasons why benzoic acid associates in benzene.

Experiment 9.7   Determine the dissociation constant of picric acid from its distribution between benzene and water

Requirements: Glass-stoppered bottles, thermostat at 25 °C, saturated solution of picric acid in water, benzene, standard 0.02 mol dm^{-3} sodium hydroxide solution.

Procedure: Prepare the following mixtures in the stoppered bottles:

Volume of saturated picric acid solution/cm^3	Volume of water/cm^3	Volume of benzene/cm^3
100	0	50
50	0	50
50	50	50
25	50	50

Stopper each bottle and shake vigorously in a thermostat to attain distribution equilibrium at 25 °C. Remove a 25-cm^3 aliquot of the aqueous layer from each bottle and titrate with the standard sodium hydroxide solution using phenolphthalein as indicator. The colour change is from greenish-yellow to orange-yellow. Titrate a 10-cm^3 sample of the benzene layer to the same end-point. This titration can only be achieved by shaking vigorously in a glass-stoppered bottle. Owing to the formation of a white benzene-water emulsion, it is advisable to keep a sample of the solution, before the colour change has occurred for comparison.

Treatment of experimental data and discussion

1. From the titration figures, calculate the concentrations of picric acid (mol dm^{-3}) in both phases of each mixture. If c_w' and c_o' are the concen-

trations in water and benzene respectively for a second determination, then equation (9.13) can be written:

$$K_c = \frac{(c_w - c_o/D)^2}{c_o/D} = \frac{(c'_w - c'_o/D)^2}{c'_o/D} \tag{9.13a}$$

Using the experimental results in pairs, calculate several values of D, (equation (9.13a)). With an average value of D use the same equation to calculate K_c for each set of readings.

2. Explain why it is important that none of the benzene layer contaminates the aqueous layer taken for titration.
3. Is it justified to neglect the H_3O^+ due to the ionization of water?
4. Calculate the pH of a 0.1 mol dm^{-3} solution of picric acid.

Experiment 9.8   Determine the stability constant of I_3^- by the distribution method

Theory: From the distribution of iodine between a solution of potassium iodide and carbon tetrachloride it is possible to determine the equilibrium constant for the reaction:

$$I_2 + I^- \rightleftharpoons I_3^-$$

$$K_{I_3^-} = \frac{[I_3^-]_w}{[I_2]_w[I^-]_w} \tag{9.14}$$

Requirements: Glass-stoppered bottles, thermostat at 25 °C, saturated solution of iodine in carbon tetrachloride, pure carbon tetrachloride, solid potassium iodide, standard 0.01 mol dm^{-3} sodium thiosulphate solution, starch solution or iodine indicator.

Procedure: (a) Determine the distribution coefficient for iodine between carbon tetrachloride and water. Prepare the following mixtures in stoppered bottles:

Volume of saturated I_2 (in CCl_4) solution/cm^3	Volume of CCl_4/cm^3	Volume of water/cm^3
25	0	200
10	0	200
5	5	200
5	10	200
5	15	200

Stopper each bottle and shake vigorously for 20 min, clamp in the thermostat for 15 min, shake for a further 5 min and replace in the thermostat to attain equilibrium at 25 °C.

Sample the carbon tetrachloride layer by inserting a 2-cm^3 pipette into the lower layer. The aqueous layer can be excluded by closing the top of the pipette with a finger whilst dipping it in. Transfer the measured sample to a flask, add 2 g solid potassium iodide and 10 cm^3 of distilled water. Acidify with 2 drops of dilute sulphuric acid, and titrate with the standard sodium thiosulphate solution, using an indicator. Repeat the determination on another sample. With a 50-cm^3 pipette, sample the aqueous phase and again titrate the iodine with the sodium thiosulphate solution.

The ratio of the titres of the two layers, in terms of the volume of sodium thiosulphate solution required to titrate the same volume of both layers, gives the distribution coefficient, provided equilibrium has been attained. As a guarantee of this, subject each bottle to further shaking before repeating the titrations.

(b) Study the tri-iodide equilibrium. Prepare 250 cm^3 of 0.05 mol dm^{-3} potassium iodide solution by weighing out the oven-dried salt. Prepare the following mixtures in glass-stoppered bottles:

Volume of KI solution/cm^3	Volume of water/cm^3	Volume of saturated I$_2$ solution/cm^3	Final concentration of KI/mol dm^{-3}
100	0	25	0.05
50	50	25	0.025
25	75	25	0.0125

and shake at 25 °C to attain distribution equilibrium as in (a). Remove 5-cm^3 samples of the carbon tetrachloride layers and 25 cm^3 of the aqueous layers and titrate as before with the standard sodium thiosulphate solution.

If possible repeat the above determinations at 35 and 45 °C.

Treatment of experimental data and discussion

1.   From the titration figures of (a) obtain values of D (equation (9.7b)) at four or five different concentrations of iodine. The results may show a slight drift with concentration but may be averaged for use in section (b). Convert all the titration figures of (b) to give concentrations of iodine/mol dm^{-3}. The titration of the aqueous layer in (b) gives $[\Sigma I_2]$, the total titratable iodine; the titration of the solvent layer gives $[I_2]_0$, and the total potassium iodide [KI] is known. Since there is no ionic form of iodine in the solvent layer, the following relationships are valid:

$$[\Sigma I_2] = [I_2]_w + [I_3^-] \tag{9.15}$$

$$[KI] = [I^-] + [I_3^-] \tag{9.16}$$

$$[I_2]_w = [I_2]_o/D \tag{9.7b}$$

Knowing the value of D, calculate $[I_2]_w$ (equation (9.7b)). From equation (9.15) knowing $[\Sigma I_2]$ and $[I_2]_w$, calculate $[I_3^-]$. Using this value and [KI], calculate $[I^-]$. Hence calculate $K_{I_3^-}$ from equation (9.14) for the three different mixtures.

2.   Calculate $\Delta G^{\ominus}$ for the equilibrium (equation (9.2)) at each temperature and, if $K_{I_3^-}$ has been determined at different temperatures, $\Delta H^{\ominus}$ (equation (9.5)), and hence $\Delta S^{\ominus}$. Compare the experimental values of the thermodynamic quantities with those calculated from tabulated value of thermodynamic data. Comment on the sign of $\Delta S^{\ominus}$.

3.   Discuss the main sources of error in this experiment and indicate how these could be minimized.

Experiment 9.9   Determine the stability constant of I_3^- by the distribution method (spectroscopic analysis)

Requirements: Glass stoppered bottles, thermostat at 25 °C, Winchester quart bottles, separating funnels, 0.002 mol dm^{-3} solution of iodine in carbon tetrachloride, 0.1 mol dm^{-3} potassium iodide solution, pure carbon tetrachloride. Spectrophotometer with 1 cm glass cells.

Procedure:

(*a*) Determine the partition coefficient of iodine between carbon tetrachloride and water. Dilute a 5 cm^3 portion of the stock 0.002 mol dm^{-3} iodine solution with 15 cm^3 of carbon tetrachloride; store in a stoppered bottle. Measure the absorption spectrum of this solution (1 cm glass cell) using carbon tetrachloride as the reference solvent, over the range 500—540 nm; select the wavelength of maximum absorbance. Take a fresh sample of the diluted solution and measure the absorbance at the wavelength of maximum absorbance, and hence calculate the absorbance (A_1) of the stock solution before dilution.

Transfer 10 cm^3 of the stock iodine solution, 10 cm^3 of carbon tetrachloride and 1500 cm^3 of water to a Winchester quart bottle and shake the mixture mechanically for 20 min. Allow the layers to separate, decant off most of the aqueous layer and transfer the remaining phases to a separating funnel. Separate off the carbon tetrachloride layer and determine, in duplicate, the absorbance at the wavelength of maximum absorbance, A_2 (it may be necessary to dilute the solution first). Record the temperature.

(*b*) Determine the distribution of iodine between carbon tetrachloride and potassium iodide solution. Transfer 10 cm^3 of the stock iodine solution and 10 cm^3 of carbon tetrachloride to each of 4 stoppered bottles (labelled 1 to 4) and add to successive bottles 40, 60, 80 and 100 cm^3 of the potassium iodide solution. Shake the bottles mechanically for 10 min. Allow the two layers to separate and determine the absorbance of the iodine in the carbon tetrachloride layer, as before.

Treatment of experimental data and discussion

1.   If V_o and V_w are the volumes of the carbon tetrachloride (20 cm^3) and aqueous (1500 cm^3) layers respectively; then since the concentrations of iodine in the two layers are proportional to the absorbance by the Beer—Lambert law (equation (5.10)) it follows that the partition coefficient is:

$$D = \frac{[I_2]_o}{[I_2]_w} = \frac{A_2 V_w}{(A_1/2 - A_2) V_o} \tag{9.17}$$

From the tabulated results calculate D for the distribution of iodine between carbon tetrachloride and water.

2.   The apparent partition coefficient of iodine between carbon tetrachloride and potassium iodide, D' may be defined as:

$$D' = \frac{[I_2]_o}{[I_2]_w + [I_3^-]_w} \quad\quad\quad (9.18)$$

where the denominator represents the total concentration of iodine in the aqueous layer. Calculate D' for each of the bottles (1 to 4) using equation (9.17), where A_1 is, as before, the absorbance of the stock solution and A_2 the absorbance after shaking with the aqueous solution, V_w varies for the four bottles.

3.   Combining equations (9.17) and (9.18) and rearranging gives:

$$\frac{D}{D'} = \frac{[I_2]_o([I_2]_w + [I_3^-]_w)}{[I_2]_w[I_2]_o}$$

$$= 1 + \frac{[I_3^-]_w}{[I_2]_w} \qu\quad\quad (9.19)$$

Substituting the value of $[I_3^-]_w/[I_2]_w$ from equation (9.19) into equation (9.14) and rearranging gives:

$$K_{I_3^-} = \frac{(D/D' - 1)}{[I^-]_w}$$

Using the values of D and D' calculate $K_{I_3^-}$ for each concentration of I^-.

4.   Calculate the value of $\Delta G^\ominus$ for the formation of I_3^- (equation (9.2)) and compare it with that calculated from tabulated thermodynamic data.

5.   Discuss the main sources of error in this method.

Experiment 9.10   Determine the formula of the complex formed between the copper(II)ion and ammonia

Theory: The distribution coefficient of ammonia between chloroform and water is given by:

$$D = [NH_3]_o/[NH_3]_w \quad\quad\quad (9.20)$$

Although in the presence of copper sulphate solution some of the ammonia is present as the complex cuprammonium ion, $Cu(NH_3)_n^{++}$, the relationship (9.20) still applies to the concentration of 'free' ammonia. Thus, from a study of the distribution of ammonia between water and chloroform and an aqueous solution of copper sulphate and chloroform, the formula of the complex ion can be determined, assuming that the equilibrium:

$$Cu^{++} + nNH_3 \rightleftharpoons Cu(NH_3)_n^{++}$$

is completely to the right.

Requirements: 250 cm^3 separating funnels, pure chloroform, approx. 1 mol dm^{-3}

solution of aqueous ammonia, standard 0.1 mol dm^{-3} copper sulphate solution and standard 0.05 mol dm^{-3} hydrochloric acid.

Procedure: Standardize the ammonia solution by diluting 25 cm^3 to 250 cm^3; titrate 25 cm^3 of the diluted solution with the standard acid, using bromocresol green as indicator. From the standardized solution, prepare a standard two-fold dilution (ie. about 0.5 mol dm^{-3}).

(a) Determine the distribution coefficient for ammonia between chloroform and water. Prepare, in duplicate, the following mixtures in the stoppered funnels: (i) 25 cm^3 of chloroform and 25 cm^3 of 1 mol dm^{-3} aqueous ammonia and (ii) 25 cm^3 of chloroform and 25 cm^3 of 0.5 mol dm^{-3} aqueous ammonia solution. Shake each funnel thoroughly for 10 min and allow the layers to separate. Run off the lower chloroform layer into a clean dry flask and transfer a 20 cm^3 aliquot into a 500 cm^3 conical flask. Add about 200 cm^3 of water and titrate with the standard acid using bromocresol green as indicator. Shake the flask frequently during the titration to obtain an accurate end-point (use more indicator than normal since this has a high solubility in chloroform).

(b) Study the formation of the cuprammonium ion. Prepare in the funnel a mixture of 75 cm^3 of chloroform, 25 cm^3 of 1 mol dm^{-3} aqueous ammonia and 25 cm^3 of copper sulphate solution; set up a duplicate mixture. Shake thoroughly for 10 min and allow the layers to separate. Run off the chloroform layer and transfer 50 cm^3 into a 500 cm^3 conical flask and determine the concentration of ammonia in this layer by titration as before.

Treatment of experimental data and discussion

1. From the titration figures of (a) calculate the concentration of ammonia in the chloroform layer and, by difference from the total amount of ammonia added originally, the ammonia concentration in the aqueous layer. Hence calculate the distribution coefficient (equation (9.20)) and show that it is independent of the ammonia concentration. From the titration figures of (b) calculate the concentration of ammonia in the chloroform layer and hence the total number of moles of ammonia present in the 75 cm^3 of chloroform (X). Using the value of D obtained previously, calculate the total concentration of 'free' ammonia in the copper sulphate solution and hence the total number of moles of ammonia in the 50 cm^3 of aqueous solution (Y). The difference between the number of moles of ammonia in the original 25 cm^3 of ammonia solution (Z) and the ammonia present in the 'free' state and in the chloroform layer (ie. Z−X−Y) is the number of moles of ammonia associated with the number of moles of Cu^{++} in the 25 cm^3 of copper sulphate solution. Hence deduce the formula of the complex.
2. Calculate the stability constant for the complex ion.
3. Discuss the general applicability of the distribution method for the determination of stability constants.

3 Potentiometric methods

Acid and base dissociation constants can be calculated from a measured hydrogen ion concentration in a solution containing known amounts of a weak acid (base) and its salt with a strong base (acid). This is best accomplished by a potentiometric titration as in Expt. 6.2, 6.3 and 6.4. The solubility product can be determined from the results of a potentiometric titration (Expt. 6.5), or by studies of the emf. of concentration cells (Expt. 4.7).

Experiment 9.11   Plot the titration curve of an amino-acid and evaluate the acid and base dissociation constants

Theory: An amino-acid ionizes according to the pH of the solution. The following stages can be recognized:

$$R . CH . \overset{+}{N}H_3 . COOH \qquad R . CH . \overset{+}{N}H_3 . COO^- \qquad R . CH . NH_2 . COO^-$$

	acidic	isoelectric	basic
Net charge	+1	0	−1

The isoelectric species, which bears no charge, is a dipolar ion. Amino-acids possessing two ionizing groups are characterized by two pK_a values (numbered from the acid to the alkaline region): pK_{a_1} due to the carboxyl group, and pK_{a_2} due to the amino group. The pK_{a_2}, eg. for glycine (9.77) can be shown to be due to the amino group by adding formaldehyde to the amino-acid and titrating with sodium hydroxide.

The isoelectric point, pI, of an amino-acid with two pK_a values is defined as:

$$pI = \frac{pK_{a_1} + pK_{a_2}}{2} \tag{9.21}$$

For amino-acids with three dissociating groups, only the predominating pK_a values are taken into account, eg. pK_{a_1} and pK_{a_2} for monoamino-dicarboxylic acids.

For moderately weak acids (eg. acetic acid) and bases, the concentration of salt formed during a titration is equivalent to the volume of titrant added. For very weak acids and bases, where the salts are strongly hydrolysed, this is not correct. In these cases, the salt concentration can be calculated from a knowledge of the amount of titrant added and the pH, since the pH is set up solely by the amount of titrant that has failed to react. The volume of titrant that has failed to react is obtained from the pH of the solution; thus, the volume actually combined can be calculated.

Requirements: pH meter, glass electrode for use to pH 13.0 and calomel electrode. Standard 0.1 mol dm^{-3} hydrochloric acid and sodium hydroxide solutions, 0.1 mol dm^{-3} solutions of amino-acids (eg. glycine, glutamic acid and histidine).

Procedure: Set up and standardize the pH meter. Pipette 25 cm^3 of the amino-acid solution into a 250-cm^3 beaker and determine the pH. Add standard hydrochloric acid from a 50-cm^3 burette, in small amounts at the start, stir with a magnetic stirrer (or by bubbling through a stream of pure nitrogen) and determine the pH after each addition. Continue adding the acid, now in larger amounts, until the pH falls to about 1.3 (150—200 cm^3 of acid may be required depending on the amino-acid).

Carefully wash the electrodes, restandardize the pH meter, and titrate another 25-cm^3 aliquot of the amino-acid with the standard sodium hydroxide solution until the pH reaches 12.5 (150—200 cm^3 may be required).

Repeat the titration with sodium hydroxide, having previously added 5 cm^3 of formaldehyde to the 25-cm^3 sample of the amino-acid solution.

Treatment of experimental data and discussion

1. The excess titrant added must first be calculated; the correction need only be applied for solutions in which the pH is less than 3.0 or greater than 11.0. Plot a curve showing the relationship between the volume of standard acid or alkali added to water to give a constant volume of 100 cm^3. This can be determined experimentally or calculated from the equation:

$$pH = -\log \frac{gx}{100} \quad \text{(for acids)}$$

or

$$pH = 14 - pOH = 14 + \log \frac{gx}{100} \quad \text{(for alkalies)}$$

where g is the volume of acid or alkali added, of concentration x mol dm^{-3}. From this curve, it is possible to calculate the volume of acid or alkali (g) necessary to produce a given pH when the total volume of the solution is 100 cm^3. The excess titrant at this pH for the total volume (v) during a titration is given by $gv/100$, which when subtracted from the volume of titrant added gives the 'corrected volume' — ie. the volume of titrant in combination with the amino-acid.

Tabulate your results in the following columns: volume of titrant, total volume of solution, pH, volume of excess titrant, corrected volume of titrant. Plot the titration curves of pH against corrected volume of acid or alkali and also the graphs of pH against log c_{A^-}/c_{HA} (equations (6.25) and (6.26)). Hence deduce the pK_a and pI of the acid.

2. Explain the effect of formaldehyde on the titration curve. Such a titration in the presence of formaldehyde, using phenolphthalein as indicator, can be used in the determination of amino groups in polypeptides; explain why a potentiometric titration is not necessary.

3. Are there any advantages to be gained by bubbling pure nitrogen through the solution instead of stirring mechanically?

Experiment 9.12 Determine the chelate stability constant of the cobalt(II)ion with glutamic acid

Theory: A dibasic amino acid dissociates according to the equations:

$$H_2A + H_2O \rightleftharpoons H_3O^+ + HA^-$$
$$HA^- + H_2O \rightleftharpoons H_3O^+ + A^{--}$$

for which the equilibrium constants, neglecting the activity coefficients are respectively:

$$K_1 = \frac{[H^+]\,[HA^-]}{[H_2A]} \tag{9.22}$$

$$K_2 = \frac{[H^+]\,[A^{--}]}{[HA^-]} \tag{9.23}$$

In the presence of a divalent metal ion, chelate formation occurs:

$$M^{++} + A^{--} \rightleftharpoons MA$$

and

$$MA + A^{--} \rightleftharpoons MA_2^{--}$$

for which the formation constants are respectively:

$$K_3 = \frac{[MA]}{[M^{++}]\,[A_2^{--}]} \tag{9.24}$$

and

$$K_4 = \frac{[MA^{--}]}{[MA]\,[A^{--}]} \tag{9.25}$$

As a direct consequence of chelate formation, the high pH buffer region of the potentiometric titration curve of the acid is greatly reduced in the presence of the metal ions (Fig. 9.1), corresponding to the equilibrium:

$$M^{++} + HA^- \rightleftharpoons MA + H^+$$

If c_M and c_A are the total concentrations of metal ion and amino acid species respectively, and a is the number of moles of base added per mole of acid present

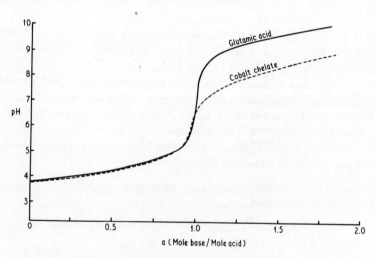

Fig. 9.1   Titration curves for glutamic acid and for the cobalt chelate of glutamic acid

during the titration, then the following relationships hold for solutions of amino acid and metal ion:

$$c_A = [H_2A] + [HA^-] + [A^{--}] + [MA] + 2[MA_2^{--}] \qquad (9.26)$$

$$c_M = [M^{++}] + [MA] + [MA_2^{--}] \qquad (9.27)$$

$$ac_A + [H^+] - [OH^-] = [HA^-] + 2[A^{--}] + 2[MA] + 4[MA_2^{--}] \qquad (9.28)$$

$[A^{--}]$ is calculated by subtracting equation (9.28) from twice equation (9.26) and eliminating $[H_2A]$ and $[HA^-]$, using equations (9.22) and (9.23):

$$[A^{--}] = \frac{(2-a)c_A - [H^+] + [OH^-]}{\dfrac{2[H^+]^2}{K_1 K_2} + \dfrac{[H^+]}{K_2}} \qquad (9.29)$$

The average number of moles of ligand, A^{--}, bound per mole of metal ion is given by:

$$n = \frac{[MA] + 2[MA^{--}]}{c_M} \qquad (9.30)$$

which from equations (9.22), (9.23) and (9.26), becomes:

$$n = \frac{1}{c_M}\left\{ c_A - \left(\frac{[H^+]^2}{K_1 K_2} + \frac{[H^+]}{K_2} + 1\right)[A^{--}]\right\} \qquad (9.31)$$

The values of n for different values of $[A^{--}]$ obtained from the high pH region of the graph are plotted against pA (ie. $-\log[A^{--}]$); the value of pA at $n = 0.5$ is pK_3 and at $n = 1.5$ is pK_4.

Requirements: pH meter, a recording type is an advantage, glass and calomel electrode system, titration vessel in which the solution can be stirred during the titration in an atmosphere of nitrogen (purified by passage through soda lime, sulphuric acid and water). 1.0 cm^3 microburette. 0.001 mol dm^{-3} solution of glutamic acid made up in 0.1 mol dm^{-3} potassium chloride solution, standard carbonate-free 0.1 mol dm^{-3} sodium hydroxide solution, 0.025 mol dm^{-3} solution of cobalt nitrate made up in 0.1 mol dm^{-3} potassium chloride solution.

Procedure: Set up and standardize the pH meter according to the operational manual. Pipette 25 cm^3 of the glutamic acid solution into the titration cell (thermostat if possible) and thoroughly displace all acidic and basic gases by bubbling the purified nitrogen through for about 30 min; the pH will eventually become constant at about pH 3.8. Record the pH value, and slowly add the standard sodium hydroxide solution from the microburette, keep the solution well stirred in a carbon dioxide-free atmosphere and determine the pH after the addition of each aliquot (*c.* 0.05 cm^3). Continue the titration until the number of moles of base added is twice the number of moles of acid originally present (ie. about 1 cm^3). The titration is best accomplished with a recording pH meter, although satisfactory results can be obtained with a manual instrument.

Repeat the experiment, but now add 1 cm^3 of the cobalt nitrate solution to the 25 cm^3 of acid before commencing the titration.

Treatment of experimental data and discussion

1. For the first titration, in the absence of metal ions, plot the curve of pH against the volume of sodium hydroxide solution and from the volume added at the equivalence point, mark off on the abscissa the corresponding values of a (ie. moles of base added per mole of acid). a is a measure of the salt concentration, $(1 - a)$ is the acid concentration remaining. From the graph, read off, and tabulate, the values of pH at which $a = 0.05, 0.10, 0.15$, etc. Plot graphs of log [salt]/[acid] ie. log $(a/1 - a)$ against pH over the two pH ranges and hence determine pK_1 and pK_2. Correct these values by including the appropriate activity coefficient(s).

2. For the titration in the presence of the metal ions, again plot the curve of pH against a. Read off, and tabulate, the pH values at $a = 1.20, 1.25$, etc. and convert these to actual hydrogen ion concentrations ($\gamma_{H^+} = 0.83$ at $I = 10^{-1}$ mol dm^{-3}). For each value of a, calculate $[A^{--}]$ using equation (9.29), and n, using equation (9.31). Plot n against pA (ie. $-\log[A^{--}]$) and read off the values of pK_3 and pK_4 when $n = 0.5$ and 1.5 respectively.

3. Comment on the values of the stability constants and discuss the effect on the values of changing the metal ion or the amino acid.

Experiment 9.13 Determine the hydrolysis constant of aniline hydrochloride in water electrometrically (see also Expt. 9.17)

Theory: The hydrolysis constant, for the reaction:

$$C_6H_5 . NH_3^+ + H_2O \rightleftharpoons C_6H_5NH_2 + H_3O^+$$

is given by:

$$K_h = \frac{c_{H_3O^+} \, c_{C_6H_5NH_2}}{c_{C_6H_5NH_3^+}} = \frac{c\alpha^2}{1 - \alpha} \qquad (9.32)$$

where α is the degree of hydrolysis.

Thus if $c_{H_3O^+}(= c\alpha)$ can be measured, α, and hence K_h can be determined.

Requirements: pH meter and glass electrode (or potentiometer and hydrogen electrode), calomel electrode, solid aniline hydrochloride.

Procedure: Set up the pH meter or potentiometer and calibrate. Prepare a standard 0.1 mol dm^{-3} solution of aniline hydrochloride and dilute to give solutions of concentrations: $0.01, 0.001$, and 0.0001 mol dm^{-3}. Determine the pH of each solution.

Treatment of experimental data and discussion

1. From the pH, determine $c_{H_3O^+}$, using the expression pH $= -\log c_{H_3O^+}\gamma_{H_3O^+}$,

and the value of $\gamma_{H_3O^+}$ from the simple Debye–Hückel equation (6.23*b*). Hence calculate α and K_h for each solution.

2. Suggest another, non-electrical, method for determining K_h.

3. Prove that $K_h = K_w/K_b$ where K_w is the ionic product for water. Hence calculate K_b, the basic dissociation constant, of aniline.

4. Discuss the errors involved in the use of the Debye–Hückel equation for the activity coefficient of H_3O^+.

Experiment 9.14   Determine the ionic product of water

Theory: The ionic product of water defined by:

$$K_w = a_{H_3O^+} a_{OH^-} = m_{H_3O^+} m_{OH^-} \gamma_{H_3O^+} \gamma_{OH^-} \tag{9.33}$$

can be determined from conductance measurements on very pure water or from the emf. of the cell:

$$\ominus \ \text{Pt, } H_2 | KOH \cdot KCl | AgCl | Ag \ \oplus$$

The emf. of this cell is given by:

$$E = E^\ominus - \frac{RT}{\mathscr{F}} \ln m_{H_3O^+} m_{Cl^-} \gamma_{H_3O^+} \gamma_{Cl}$$

Substituting from equation (9.33) gives:

$$E = E^\ominus - \frac{RT}{\mathscr{F}} \ln K_w - \frac{RT}{\mathscr{F}} \ln \frac{m_{Cl^-}}{m_{OH^-}} - \frac{RT}{\mathscr{F}} \ln \frac{\gamma_{Cl^-}}{\gamma_{OH^-}}$$

or

$$E - E^\ominus + \frac{RT}{\mathscr{F}} \ln \frac{m_{Cl^-}}{m_{OH^-}} = -\frac{RT}{\mathscr{F}} \ln K_w - \frac{RT}{\mathscr{F}} \ln \frac{\gamma_{Cl^-}}{\gamma_{OH^-}} \tag{9.34}$$

Thus, from measurements of E at different m_{Cl^-} and m_{OH^-} and, knowing $E^\ominus$ for the cell, ie. the standard electrode potential of the Ag, AgCl electrode, all the terms on the left-hand side of the equation are known. The graph of $\sqrt{I}$ against these values is linear and of intercept $-RT/\mathscr{F} \ln K_w$.

Requirements: Potentiometer and galvanometer or digital voltmeter, hydrogen electrode, silver, silver chloride electrode, standard 0.1 mol dm^{-3} potassium hydroxide and 3.5 mol dm^{-3} potassium chloride solutions.

Procedure: Set up and standardize the potentiometer circuit. From the standard solutions, prepare mixtures which are all 0.01 mol dm^{-3} with respect to potassium hydroxide, and in which the potassium chloride is in the range 0.005 to about 3.0 mol dm^{-3}. Connect the electrodes to the potentiometer, taking the usual precautions for the hydrogen electrode, and measure the emf. between the electrodes in each solution. Record the polarity of the electrodes and the temperature of measurement.

Treatment of experimental data and discussion

1. From the known value of $E^{\ominus}(\mathrm{AgCl, Ag, Cl^-})$, the measured value of E at known m_{OH^-} and m_{Cl^-} calculate the value of the left-hand side of equation (9.34). Plot a graph of these values against $\sqrt{I}$ and determine the intercept and hence K_w. Calculate $\Delta G^{\ominus}$ for the ionization of water.

2. The above calculation has been made using concentrations expressed as mol dm^{-3} instead of molality (ie. mol kg^{-1}); comment on the errors involved in the assumption that the two concentration scales are the same.

3. Write down the electrode and cell reactions and derive equation (9.34) from first principles.

4. Comment on the use of equation (9.34) when the emf. is measured at only one set of concentrations.

5. What additional information can be obtained from a knowledge of the variation of K_w with temperature?

Experiment 9.15   Determine the equilibrium constant of the reaction:
$$\mathrm{Ag(NH_3)_2^+ \rightleftharpoons Ag^+ + 2NH_3}$$

Theory: The instability or dissociation constant of this reaction is:

$$K_i = \frac{a_{\mathrm{Ag}^+}\, a_{\mathrm{NH_3}}^2}{a_{\mathrm{Ag(NH_3)_2^+}}} \tag{9.35}$$

The electrode potential of a silver electrode dipping in a solution of the complex gives a measure of a_{Ag^+}, since from equation (6.35)

$$E(\mathrm{Ag^+, Ag}) = E^{\ominus}(\mathrm{Ag^+, Ag}) + \frac{RT}{\mathscr{F}} \ln a_{\mathrm{Ag}^+}$$

Experimentally, this is most easily achieved by setting up the concentration cell without transport:

$$\mathrm{Ag} \left| \begin{array}{c} 0.025\ \mathrm{mol\ dm^{-3}\ AgNO_3\ in} \\ 1\ \mathrm{mol\ dm^{-3}\ ammonia} \\ (a_{\mathrm{Ag}^+})_1 \end{array} \right| \begin{array}{c} \mathrm{Satd.} \\ \mathrm{potassium\ nitrate} \end{array} \left| \begin{array}{c} 0.01\ \mathrm{mol\ dm^{-3}\ AgNO_3} \\ \\ (a_{\mathrm{Ag}^+})_2 \end{array} \right| \mathrm{Ag}$$

the emf. of which is:

$$E = \frac{RT}{\mathscr{F}} \ln \frac{(a_{\mathrm{Ag}^+})_2}{(a_{\mathrm{Ag}^+})_1} \tag{9.36}$$

Requirements: Potentiometer and galvanometer or digital voltmeter, two silver electrodes, half cells (eg. Fig. 7.11), standard 0.05 mol dm^{-3} silver nitrate solution, 2 mol dm^{-3} ammonia solution, saturated potassium nitrate solution for bridge.

Procedure: Set up the potentiometer circuit and calibrate. Clean and electroplate the two silver electrodes (p. 306) and check that when they are both immersed in the same silver nitrate solution there is no emf. Dilute two portions of the silver nitrate solution (*a*) with an equal volume of 2 mol dm^{-3} ammonia solution, and (*b*) with 4 times the volume of water. Set up the cell, and taking the usual precautions of washing and filling the half cells, measure the emf. Record the polarity of the electrodes. Repeat the determination of emf. for other cells containing different dilutions of the silver nitrate.

Treatment of experimental data and discussion

1. From the emf. of the cell, calculate the activity of silver ions in the solution containing the complex $(a_{Ag})_1$ (equation (9.36)) and from the activity coefficient (p. 316), determine the concentration of free silver ions in solution. By difference calculate the concentration of $Ag(NH_3)_2^+$ and hence NH_3. Since $\gamma_{Ag^+} = \gamma_{Ag(NH_3)_2^+}$ and γ_{NH_3} is unity, concentrations can be used directly instead of activities in equation (9.35).

2. Write down the electrode reactions and derive equation (9.36) from first principles.

3. Comment on the accuracy of the method when the concentration of free silver ions in the two half cells are approaching one another.

4 Conductometric methods

In solids, electricity is carried by electrons and, in solution, by ions. Both types of conductor obey Ohm's law:

$$I = E/R \tag{9.37}$$

where E/V is the pd. between the ends of a conductor of resistance R/Ω carrying a current I/A.

The resistivity, ρ, is defined by:

$$R = \frac{\rho L}{A} = \frac{1}{\kappa} \frac{L}{A} \tag{9.38}$$

where L/m is the length and A/m^2 the cross-sectional area of the conductor.

The conductance (G) is the reciprocal resistance and the conductivity $\kappa/\Omega^{-1} m^{-1}$ is the reciprocal resistivity.

When comparing the conducting properties of different substances it is preferable to use chemically comparable quantities, eg. equimolar quantities. The molar conductivity defined as:

$$(\Lambda/\Omega^{-1} m^2 mol^{-1}) = \frac{(\kappa/\Omega^{-1} m^{-1})}{(c/mol\ m^{-3})} \tag{9.39}$$

is a measure of the number and rate of migration of all the ions from 1 mole of the solute. The value of Λ of all electrolytes increases with dilution; this can be accounted for in terms of an increase in the number of ions and an increase in the mobility of the ions. Arrhenius, making the tacit assumption that the mobilities of the ions are independent of concentration, suggested that increased dissociation occurs on dilution, and that only at infinite dilution is dissociation complete. On this basis, the degree of dissociation, α, can be expressed as:

$$\alpha = \Lambda/\Lambda_0 \tag{9.40}$$

where Λ_0, the molar conductivity at infinite dilution, is a measure of the capacity of the completely dissociated solute to carry a current. This theory is approximately correct for weak electrolytes. If c is the concentration of the weak electrolyte AB and α the degree of dissociation:

$$AB \rightleftharpoons A^+ + B^-$$

then

$$K_c = \frac{[A^+][B^-]}{[AB]} = \frac{c\alpha^2}{1 - \alpha} \tag{9.41a}$$

$$= \frac{\Lambda^2 c}{\Lambda_0(\Lambda_0 - \Lambda)} \tag{9.41b}$$

Thus, K_c can be determined from conductance measurements. Using activities in place of concentrations:

$$K_{\text{therm.}} = \frac{a_{A^+} a_{B^-}}{a_{AB}} = K_c \frac{\gamma_{A^+} \gamma_{B^-}}{\gamma_{AB}} = K_c \frac{\gamma_\pm^2}{\gamma_{AB}}$$

But

$$-\log \gamma_\pm = A\sqrt{I}$$

Hence

$$\log K_{\text{therm.}} = \log K_c - 2A\sqrt{I} \tag{9.42}$$

since $\gamma_{AB} = 1$.

Strong electrolytes, however, do not obey this Arrhenius theory, even approximately. All strong electrolytes are ionized even in the solid state, so the increased conductivity on dilution cannot be attributed to increased dissociation. Strong coulombic forces exist between ions, and it was on this basis that Debye, Hückel and Onsager put forward their equation for the conductivity of very dilute solutions of strong electrolytes:

$$\Lambda = \Lambda_0 - (A + B\Lambda_0)\sqrt{c} \tag{9.43}$$

where A and B are constants. For strong electrolytes, the $\Lambda - \sqrt{c}$ graphs are approximately linear and of the correct slope as the concentration approaches zero. Since A and B are constants, it is theoretically possible to calculate Λ_0 from Λ measured at any concentration. Owing to slight departures from the theoretical behaviour, the values of Λ_0 so obtained are a function of c; the graph of these values of Λ_0 plotted against c and extrapolated to $c = 0$ gives a correct value of Λ_0.

For weak electrolytes, however, this method cannot be used to determine Λ_0. Λ_0 is calculated from the ion conductivity (Λ_0) or the mobilities (u) of the anion and cation at infinite dilution, using Kohlrausch's law of independent migration of ions:

$$\Lambda_0 = (\Lambda_{0+} + \Lambda_{0-}) = \mathscr{F}(u_+ + u_-) \tag{9.44}$$

For acetic acid (using the values of ion conductivities (p. 305)).

$$\Lambda_0(CH_3COOH) = \Lambda_0(H^+) + \Lambda_0(CH_3COO^-) = [349.7 + 40.9] \times 10^{-4}$$
$$= 390.6 \times 10^{-4} \, \Omega^{-1} \, m^2 \, mol^{-1}$$

An alternative method based on the same principle for acetic acid is:

$$\Lambda_0(CH_3COOH) = \Lambda_0(HCl) + \Lambda_0(CH_3COONa) - \Lambda_0(NaCl)$$

the values of Λ_0 for the strong electrolytes can be obtained by extrapolation of the appropriate $\Lambda - \sqrt{c}$ curve.

Measurement of conductance of solutions

The essential requirements for measuring conductances are a suitable cell containing metal electrodes, between which the current passes through the solution, and a means of measuring the resistance of the cell and solution. Polarization, which occurs when a direct current is applied to a solution is overcome by using an alternating current; in this way polarization occurring in the first half cycle is reversed during the second. In the a.c. bridge the moving coil galvanometer, used in a d.c. Wheatstone bridge, is replaced by an instrument capable of detecting a.c. voltages of the order of 1 mV.

An a.c. bridge responds to inductance and capacitance as well as resistance and a perfect null point can only be obtained if the bridge is balanced for both reactance and resistance. In commercial bridges, capable of working over a wide frequency range, a sensitive, null-balance detector is used to obtain a perfect null point.

Various types of conductance cell have been described (Fig. 9.2).

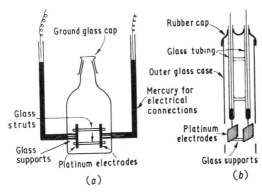

Fig. 9.2   Conductance cells. (a) Bottle type. (b) Dip type

For accurate conductance measurements capped bottle cells are recommended; these can be of different cell constant (L/A) depending on the separation and area of the two platinum plates. The cells are constructed of Pyrex glass; thick platinum electrodes are used to ensure rigidity. The dip type cell is very robust and satisfactory for conductometric titrations. The electrodes are either left bright or greyed.

Platinize the electrodes (p. 306). Before use, the glass and electrodes must be completely freed from electrolytes. Thoroughly wash the cell with distilled water and then pass steam into it in the inverted position for 20—24 hours. Fill with conductivity water and measure the conductance, leave aside for several days and again measure the conductance. Repeat this until there is no further change in the conductance of the water on storage. When not in use, leave the cell filled with conductance water. Similar steaming out and storage applies to all glassware used in conductance measurements.

A cell holder, to support the cell, immersed as far as possible in a thermostat, is readily constructed from Perspex.

It is not possible to determine the cell constant ($J = L/A$) accurately from the geometric dimensions of the cell; it is obtained by calibration with a solution of known conductivity κ; if this solution has a resistance R, then $J = \kappa R$ (equation (9.38)), or $J = \kappa/G$. If the resistance (conductance) of an unknown solution in this cell is R' (G') then its conductivity, $\kappa' = J/R' = JG'$.

A solution of potassium chloride is the accepted reference solution for the calibration of conductance cells. The molar conductivity/Ω^{-1} m^2 mol^{-1} of such a solution at 25 °C of concentration c/mol m^{-3}, is calculated from the equation:

$$\Lambda(\text{KCl}, 298\ \text{K}) = 149.93 \times 10^{-4} - 2.992 \times 10^{-4}\, c^{1/2} + 58.74 \times 10^{-7}\, c \log c +$$
$$22.18 \times 10^{-7}\, c \quad (9.45)$$

(Lind, Zwolenik and Fuoss, *J. Am. chem. Soc.*, 1959, **81**, 1557).

Purify the sample of potassium chloride by recrystallizing from conductance water three times and fusing in a platinum crucible. Allow to cool in a desiccator. Prepare a solution by dissolving about 1 g (apply the buoyancy correction) in conductance water and making up to 1 dm^3. The conductivity of the water must be known; it can be obtained from measurements in a cell of known constant, or, by a series of successive approximations, using first an estimated constant from the geometric measurements of the cell. A more accurate value of J, obtained from the measurements to be described, can then be used to give a better value for the conductivity of the water and hence a better value of J.

Wash out the cell twice with the potassium chloride solution and completely fill with the solution. Stopper the cell and allow it to attain temperature equilibrium in the thermostat at 25.00 ± 0.02 °C. Now measure the resistance or conductance. Repeat with other fillings of the cell with the same solution until concordant results are obtained.

For the solution calculate Λ (equation (9.45)) and hence κ (equation (9.39)) and finally J:

$$J = \frac{L}{A} = \frac{\kappa(\text{KCl}) + \kappa(\text{H}_2\text{O})}{\text{measured conductance}} \quad (9.46)$$

Normal laboratory distilled water generally has a conductivity of $3-5 \times 10^{-4}\ \Omega^{-1}\ \text{m}^{-1}$. Since by extreme purification it has been reduced to $0.04 \times 10^{-4}\ \Omega^{-1}\ \text{m}^{-1}$, it follows that most of this conductivity is due to such impurities as carbon dioxide, ammonia, salts, etc. Very elaborate precautions are required to reduce the conductance to less than $0.1 \times 10^{-4}\,\Omega^{-1}\,m^{-1}$, since in the presence of atmospheric carbon dioxide it equilibrates at a value of $0.8 \times 10^{-4}\,\Omega^{-1}\,m^{-1}$. It is thus necessary to collect, store and handle the water in an atmosphere free of carbon dioxide.

The principle of the method is to redistil laboratory distilled water, to which alkali and potassium permanganate have been added, from an electrically heated copper steam-can. A spray trap eliminates droplets of water passing over. Part of the steam is condensed in a tin-plated condenser and the water collected in a glass vessel fitted with a soda-lime tube to exclude carbon dioxide. A distillate of $\kappa \leqslant 1 \times 10^{-4}\,\Omega^{-1}\,m^{-1}$ is rejected. The condensate from the remaining steam, of inferior quality, can be used for washing (Vogel and Jeffery. *J. Chem. Soc.* (1931) 1201).

Experiment 9.16   Determine the molar conductivity of benzoic acid over a range of concentrations and hence its dissociation constant

Requirements: Conductance bridge, bottle type conductance cell, thermostat at $25.00 \pm 0.02\ ^{\circ}C$, Pyrex glass flasks ($750\ cm^3$) with ground glass caps, which have been steamed out, conductance water, pure solid potassium chloride and benzoic acid (or phenylacetic acid).

Procedure: Set up the conductance bridge according to the instructions and determine the constant of the conductance cell, using a solution of potassium chloride (equation (9.46)). Prepare a solution, by weight, of benzoic acid in conductance water. Accurately weigh out about $0.65\,g$ of purified benzoic acid into a dry steamed-out flask, add about $500\ cm^3$ of conductance water and reweigh; this solution is approximately $0.01\ mol\ dm^{-3}$. Prepare more dilute solutions by repeated dilution by weight, using a weight burette, in other Pyrex flasks to give solutions of about $0.005, 0.0025, 0.00125, 0.0006$ and 0.0003 m. Carefully cap each flask and suspend in the thermostat at $25\ ^{\circ}C$.

Rinse out the cell with conductance water, then with the given solution, and finally fill with this solution. Replace the cell in the thermostat and connect to the conductance bridge. Obtain a balance on the bridge (balancing both resistance and capacitance), leave the cell for 15–20 min and again record the resistance (or conductance) at balance. Continue in this way until there is no further change of balance point; record this value. Rinse out the cell and again fill with the same solution; this check is most essential, particularly with the very dilute solutions, where appreciable amounts of material adsorbed by the glass and platinum electrodes may be desorbed.

Using this technique, measure the resistance of each solution, for preference, starting with the most dilute. Finally, determine the conductance of the water used throughout the experiment.

Treatment of experimental data and discussion

1.   From the measured conductances of the KCl solutions determine the cell constant (equation (9.46)).

2.   Express all the concentrations of the acid in mol m^{-3}. From the measured conductance of each solution calculate the conductivity (equation (9.38)). Subtract the conductivity of water and evaluate the molar conductivity of the electrolyte (equation (9.39)) at the different concentrations.

3.   Plot a graph of Λ against $\sqrt{c}$ and show the impossibility of determining Λ_0 by extrapolation. Use the law of independent migration of ions to calculate Λ_0 from the known molar conductivities of the ions (p. 305). From this value of Λ_0 calculate α (equation (9.40)) and hence K_c (equation (9.41a)) at each concentration. Plot log K_c against $\sqrt{c}$ and from the intercept determine K_{therm} (equation (9.42)).

4.   Test the validity of Ostwald's dilution law (equation (9.41b)), which may be written in the form:

$$c\Lambda = K_c \frac{\Lambda_0^2}{\Lambda} - K_c\Lambda_0$$

by plotting $c\Lambda$ against $1/\Lambda$. Determine the slope ($K_c\Lambda_0^2$) and the intercept ($-K_c\Lambda_0$) and hence deduce K_c and Λ_0. Compare the value of Λ_0 with that calculated above.

5.   Explain why the conductivity of the water, used to prepare the solutions, must be as small as possible.

6.   Discuss the relative advantages and disadvantages of using platinum black electrodes in conductance cells.

Experiment 9.17   Determine the degree of hydrolysis and the hydrolysis constant of aniline hydrochloride in water conductometrically (see also Expt. 9.13)

Theory: Aniline hydrochloride, the salt of a strong acid and a weak base, is hydrolysed in water giving $(1 - \alpha)$ moles of unhydrolysed salt and α moles of acid and base, where α is the degree of hydrolysis:

$$C_6H_5 . NH_3Cl \rightleftharpoons C_6H_5 . NH_3^+ + Cl^-$$

$$C_6H_5 . NH_3^+ + H_2O \rightleftharpoons C_6H_5 . NH_2 + H_3O^+$$

The weak, undissociated base, aniline, will not contribute to the molar conductivity of the solution (Λ) which may therefore be written:

$$\Lambda = (1 - \alpha)\Lambda_c + \alpha\Lambda_{\text{HCl}} \tag{9.47}$$

whence

$$\alpha = \frac{\Lambda - \Lambda_c}{\Lambda_{\text{HCl}} - \Lambda_c} \tag{9.48}$$

where Λ_c and Λ_{HCl} are the molar conductivities of the unhydrolysed salt and free acid respectively. Λ_{HCl} is generally taken as the molar conductivity at infinite dilution; it is more correct however to use the value at the same total ionic strength. Λ_c is obtained by adding free aniline to the salt solution which suppresses the hydrolysis of the salt. The conductance of such a mixture is virtually that of the unhydrolysed salt.

The hydrolysis constant, K_h, is given by:

$$K_h = \frac{c_{H_3O^+} c_{C_6H_5NH_2}}{c_{C_6H_5 . NH_3^+}} = \frac{c\alpha^2}{1 - \alpha} \qquad (9.32)$$

where c/mol dm^{-3} is the concentration of the aniline hydrochloride in solution. The thermodynamic hydrolysis constant is practically the same as the classical expression, since $\gamma_{C_6H_5 . NH_2} \sim 1$ and $\gamma_{H_3O^+} = \gamma_{C_6H_5 . NH_3^+}$.

Requirements: Conductance bridge, bottle type conductance cell, thermostat at 25 °C; Pyrex glass (capped) flasks which have been steamed out, conductance water, pure solid potassium chloride, aniline hydrochloride and aniline. (Aniline hydrochloride is best prepared fresh by passing dry hydrogen chloride gas through a 10% solution of aniline in dry ether. Filter off the white precipitate of the hydrochloride, wash with dry ether and store in a desiccator over caustic potash.)

Procedure: Set up the conductance bridge according to the instructions and determine the constant of the conductance cell, using a solution of potassium chloride (p. 242). Prepare 0.01 and 0.05 mol dm^{-3} solutions of aniline hydrochloride in conductance water. Taking all the usual precautions (Expt. 9.16) of rinsing the cell and allowing temperature equilibrium to be established, measure the resistance of one solution. When a constant balance point has been attained, record this, and add a few small drops of aniline, and again determine the resistance. Add further drops of aniline to check that there is no further change of resistance. Repeat these determinations with the other solution of aniline hydrochloride. Finally, determine the conductance of the water.

Treatment of experimental data and discussion

1. From the measured conductances, the cell constant and the concentrations calculate the conductivities (equation (9.38)) and the molar conductances Λ and Λ_c (equation (9.39)). Determine the degree of hydrolysis, α (equation (9.48)), and the hydrolysis constant (equation (9.32)) at each concentration. (Assume Λ_0 for HCl = 426.0 Ω^{-1} cm^2 mol^{-1}).
2. Comment on the variation of K_h with concentration.
3. Discuss the errors involved in using the limiting molar conductance of HCl at infinite dilution.
4. Prove that $K_h = K_w/K_b$ and hence calculate the value of K_b, the basic dissociation constant, of aniline.

5   Spectrophotometric methods

Spectrophotometric methods can be used to measure dissociation or ionization constants, provided that there is a pronounced difference in the absorption between the molecular and ionic forms of the substance. Indicator constants can be determined in the visible range of the spectrum and, dissociation constants of such substances as asprin, salicylamide and p-nitrophenol, in the ultraviolet region.

If, for the dissociation of the weak acid, represented classically:

$$HA \rightleftharpoons H^+ + A^-$$

$$c_{HA} \qquad c_{H^+} \quad c_{A^-}$$

the molar extinction coefficients ($\epsilon = A/cl$, equation (5.12)), at the chosen wavelength, of HA and A^- are ϵ_{HA} and ϵ_{A^-} respectively, and, if the total acid concentration is c/mol dm^{-3} and the measured extinction coefficient ϵ, then the following identities hold:

$$\epsilon c = c_{HA} \epsilon_{HA} + c_{A^-} \epsilon_{A^-}$$

and

$$c = c_{HA} + c_{A^-}$$

whence it can be shown that:

$$\frac{c_{A^-}}{c_{HA}} = \frac{\epsilon_{HA} - \epsilon}{\epsilon - \epsilon_{A^-}} \tag{9.49}$$

Since, for the weak acid:

$$pH = pK_a + \log \frac{c_{A^-}}{c_{HA}} + \log \gamma_{A^-} \tag{6.22}$$

It follows that:

$$pH = pK_a + \log \frac{\epsilon_{HA} - \epsilon}{\epsilon - \epsilon_{A^-}} + \log \gamma_{A^-} \tag{9.50}$$

since the activity coefficient of the undissociated acid is unity.

Alternatively if A_{HA}, A_{A^-} and A are the measured absorbances (all at the same wavelength) of the acid solution, the alkaline solution and a solution of intermediate pH respectively, and if α is the degree of dissociation, then:

$$\frac{c_{A^-}}{c_{HA}} = \frac{\alpha}{1 - \alpha} \tag{9.51}$$

and

$$A = (1 - \alpha)A_{HA} + \alpha A_{A^-}$$

or

$$\alpha = (A_{HA} - A)/(A_{HA} - A_{A^-}) \tag{9.52}$$

Combining equations (6.22), (9.51) and (9.52) and rearranging gives:

$$pH = pK_a + \log \frac{(A_{HA} - A)}{(A - A_{A^-})} + \log \gamma_{A^-} \tag{9.53}$$

γ_{A^-} can be calculated from the equation:

$$-\log \gamma_{A^-} = \frac{0.505z^2\sqrt{I}}{1 + B\sqrt{I}} \qquad (6.23a)$$

whence:

$$pH = pK_a + \log \frac{(A_{HA} - A)}{(A - A_{A^-})} - \frac{0.505\sqrt{I}}{1 + B\sqrt{I}} \qquad (9.54)$$

Experiment 9.18   Determine the ionization constant of bromophenol blue

Requirements: Spectrophotometer with thermostatted 1-cm glass cells, pH meter and glass electrode assembly solid bromophenol blue, 1 mol dm^{-3} solutions of sodium acetate, hydrochloride acid and sodium hydroxide.

Procedure: Prepare the stock indicator solution: moisten 0.01 g of powdered bromophenol blue with 2–3 drops of ethanol, dissolve completely in 0.15 cm^3 of 1 mol dm^{-3} sodium hydroxide solution and make up to 100 cm^3 with distilled water.

    Investigate the pH values of various buffer solutions containing 20 cm^3 of sodium acetate and various quantities (not more than 19.5 cm^3) of hydrochloric acid. Then, still using 20 cm^3 of sodium acetate, make up a set of six buffer solutions in standard 50 cm^3 volumetric flasks, having pH values evenly distributed in the range 3.2 to 5.0. To each flask add 5 cm^3 of indicator solution and distilled water to make up to 50 cm^3. In addition make up a solution containing 20 cm^3 of hydrochloric acid, 5 cm^3 of indicator and 25 cm^3 of water and another with 20 cm^3 sodium acetate solution, 5 cm^3 of indicator and 25 cm^3 of water. Measure and record the pH value of each solution.

    Using a recording spectrophotometer record the spectra of these solutions (against a blank of distilled water) in the range 450–650 nm on one piece of chart paper. For each solution read the value of the absorbance at λ_{max}, from the scale of the instrument or from the chart, whichever is the more accurate.

Treatment of experimental data and discussion

1. For each solution evaluate the expression $\log (A_{HA} - A)/(A - A_{A^-})$ at λ_{max} and plot against the pH of the solution. From the intercept calculate the ionization constant of the indicator, making allowance for the ionic strength of the solution (equation (9.54)). Does the indicator contribute to the ionic strength of the solution?
2. Discuss the conditions necessary for the occurrence of an isosbestic point.
3. Plot the absorbance of each solution against the pH and show that it is possible to determine the pH of an unknown solution by this method.
4. Explain why the pH range for study is limited.
5. Discuss the possible limitations of the method.

Experiment 9.19   Determine the dissociation constant of *p*-nitrophenol

Requirements: Spectrophotometer, with thermostatted 1-cm quartz cells, pH meter and glass electrode assembly, pure p-nitrophenol, eight buffer solutions in the range pH 5—10 (p. 310).

Procedure: To a sample of the buffer solution, add 1 cm^3 of a solution of p-nitrophenol (0.05 g in 100 cm^3) in water and make up to 100 cm^3 with the buffer solution. The final concentration of p-nitrophenol is approximately 3.6×10^{-5} mol dm^{-3}.

Measure the pH of an aliquot of each solution. Using a recording spectrophotometer, record the spectra of these solutions (against the buffer solution as blank) in the range 300—440 nm on one piece of chart paper. Alternatively, determine the absorbance of each solution against the buffer solution as the blank, at 10 nm intervals in the range 300—440 nm. Smaller intervals are required to locate the maxima.

Treatment of experimental data and discussion

1. Two absorption maxima are obtained at approximately 317 nm, due to the acid form of the nitrophenol, and 407 nm, due to the ionized salt form, with an isosbestic point at about 350 nm. Calculate the extinction coefficients (equation (5.12)) of the p-nitrophenol in all the solutions at the recorded wavelengths of maximum absorption.

2. Two independent estimates of the dissociation constant of p-nitrophenol can be made. At 407 nm in acid solution, there is no background absorption due to the peak at 317 nm, ie. the acid form does not absorb at all at this wavelength. If ϵ_B is the extinction coefficient when the p-nitrophenol is completely in the salt-like form (ie. at pH 10), then at any lower pH the observed extinction coefficient $\epsilon_{obs.} = \alpha \epsilon_B$, where α is the fraction of p-nitrophenol in the salt form. Absorption at 317 nm is mainly due to the acid, but even at pH values less than 5 there is a small amount of background due to the salt form; hence, the extinction coefficient for this must be subtracted from all the observed values before α is calculated.

    From the data determine the value of α at each pH, and, knowing the concentration of p-nitrophenol, determine K_c. Make the appropriate correction for activity coefficients and determine K_{therm}.

3. Discuss the conditions necessary for the occurrence of an isosbestic point.

Experiment 9.20   Determine the strength of the base 2-nitro-4-methyl-aniline

Theory: The strength of a base may be defined as its ability to accept a proton; thus the equilibrium constant, K, of the reaction:

$$B + H_3O^+ \rightleftharpoons BH^+ + H_2O$$

is a measure of the strength of the base B, as is also the value of K_a for the reverse reaction, ie. the ionization of the conjugate acid BH$^+$.

$$K_a = \frac{1}{K} = \frac{a_B a_{H_3O^+}}{a_{BH^+} a_{H_2O}} \tag{9.55}$$

Thus the greater K_a for BH$^+$, the weaker is the base B. Since the neutral

2-nitro-4-methyl-aniline is orange and the protonated form is colourless, the ratio c_B/c_{BH^+} may be obtained spectrophotometrically.

Requirements: Spectrophotometer with thermostatted 1-cm glass cells, solid 2-nitro-4-methyl-aniline, 3 mol dm^{-3} hydrochloric acid and 5 mol dm^{-3} sodium hydroxide solution.

Procedure: Prepare an aqueous solution (stock) of 2-nitro-4-methyl-aniline of concentration about 8×10^{-4} mol dm^{-3}.

    Make up the following solutions in 25 cm^3 volumetric flasks:
(i)   10 cm^3 stock solution diluted to 25 cm^3 with distilled water,
(ii)   10 cm^3 stock solution 10 cm^3 concentrated hydrochloric acid, diluted to 25 cm^3 with water,
(iii) 10 cm^3 stock solution, 10 cm^3 5 mol dm^{-3} sodium hydroxide solution diluted to 25 cm^3 with water.

In solution (ii), the base will be almost completely protonated (BH$^+$), while in (iii) it will be in the neutral state (B). Record the absorption spectrum of each solution against water as a blank, over the wavelength range 360–560 nm. From these spectra, select a suitable wavelength (c. 420 nm) at which to measure the concentration of the neutral molecule in solutions containing different concentrations of hydrochloric acid.

    Check the validity of the Beer–Lambert law (p. 94) for the neutral molecule by accurately preparing three or four solutions of different concentration in the range 5×10^{-4} to 5×10^{-5} mol dm^{-3} and measuring the absorbance of each of the solutions at the wavelength of maximum absorption.

    Prepare a series of solutions (at least 5) in 25 cm^3 volumetric flasks, each containing 10 cm^3 of stock solution of base and x cm^3 of standard 3 mol dm^{-3} hydrochloric acid and made up to a total volume of 25 cm^3 with water, where x is varied so that the final acid concentrations cover the range 0.04 to 2 mol dm^{-3}. Measure the absorbance of each solution at the selected wavelength.

Treatment of experimental data and discussion

1.   Combining equations (9.51) and (9.52) for the system under study gives:

$$\frac{c_B}{c_{BH^+}} = \frac{A_{BH^+} - A}{A - A_B} \tag{9.56}$$

where A_{BH^+} and A_B are the absorbances of solutions (ii) and (iii) respectively, and A the absorbance in the presence of different concentrations of hydrochloric acid (all measured at the same wavelength). Thus neglecting activity coefficients and combining equations (9.55) and (9.56):

$$K_a = \frac{A_{BH^+} - A}{A - A_B} c_{H_3O^+} \tag{9.57}$$

In this experiment, unlike Expt. 9.18 and 9.19, it is not possible to measure A_{BH^+} experimentally. On rearranging, equation (9.57) becomes:

$$\frac{A - A_B}{c_{H_3O^+}} = \frac{A_{BH^+}}{K_a} - \frac{A}{K_a}$$

Plot a graph of A against $(A - A_B)/c_{H_3O^+}$ and extrapolate to $(A - A_B)/c_{H_3O^+} = 0$, when the intercept is A_{BH^+}. Using this extrapolated value of A_{BH^+} calculate K_a (equation (9.57)) and hence pK_a at each acid concentration. The value of K_a corrected for non-ideal behaviour may be obtained by an approximate method: plot the pK_a against the acid concentration and extrapolate to zero acid concentration.

2.   Explain why it is not possible to measure A_{BH^+}.

Spectrophotometric methods are also applicable to the study of complex ions. Job's method of continuous variation is a simple method for determining the formula and the stability of a single compound or ion-pair formed by two components in solution, and may be extended to systems in which more than one compound is formed from a pair of components. The method is applicable to the formation of complex ions represented by:

$$A + nB \rightleftharpoons AB_n$$

where A is a metallic ion and B either a neutral molecule (eg. ammonia), or an anion. The determination of n involves, in principle, the measurement of a suitable property (eg. absorbance) of a series of mixtures of A and B of identical molar concentration. A curve is then plotted of composition against the difference between each observed value and the corresponding value of the property calculated on the assumption that there is no reaction. The curve passes through a maximum (minimum) value if the measured property has a larger (smaller) value for the complex ion than for A or B. The composition of the maximum or minimum bears a simple relationship to n and is independent of the equilibrium constant. A smooth curve in the region of the maximum (minimum) is characteristic of an equilibrium reaction. In contrast, the maximum (minimum) lies at the intersection of two straight lines, when the reaction is complete.

The absorption of monochromatic radiation is a very suitable property for this method, since complex ions often have absorption maxima at different wavelengths to their components.

If solutions of A and B of identical concentration, $M/\text{mol dm}^{-3}$ are mixed in varying proportions by the addition of $(1 - x)$ parts of A to x parts of B $(x < 1)$, with no appreciable volume change on mixing, then the concentrations of A, B and AB_n and their interrelationship are given by:

$$c_A = M(1 - x) - c_{AB_n} \tag{9.58}$$

$$c_B = Mx - nc_{AB_n} \tag{9.59}$$

$$c_A c_B^n = Kc_{AB_n} \tag{9.60}$$

where K is the instability constant.

The condition for a maximum in the curve of c_{AB_n} against x is that:

$$\frac{dc_{AB_n}}{dx} = 0 \tag{9.61}$$

If equations (9.58), (9.59) and (9.60) are differentiated with respect to x and combined with equations (9.58), (9.59), (9.60) and (9.61) it can be shown that:

$$n = \frac{x}{1 - x} \qquad (9.62)$$

thus n can be calculated from the value of x for which c_{ABn} is a maximum.

If ϵ_A, ϵ_B and ϵ_{ABn} are the molar extinction coefficients respectively at a given wavelength, then for a cell of path length, l, the measured absorbance of the mixture is given by:

$$A = l(\epsilon_A c_A + \epsilon_B c_B + \epsilon_{ABn} c_{ABn}) \qquad (9.63)$$

and the difference between A (equation (9.63)) and the absorbance which the mixture would have had in the absence of reaction (A') is:

$$A - A' = l(\epsilon_A c_A + \epsilon_B c_B + \epsilon_{ABn} c_{ABn} - \epsilon_A M(1 - x) - \epsilon_B Mx) \qquad (9.64)$$

which on substitution for c_A and c_B from equations (9.58) and (9.59) becomes:

$$A - A' = l c_{ABn}(\epsilon_{ABn} - \epsilon_A - n\epsilon_B)$$

or
$$\frac{d(A - A')}{dx} = l(\epsilon_{ABn} - \epsilon_A - n\epsilon_B)\frac{dc_{ABn}}{dx} \qquad (9.65)$$

ie. $A - A'$ passes through a maximum (minimum) when c_{ABn} is a maximum if ϵ_B is greater (less) than $\epsilon_A + n\epsilon_B$. For the situation when $\epsilon_B = 0$, as is often true in complex ion formation, $A - A'$ is a maximum (minimum) when c_{ABn} is a maximum, and when ϵ_{ABn} is greater (less) than ϵ_A. Thus if the graph of $A - A'$ against x is constructed the value of x at the maximum can be determined and hence n (equation (9.62)).

If two or more compounds can be formed from A and B, the value of x at which $A - A'$ is a maximum varies with the wavelength of radiation used. If only a single compound is formed, then $A - A'$ is a maximum at a fixed value of x regardless of the wavelength used, provided only that the extinction coefficient of the complex differs from those of the components. Thus by studying the variation of $A - A'$ with x at different wavelengths it is possible to determine whether one or more complexes are formed.

Experiment 9.21   Study the equilibrium reaction between the iron(III)ion and the thiocyanate ion

Theory: The reaction:

$$Fe^{+++} + CNS^- \rightleftharpoons Fe(CNS)^{++}$$

provides a good example of the application of Job's method of continuous variations. Further it is a very simple experiment which can be studied with a filter absorptiometer.

Requirements: Filter absorptiometer and filters, or spectrophotometer for use in visible region, 1-cm glass cells. Standard 0.020 mol dm^{-3} nitric acid, standard

0.030 mol dm^{-3} hydrochloric acid, solid iron(III)nitrate and potassium thio-cyanate.

Procedure: Prepare the following solutions, each of total ionic strength, $I = 0.032$ mol dm^{-3}:

Solution X: Fe(NO$_3$)$_3$ 0.0020 mol dm^{-3}; HNO$_3$ 0.020 mol dm^{-3}.
Solution Y: KCNS 0.0020 mol dm^{-3}; HCl 0.030 mol dm^{-3}.

The nitric acid in X suppresses hydrolysis of the iron(III)nitrate and is not sufficient to oxidize the thiocyanate when X and Y are mixed. The hydrochloric acid is added to give a constant ionic strength in both solutions, while maintaining the concentrations of the Fe^{+++} and thiocyanate ions constant. With the solutions in two burettes prepare 10 cm^3 mixtures of the solutions in the ratios 0.5 : 9.5; 1 : 9 ... 9 : 1, 9.5 : 0.5.

If a spectrophotometer is available, plot the absorption spectrum of the 5 : 5 mixture against water as blank, over the range 380–540 nm, and hence determine the wavelength of maximum absorption (*c.* 460 nm). Finally, determine the absorbance of each of the mixtures, and the two original solutions, at this wavelength.

With an absorptiometer, select the appropriate filter transmitting in the range 430–480 nm and determine the absorbance of each mixture in a 1 cm cell against water as blank.

Treatment of experimental data and discussion

1. Since the original solutions do not absorb in this region A' is zero and $(A - A')$, in equation (9.65) is the measured absorbance of the solution. Plot a graph of A against x (0 to 1). Determine the formula of the complex (equation (9.62)). Comment on the shape of the curve.
2. Why is it necessary to keep the ionic strength of the solutions constant?
3. What additional information is required to prove that other complexes are not formed between Fe^{+++} and CNS$^-$?

Experiment 9.22   Study the formation of dichromate ion

Theory: In a solution of potassium chromate and hydrochloric acid in which there is a negligible amount of undissociated chromic acid, the only reaction is:

$$n \, CrO_4^{--} + H_3O^+ \rightleftharpoons H(CrO_4)^{1-2n} + H_2O$$

The formation of the dichromate ion will not interfere, its maximum concentration must occur at the same value of x as the maximum concentration of H(CrO$_4$)$^{1-2n}$. The maximum in $A - A'$ will coincide with the composition maximum.

Requirements: Spectrophotometer, 1-cm glass cells, 0.01 mol dm^{-3} potassium chromate and 0.01 mol dm^{-3} hydrochloric acid solutions.

Procedure: Plot the absorption spectrum of the potassium chromate solution against a blank of distilled water, over the wavelength range 400–560 nm and hence determine the molar extinction coefficient (equation (5.12)) of potassium chromate at the different wavelengths.

Prepare 100 cm^3 mixtures of the chromate and acid containing 100 x cm^3 of acid and 100(1 − x) cm^3 of potassium chromate solution, where x = 0.1, 0.2 ... 0.9. Measure the absorbance of each solution, against a blank of distilled water, at the following wavelengths: 410, 470, 505, 525, 535, 545 and 555 nm.

Treatment of experimental data and discussion
1. From the extinction coefficients of the pure chromate solution, calculate the absorbance, A', at each wavelength assuming that there is no reaction on mixing with the acid. Hence calculate $(A - A')$ and plot against x, for each wavelength.
2. Determine the formula of the compound formed during the reaction and comment on the shape of the curves.

Experiment 9.23 Study the formation of complex ions between the nickel ion and o-phenanthroline

Requirements: Spectrophotometer and 4-cm glass cells; 0.1 mol dm^{-3} nickel sulphate solution (free from cobalt), solid o-phenanthroline.

Procedure: Since o-phenanthroline is of limited solubility, prepare the mixtures of nickel sulphate and the base as follows: dilute 50(1 − x) cm^3 of nickel sulphate with 50 x cm^3 of water and add the calculated amount of solid o-phenanthroline required for 50 x cm^3 of 0.1 mol dm^{-3} solution, where x = 0, 0.1, 0.2, 0.3, 0.4, 0.5, 0.6, 0.67, 0.75, 0.8, 0.9. Plot the absorption spectrum of each solution against a blank of distilled water, over the range 500–650 nm.

Treatment of experimental data and discussion
1. If it is tentatively assumed that relatively stable compounds are formed when the nickel/o-phenanthroline molar ratios are 1:1, 1:2 and 1:3, then it is possible to select wavelengths at which the assumption can be verified. From the absorption curves it will be apparent that for the first compound the most suitable wavelength is 620 nm, for the second 580 nm and for the third 528 nm.

   At each wavelength plot the graph of the absorbance against x and also against the calculated curve assuming there is no reaction. Finally plot the differences in the ordinates of these two curves and locate the values of x corresponding to the maxima of the difference curves.
2. From the shapes of the $(A - A') - x$ curves determine the formulae of the complex ions (equation (9.62)).
3. What are the co-ordination numbers of nickel in the different complex ions?
4. List other complex ions which could be studied by this method.

Experiment 9.24 Investigate the formation of lead nitrate complexes in solution and determine the dissociation constant of the complex

Theory: For many electrolytes, particularly those of high or asymmetrical valency type, the graph of the molar conductivity of the solution against $c^{1/2}$ is either linear or slightly concave, but the experimental slopes are greater than those predicted theoretically and the value of the molar conductivity is less than that required by the theory. The general explanation offered for the discrepancy is that the dissociation of the electrolyte is incomplete, ie. the number of ions available for the transport of electricity is less than that expected from the stoichiometric concentration. The correct form of the limiting equation is:

$$\Lambda = \alpha[\Lambda_0 - (A + B\Lambda_0)\sqrt{\alpha c}]$$

where αc is the actual ionic concentration.

The quantity α represents the fraction of solute which is free to conduct at a given concentration. The departure of α from unity may be due to either ion association or incomplete ionization; these are indistinguishable by conductance studies.

From a study of the absorbances of various mixtures of lead perchlorate and sodium nitrate of equal concentrations, the formula of the ion pair will be established as $PbNO_3^+$. From a subsequent study of the variation of the absorbance of mixtures of lead perchlorate (at a range of concentrations) and sodium nitrate (of constant concentration) each adjusted to a fixed ionic strength, the dissociation constant, K_i of the complex may be determined.

For the system:

$$PbNO_3^+ \rightleftharpoons Pb^{++} + NO_3^-$$

$$K_i = \frac{c_{Pb^{++}} c_{NO_3^-}}{c_{PbNO_3^+}} \tag{9.66}$$

If the concentration of sodium nitrate in the mixtures is kept constant, a, while $c_{Pb^{++}}$ is varied then:

$$c_{NO_3^-} = a - c_{PbNO_3^+}$$

which on substitution into equation (9.66) and rearrangement gives:

$$c_{PbNO_3^+} = \frac{a c_{Pb^{++}}}{K_i + c_{Pb^{++}}} \tag{9.67}$$

The absorbance, A, of the solution for l m path length is given by:

$$A = [\epsilon_{NO_3^-} c_{NO_3^-} + \epsilon_{PbNO_3^+} c_{PbNO_3^+}] l$$
$$= [\epsilon_{NO_3^-}(a - c_{PbNO_3^+}) + \epsilon_{PbNO_3^+} c_{PbNO_3^+}] l \tag{9.68}$$

in the region where absorption is due solely to NO_3^- and $PbNO_3^+$.

Eliminating $c_{PbNO_3^+}$ from equations (9.67) and (9.68) gives:

$$A - la\epsilon_{NO_3^-} = \frac{la c_{Pb^{++}}}{K_i + c_{Pb^{++}}} (\epsilon_{PbNO_3^+} - \epsilon_{NO_3^-})$$

which on rearrangement becomes:

$$\frac{la c_{Pb^{++}}}{A - la\epsilon_{NO_3^-}} = \frac{K_i}{(\epsilon_{PbNO_3^+} - \epsilon_{NO_3^-})} + \frac{c_{Pb^{++}}}{(\epsilon_{PbNO_3^+} - \epsilon_{NO_3^-})} \tag{9.69}$$

Requirements: Spectrophotometer and 1-cm quartz cells. Solid yellow le. sodium nitrate, sodium perchlorate, 72% perchloric acid, ion-exchange packed with Zeocarb 225, and standard 0.1 mol dm^{-3} sodium hydroxide sol.

Procedure:

(*a*) *Preparation and analysis of lead perchlorate solution* (final solution should be approximately 2 mol dm^{-3} Pb^{++} and 5 mol dm^{-3} ClO_4^-): Add 140 g of pure yellow lead oxide to 120 cm^3 of 72% perchloric acid, ie. excess perchloric acid, which prevents the formation of basic salts. *Care*, the reaction is exothermic. Dilute to about 250 cm^3 with distilled water and filter through a clean glass sintered crucible; reject any residue. This 'stock' solution must be analysed for all constituents, ie. Pb^{++}, H^+, ClO_4^-.

Accurately prepare a 50-fold dilution of the stock solution. To 25 cm^3 of this solution add 0.05 mol dm^{-3} sodium sulphate and determine the weight of the lead sulphate precipitate by the usual gravimetric method. Hence calculate the concentration of the solution with respect to lead.

Transfer 25 cm^3 of the 50-fold dilution to the top of a prepared column of Zeocarb 225 (p. 281) in the H^+ form. Thoroughly wash with distilled water (150–200 cm^3) and titrate the eluate with 0.1 mol dm^{-3} sodium hydroxide solution. Hence calculate the concentration of the solution with respect to perchlorate and, by difference, determine the concentration of hydrogen ion.

(*b*) *Determination of formula of the complex:* Dilute the 'stock' solution accurately to give an approximately 0.1 mol dm^{-3} Pb^{++} solution, and prepare a solution of sodium nitrate of identical concentration.

Using standard 10 cm^3 volumetric flasks, prepare mixtures of these two solutions containing 10 x cm^3 of sodium nitrate and 10 $(1-x)$ cm^3 of lead perchlorate solution, where x is 0.1, 0.2, 0.3 . . . 0.9. Measure the absorbance of (*a*) the sodium nitrate solution, (*b*) the lead perchlorate solution and (*c*) the mixture containing equal parts of both solutions, against a blank of distilled water, from 260–380 nm and locate the wavelength of maximum absorption. For the solution containing nitrate this is at about 300 nm; the absorption of the lead perchlorate over the region is negligible. Determine the molecular extinction coefficient ($\epsilon_{NO_3^-}$) of the nitrate ion at 300 nm.

Now measure the absorbance of each of the prepared mixtures of sodium nitrate and lead perchlorate, against distilled water, at 300 nm.

(*c*) *Determination of the dissociation constant of the complex:* Prepare a series of solutions (50 cm^3) each 0.05 mol dm^{-3} with respect to sodium nitrate, but containing different concentrations of lead perchlorate; the range 0.1 to 0.55 mol dm^{-3} is adequate. Calculate the ionic strength of each solution (including the contribution of hydrogen ions) and add the calculated amount of sodium perchlorate to give a constant ionic strength of 2 mol dm^{-3}. If time permits, repeat over a range of ionic strengths (0.5–6.0 mol dm^{-3}) in which the final ionic strength is made up either with perchloric acid or sodium perchlorate. This eliminates the possibility of any changes which can be attributed to changes in the activity coefficients. Determine the absorbance of each solution against a blank of distilled water at 300 nm.

Treatment of experimental data and discussion

1. From the data of (b) plot a graph of $(A - A')$ against x; in equation (9.64) ϵ_A (ie. $\epsilon_{Pb^{++}}$) is negligible. Determine the value of x for which $(A - A')$ is a maximum and hence deduce n (equation (9.62)). Does the shape of the graph indicate the presence of more than one complex ion or ion pair? If so establish the formulae of each.

2. From the data of (c) and the known extinction coefficient of NO_3^-, calculate the value of $lac_{Pb^{++}}/(A - la\epsilon_{NO_3^-})$ for the different solutions and plot these values against $c_{Pb^{++}}/\text{mol m}^{-3}$ (equation (9.69)). From the measured slope and intercept calculate K_i, the instability constant of $PbNO_3^+$.

3. In (c) explain why it is necessary to keep the ionic strength of all the solutions constant.

4. Explain the effect of the variation of the ionic strength on K_i.

5. Starting from equations (9.58), (9.59), (9.60) and (9.61) deduce equation (9.62).

6. Discuss the reasons for using lead perchlorate instead of other soluble lead salts.

7. Describe any other experimental methods which are available for the study of this system.

6 Cryoscopic method

Experiment 9.25   Study the formation of complex ions by mercury(II)iodide in potassium iodide solutions by the cryoscopic method

Theory: A 1.0 mol dm^{-3} solution of potassium iodide in water freezes at a temperature below $-3\,^\circ$C. When solid mercury(II)iodide, which is insoluble in water, is added to this solution the freezing point rises to a limiting value when the solution is saturated with mercury(II)iodide. The depression of the freezing point of water by potassium iodide depends on the number of ions in solution:

$$KI \longrightarrow K^+ + I^-$$

The added mercury(II)iodide dissolves in this solution with the formation of complex ions:

$$K^+ + I^- + HgI_2 \longrightarrow K^+ + HgI_3^- \qquad (a)$$

$$2K^+ + 2I^- + HgI_2 \longrightarrow 2K^+ + HgI_4^{--} \qquad (b)$$

In both cases the concentration remains constant. If the reaction results in the formation of HgI_3^- (a) the freezing point of the potassium iodide solution will remain unchanged, since the total number of ions is constant. If, however, HgI_4^{--} is formed (b) then the freezing point must increase as the total number of ions is reduced (from 4 to 3). The maximum freezing point of the saturated solution corresponds to a solubility of 0.5 mole mercury(II)iodide/mole potassium iodide. For the formation of HgI_4^{--} alone the increase of freezing point with the addition of mercury(II)iodide should be linear.

Requirements: Beckmann thermometer, apparatus (Fig. 3.5), pure potassium iodide and mercury(II)iodide.

Procedure: Place 20 cm^3 of water in the inner tube and using the procedure described in Expt. 3.4, determine its freezing point. It will be necessary to use a freezing mixture of salt and ice in the outside container. Dissolve 3.3 g potassium iodide (0.02 mole) in this water to give a 1 mol dm^{-3} solution, and determine its freezing point (below −3 °C). Now add 1.5 g of mercury(II)iodide to this solution and determine the freezing point of the yellow solution. Add three further 1.5 g amounts of mercury(II)iodide and determine the freezing point after each addition. The last portion will probably not dissolve completely, since the solution is saturated; the suspended solid does not interfere with the freezing point determination.

Treatment of experimental data and discussion

1. Plot a graph of the depression of the freezing point of water against the number of moles of mercury(II)iodide added. Draw a horizontal line at the maximum freezing point depression found for the final saturated solution and a vertical line from the 0.01 mole mark. Join the point of intersection of these two lines to the freezing point depression of pure potassium iodide; this is the theoretical curve of ΔT against concentration if HgI_4^{--} only is formed.

2. Comment on the deviation of the experimental curve from the theoretical curve, in terms of the different complex ions that may be formed.

10

Surface chemistry and colloids

At any boundary between two phases there is an abrupt change in the inter-molecular forces, eg. at the liquid/vapour interface there exists the force of surface tension, the like of which is unknown in homogeneous systems. In the immediate neighbourhood of an interface between two phases, there is a change of concentration so that the concentration of any material is different at, and near the interface to, its bulk concentration, in either of the two phases forming the boundary. This phenomenon of *adsorption*, existing at all interfaces, is due to the abrupt change in the intermolecular forces. An electrical potential, the *electrokinetic* or ζ *potential*, existing at an interface is due, in part, to the preferential adsorption of one kind of ions on one of the phases. While these phenomena exist at any interface, they are most important and easily recognized in systems where the surface area/mass ratio is very large; eg. thin films on liquids, finely divided and porous solids and colloidal systems.

1   Adsorption by solids

The surface of a solid possesses a residual attractive force; in consequence all solids tend to adsorb on their surfaces any gas or vapour with which they come into contact. Many solids are also capable of adsorbing ions and molecules from solution; an effect which is used for the separation of substances by chromatography. Adsorption by solids is dependent upon the temperature, the nature of the adsorbent and adsorbate and on the concentration (or pressure) of the adsorbate. Adsorption from dilute solution may often be represented successfully by the empirical equation, due to Freundlich:

$$x/m = kc^{1/n} \tag{10.1}$$

where x is the amount of solute (moles) adsorbed by m grammes of adsorbent when the equilibrium concentration in solution is $c/\text{mol dm}^{-3}$, k and n are constants ($n > 1$). The graph of $\log c$ against $\log x/m$ is linear and of slope $1/n$. This classical isotherm is a special case of a more general relationship originally derived by Langmuir for the adsorption of gases on solids:

$$\frac{x}{m} = \frac{ac}{1 + bc} \tag{10.2}$$

where a and b are experimental constants. Both adsorption isotherms show that adsorption increases with concentration, while Langmuir's equation further suggests that at high concentrations the amount adsorbed approaches a limiting value, at which point it is envisaged that the solid is covered with a monolayer. Adsorption from solution decreases with increase of temperature; the forces involved in this 'physical' adsorption process are of the van der Waals type in contrast to the 'chemisorption' of gases on solids which involves the formation of bonds akin to valence bonds.

Experiment 10.1   Determine the adsorption isotherm of acetic acid from aqueous solution by charcoal

Requirements: Stoppered bottles, thermostat at 25 °C, standard 0.5 mol dm^{-3} acetic acid, 0.1 mol dm^{-3} sodium hydroxide solutions and powdered active charcoal. The charcoal should be well washed with water, to remove residual acid, and then dried.

Procedure: Weigh 5 g of charcoal into each of seven bottles containing 50, 25, 15, 7.5, 4, 2 and 0 cm^3 of 0.5 mol dm^{-3} acetic acid made up to 50 cm^3 with water. Shake, suspend in the thermostat and leave for 1 hour with intermittent shaking. Filter each solution through a dry filter paper rejecting the first 5 cm^3 of each and collect the filtrates in dry flasks. Take an aliquot of each filtrate (5 cm^3 of the first, 10 cm^3 of the second and 20 cm^3 of the remainder) and titrate with the standard alkali solution using phenolphthalein as indicator. Make duplicate titrations of each.

Treatment of experimental data and discussion
1.   From the titration figures (including the necessary allowance for the acid titration in water) calculate the total amount of acetic acid in 50 cm^3 of solution and hence by difference the amount of acetic acid (x) adsorbed by m g of charcoal. Also tabulate the equilibrium concentration (c/mol dm^{-3}) of acetic acid in each solution.
2.   Plot graphs of x/m against c; log x/m against log c and $c/(x/m)$ against c and test the validity of equations (10.1) and (10.2). Determine the values of the constants in these equations where possible.
3.   Comment on the problem of ensuring that thermodynamic equilibrium is attained in this type of experiment and describe how the establishment of equilibrium can be tested.
4.   Explain why it is necessary to discard the first 5 cm^3 of each filtrate.
5.   Give a simple derivation of the Langmuir adsorption isotherm.

Experiment 10.2   Determine the adsorption isotherm of oxalic acid from aqueous solution by charcoal

Requirements: Stoppered bottles, thermostat at 25 °C, standard 0.25 mol dm^{-3} oxalic acid and approximate 0.02 mol dm^{-3} potassium permanganate solutions, powdered active charcoal.

Procedure: Standardize the potassium permanganate solution with the oxalic acid solution. Prepare 250 cm³ portions of oxalic acid solutions by dilution, giving 0.2, 0.15, 0.10, 0.05 and 0.025 mol dm⁻³ solutions; standardize these with the potassium permanganate solution. Weigh out accurately about 2 g of active charcoal into each of six dry stoppered bottles and add 100 cm³ portions of the various oxalic acid solutions to each. Secure the stoppers, suspend in the thermostat and shake at short intervals over the period of 1—2 hours. The conditions of agitation must be the same for all samples. Filter through a small dry paper, rejecting the first runnings of each, and collect about 50 cm³ of each filtrate in a dry flask. Determine the concentration of oxalic acid in each by titrating suitable portions with the potassium permanganate solution.

Treatment of experimental data and discussion

1. Calculate the amount of oxalic acid present in 100 cm³ of solution and hence by difference the amount of oxalic acid adsorbed. Test the validity of equations (10.1) and (10.2) by plotting graphs of x/m against c; log x/m against log c and $c/(x/m)$ against c. Determine the values of the constants in the isotherms where possible.

2. Comment on the problem of ensuring that thermodynamic equilibrium is attained in this type of experiment and describe how the attainment of equilibrium may be tested.

3. Explain why different kinds of charcoal give different adsorption isotherms.

4. Explain how, by experiments of this kind, the heat of adsorption of oxalic acid on charcoal may be determined.

5. Derive the Langmuir adsorption isotherm.

2   Adsorption at liquid surfaces

The addition of a solute to water causes a change in the surface tension. Surface inactive substances (eg. inorganic electrolytes) produce a slight increase in the surface tension, while surface active substances (eg. fatty acids) cause a marked decrease in the surface tension often at very low concentrations. The change occurs as a result of the accumulation of the component with the lower surface tension at the interface. Thus, in the case of the fatty acids, it is the acid which is adsorbed at the surface. Substances adsorbed at a water/air interface have an amphipathic structure, ie. they are composed of a water-soluble, hydrophilic group (eg. —COOH) attached to a hydrophobic group (eg. hydrocarbon residue). The surface excess concentration (Γ/mol m⁻²) at a concentration (c/mol m⁻³), is expressed by the Gibbs adsorption isotherm; which for dilute solutions may be written:

$$\Gamma = -\frac{c}{RT}\left(\frac{d\gamma}{dc}\right) \tag{10.3}$$

Assuming this equation to be valid, Γ can be calculated from a study of the variation of the surface tension, γ, with concentration.

Experiment 10.3   Study the variation of the surface tension of solutions of butan-1-ol with concentration

Requirements: Capillary rise surface tension apparatus (Expt. 2.7) or du Nouy balance (Expt. 2.11). 1.0 mol dm^{-3} aqueous stock solution of butan-1-ol.

Procedure: Dilute the stock solution to give solutions of concentration: 0.8, 0.7, 0.6, 0.5, 0.4, 0.3, 0.2, 0.1, 0.05 and 0.025 mol dm^{-3} with respect to butan-1-ol. Using the appropriate technique determine the surface tension of water and each of the solutions of butan-1-ol; record the temperature of the measurement. Determine the density of each solution at the same temperature (Expt. 2.2).

Treatment of experimental data and discussion

1. Calculate the surface tension of each solution and plot a graph of $\gamma/N\ m^{-1}$ against $c/mol\ m^{-3}$. Construct tangents to this curve at definite values of c and measure the slope of each. (A better method is to calculate the best equation to the γ-c curve by the method of least squares (p. 11); this is a quadratic and the slope at given values of c can be obtained more accurately by differentiation.) Hence calculate the value of the surface excess concentration at each value of c (equation (10.3)).

2. Plot Γ against c and show that Γ approaches a limiting value at higher values of c. This limiting value gives the number of moles of butan-1-ol per square metre of surface. The area occupied by 1 mole is thus $1/\Gamma$ and the area occupied by 1 molecule is $1/\Gamma N_A$ m^2 or $10^{20}/\Gamma N_A$ Å^2. Hence calculate the limiting area occupied by 1 molecule of the alcohol.

3. At room temperature the cross-section of a saturated aliphatic chain is 18.5 Å^2. Account for the difference between this value and the experimental value.

4, Discuss the orientation of the alcohol molecules at the interface in terms of your experimental observations.

5. Predict the limiting area for 1 molecule of either pentan-1-ol or hexan-1-ol.

6. Does this experiment test the validity of equation (10.3)?

Experiment 10.4   Demonstrate that Rhodamine 6G, is adsorbed at the water/air interface

Theory: The original method used to verify the Gibbs adsorption isotherm was to pass air bubbles through a column of solution and compare the concentration of the solute in the liquid collected from the bursting bubbles (ie. the adsorbed solute at the water/air interface) with the concentration of the solution remaining in the column. This method showed conclusively that surface active solutes were adsorbed but the adsorption was much greater than that predicted by equation (10.3).

Requirements: Chromatography column (30 cm × 2 cm) fitted with a sintered glass disc and tap, Rhodamine 6G, supply of air from cylinder.

Procedure: Fit the top of the column with a rubber bung carrying a short bent tube leading to a small flask. Half fill the column with a dilute aqueous solution of

Rhodamine 6G, and pass a slow stream of air through the solution. Regulate the flow so that a column of froth rises from the surface and passes through the tube into the receiving flask. Take care that no solution as such is forced over. When sufficient liquid has been collected, pour a sample into a test tube and compare it with the original dye solution and with the solution remaining in the column (by eye).

Discussion

1. Comment on the difference in the concentration of the dye collected in the flask and that remaining in solution in the tube.
2. Explain, why in such an experiment the adsorption is found to be greater than that predicted by equation (10.3).
3. What methods are available for the experimental verification of the Gibbs adsorption isotherm?

3   Insoluble surface films

When a drop of a non-volatile insoluble liquid is placed on the surface of water, depending on the relative forces of adhesion and cohesion, it may either remain as a lens on the surface or spread as a monolayer, any excess liquid remaining as a lens in equilibrium.

Stable monolayers are only produced by substances which are insoluble in water and which contain a hydrophilic group, preferably at one end of the molecule. These conditions are fulfilled by a long hydrocarbon residue attached to a polar group (eg. $-COOH$, $-CH_2OH$). On account of their electric dipole, these groups are attracted to the water surface, thereby anchoring the molecules.

These insoluble films result in a lowering of the surface tension of the water. The molecules on the surface exhibit a two dimensional force, F, known as the 'surface pressure', analogous to the three dimensional gas pressure. It can be shown that:

$$F = \gamma_0 - \gamma \tag{10.4}$$

where γ_0 and γ are the surface tensions of water and the film respectively. For insoluble films, the variation of F with the area/molecule is illustrated in Fig. 10.1, in which the extrapolated area at zero compression for long chain acids is 20.5 Å^2, independent of the chain length. This suggests that at this point the molecules are oriented with their polar groups in the water and the hydrocarbon group pointing vertically outwards.

For an ideal film $FA = kT$; the calculated $FA-F$ curves for real films (compare $pv-p$ curves for real gases) show departure from ideal behaviour.

Experimentally, these insoluble monolayers are easily formed by adding a few drops of a solution of the substance in a volatile organic solvent to a clean water surface and allowing the solvent to evaporate. From the concentration of the solution and the volume added, usually by means of a micrometer syringe, the actual number of molecules comprising the film can be calculated. The films are easily handled on surfaces by confining them between glass or perspex strips on the surface. These strips are analogous to semipermeable membranes, in which the water molecules can pass under to the clean surface on the other side, while the film molecules are prevented.

In the experimental work, it is essential that all surfaces are scrupulously clean and free from grease and dust. Glass dishes with ground tops are washed with hot chromic acid solution, rinsed thoroughly and dried (inverted) in an oven. The glass is coated with paraffin wax by applying a solution of wax dissolved in benzene whilst still hot and allowing to dry. Strips of glass (30 × 2 × 0.2 cm thick) used as barriers are treated similarly; these must be handled only at the ends. The wax provides a non-wetting surface over which the water cannot spread. Perspex dishes and strips are more easily prepared by washing with hot soapy water and rinsing under a running tap, they may be coated with a silicone oil, applied as a solution in carbon tetrachloride.

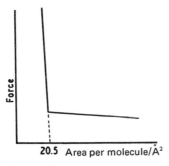

Fig. 10.1   Force—area curve for fatty acid on water

The dish is placed on a flat surface and filled with water until it overflows. A barrier placed across it will now separate the surface into two portions. To test if the surface is clean, sprinkle powdered talc, from a shaker, on the surface and then move the barrier across it. If the talc powder moves in advance of the barrier (ie. before the barrier has reached it) the surface is contaminated. It can be swept clean by pulling barriers across, always in the same direction, and always leaving one barrier on to keep back the contamination.

Experiment 10.5   Determine the area occupied by a long chain acid molecule on a water surface

Requirements: Large rectangular glass or Perspex dish with ground edges, long strips of glass or perspex (30 × 2 × 0.2 cm thick), micrometer syringe, *n*-hexane or benzene, solid octadecanoic or hexadecanoic acid.

Procedure: Prepare a solution of the acid in *n*-hexane or benzene (0.3 g dm^{-3}). Thoroughly clean the dish, stand it on a level surface in a tray, and fill the dish with distilled water until it overflows. Test the surface for contamination and renew if necessary. Now sprinkle talc uniformly over the entire surface, and with the syringe held vertically add 0.1 cm^3 of the acid solution. The solvent evaporates, leaving a circular monolayer of acid on the surface; this is revealed by the talc particles.

Carefully measure the diameter of the circle so formed using either a rule or travelling microscope.

Empty the dish, refill with clean water and repeat the above with different additions of the two acid solutions.

Treatment of experimental data and discussion

1. From the diameter of the oil film calculate its area, and hence the area occupied by 1 mole and by 1 molecule of each acid. Assuming that the density of the acid in the film is the same as in the bulk, calculate the thickness of the film.
2. Comment on the values obtained and compare the measured area with the cross-sectional area of an aliphatic hydrocarbon chain, 18.5 Å². Discuss the orientation of the acid molecules at the air-water interface.
3. How would the area be affected by the presence of a double bond in the hydrocarbon chain near the carboxyl group?

Experiment 10.6   Determine the $F-A$ curve for a long chain acid on water using a surface pressure balance

Theory: In this experiment the film is confined between a fixed and a floating barrier. The molecules of film on the surface exert a pressure (F) on the floating barrier (Fig. 10.2a), thereby displacing it in the direction of the water surface. By means of a torsion head coupled to an optical lever, the pressure can be calculated, and, by varying the area in which the film is confined, a complete $F-A$ curve may be plotted.

Requirements: Surface pressure balance, Perspex trough with ground edges (Fig. 10.2c), long Perspex strips, aluminium or mica floating barrier, Agla micrometer syringe, solution of acid in ether (0.001 g in 100 cm³). Suitable acids are hexadecanoic (palmitic), octadecanoic (stearic) and 9-octadecanoic (oleic).

Procedure: Thoroughly clean the trough and strips. The floating barrier, which contains two holes to locate the stirrup attached to the torsion wire, is about 1 cm short of the width of the trough. In this way frictional forces between the walls and float are eliminated. Film is prevented from leaking between the float and sides by attaching thin waxed-cotton threads to the extreme ends of the float. Coat the float lightly with wax and attach the thread. Set up the pressure balance on a level vibration-free bench and fill the trough with distilled water. Without fingering, place the float on the water surface and attach it to the stirrup; make sure that the threads are lying on the water surface. Attach these to the sides of the trough with wax, leaving sufficient on the surface to permit movement of the float. Place a Perspex strip (handling it only at its extremities) on the water surface near the floating barrier and gently pull it away across the surface so that it sweeps away any

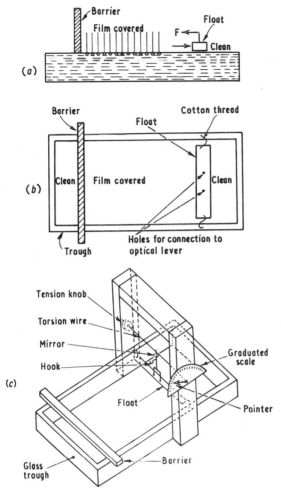

Fig. 10.2 Surface pressure balance. (*a*) Film balance diagrammatic. (*b*) Plan of trough. (*c*) Complete apparatus

contaminating film. Repeat this sweeping several times, always leaving one strip at the far end of the trough to prevent the contamination from spreading back.

Arrange a light source so that light reflected from the small mirror on the stirrup is brought to focus on a scale. Record the zero position of the spot with a clean water surface on either side of the float. Now add small weights to the hook mounted at right angles to the centre of the torsion wire; this causes a displacement of the float (similar to that produced by a film). By means of the torsion head, twist the wire and bring the light spot back to its zero position; record the number of degrees through which the torsion head is turned. Repeat for other weights. Knowing the distances of the weight and the float from the centre of the wire, the force (N per metre length of float) equivalent to one degree on the torsion head can be calculated.

In some commercial surface pressure balances there is no provision for the return of the floating barrier to the original position. Movement of the float is recorded by the displacement of the light spot along a scale; this must be calibrated in a similar manner to that described.

Again sweep the surface and record the zero position of the light spot. Now apply a known volume ($0.05-0.1$ cm^3) of the acid solution, by means of a micrometer syringe, to the water surface between the floating barrier and the Perspex strip at the end of the trough. Allow the ether to evaporate and check that the film is not escaping round the float. If the float has been displaced, return it to the zero position and record the angle on the torsion head and the distance between the barriers. Advance the strip 1 cm, return the float to the zero position, and again record the angle and the distance. Repeat at ever decreasing distances until a large increase in force for a relatively small decrease in area indicates the formation of a close packed film. Measure the width of the trough.

(Alternatively, record the displacement of the spot of light along the scale.)

Clean the apparatus and repeat the experiment using a 0.1 mol dm^{-3} solution of hydrochloric acid to support the film. Study also the effect of changing the chain length and the effect of the presence of a double bond in the hydrocarbon chain.

Treatment of experimental data and discussion

1. Calculate the number of molecules of the acid on the surface and hence A, the area available per molecule (in Å^2) for each area studied. Plot the graph of F, obtained from the angle on the torsion head (or the displacement of the light spot), against A. Extrapolate the steep part of the curve to zero compression and hence obtain the limiting area per molecule.

2. Discuss the shapes of the $F-A$ curves for the different acids in terms of the orientation of the acid molecules at the interface.

3. Explain the significance of the limiting area and discuss the factors which determine its value (in particular compare the values for stearic and oleic acids).

4. Give an account of the main sources of error in this method and suggest ways in which they could be minimized.

4   Electrical properties of interfaces

At any phase boundary there exists an electrical double layer due to the asymmetric distribution of ions, whereby one type of ion is preferentially attracted to one of the phases. Two types of surface are recognized: the *non-ionogenic surface* where the charge is due to the adsorption of foreign ions from the electrolyte (eg. oil droplets, air bubbles, finely powdered charcoal), and the *ionogenic surface* where the charge is due to the dissociation of ionogenic groups (eg. $-COOH$, $-NH_2$,

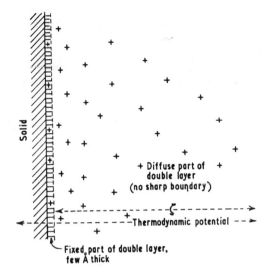

Fig. 10.3 Diagrammatic illustration of the electrical double layer

as in proteins). Whatever the origin of the double layer, particles in aqueous salt solutions possess a charge, so that on the application of an electric field they migrate through the solution; this phenomenon is known as *electrophoresis*. Conversely, if a potential is applied to a system in which the solid phase is held stationary (eg. solution in a tube) the liquid moves relative to the surface, *electro-osmosis*. The electrophoretic (-osmotic) mobility, $\bar{v}$, is the linear velocity of the particle (liquid), v/m s^{-1} for unit applied field strength X/V m^{-1}. Thus $\bar{v} = v/X$ m^2 s^{-1} V^{-1}. The determination of $\bar{v}$ requires the simultaneous measurement of the velocity and applied field strength. The mobility is related to the electrokinetic or ζ potential by the Smoluchowski equation:

$$\bar{v} = \frac{v}{X} = \frac{\epsilon \zeta}{4\pi\eta} \tag{10.5}$$

where ϵ and η are the permittivity and viscosity of the medium respectively. The ζ potential is defined as the potential between the liquid side of the fixed part of the double layer and the bulk of the solution (Fig. 10.3). Studies of the variation of ζ with ionic strength and pH of different surfaces, eg. coated particles, biological cell surfaces, are of great value in elucidating surface components and surface structure. The determination of ζ potentials will not be described owing to the complicated nature of the apparatus and the precautions which have to be taken. A commercial Particle Electrophoresis Apparatus, available from Rank Bros. (Bottisham, Cambridge), is suitable for precise measurements on colloidal solutions and suspensions.

Experiment 10.7   Demonstrate the phenomenon of electro-osmosis

Requirements: Apparatus (Fig. 10.4), capillary tube with very fine bore or tube with wide bore packed with macerated filter paper, sealed in platinum electrodes, source of 200–500 V d.c., digital voltmeter, solid potassium chloride, calcium chloride and thorium nitrate.

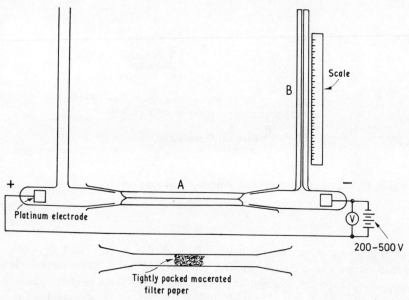

Fig. 10.4   Apparatus for electro-osmosis

Procedure: Thoroughly clean the apparatus, which consists of a capillary tube A (10 cm long) connected by ground glass joints to the electrode compartments carrying platinum electrodes. Tube B is large bore capillary tubing which magnifies the movement of liquid along A. Fill the apparatus with a 10^{-3} mol dm^{-3} solution of potassium chloride so that the meniscus is on the scale and mounted behind B. Switch on the current, record the polarity of the electrodes and the applied potential and observe the rate of migration of the liquid on the scale. (Alternatively, macerate some filter paper, soak in potassium chloride solution, and pack in the wider bore tube; study the rate of migration as before.)

Investigate the variation of the rate of migration with (*a*) the nature of the electrolyte; (*b*) its concentration; use the range 10^{-3} to 10^{-5} mol dm^{-3}, (*c*) the applied potential, and (*d*) the effect of coating the tube with gelatin.

Treatment of experimental data and discussion

1.   For each solution, calculate the applied field strength from the measured potential difference and the distance between the electrodes. Calculate the electro-osmotic mobility from the measured rate of migration and the applied field strength.

2.   For the different electrolytes plot graphs to show the variation of the

mobility (proportional to the ζ-potential, equation (10.5)) with the concentration. Discuss your results in terms of the structure of the electrical double layer.

3. Does the mobility vary with the applied field strength?
4. Discuss the main sources of error in the method and suggest ways in which they could be reduced.

5 The colloidal state

Colloidal solutions consist of particles dispersed in a homogeneous medium. The particles are smaller than coarse filterable precipitates but larger than small molecules, the generally accepted diameter range is 1—500 nm. There is, however, no sharp dividing line between colloidal solutions and true solutions on the one hand and colloidal solutions and suspensions on the other. Colloidal solutions are nevertheless easily distinguished by certain characteristic properties.

Two types of colloidal solution are recognized; the *lyophobic* (hydrophobic) sols, ie. liquid (water) fearing in which there is no affinity of the particles (disperse phase) for the dispersion medium. The stability of these sols depends mainly on the charge carried by the particles, eg. colloidal gold, silver, silver chloride, and arsenic trisulphide. *Lyophilic* (hydrophilic), ie. liquid (water) loving in which there is affinity between the disperse phase and the dispersion medium. These sols, which contain highly solvated particles, are more stable than the lyophobic sols, eg. proteins, rubber and silicic acid.

The chief difference between these two types of colloid is their behaviour towards electrolytes. Lyophobic colloids are precipitated by small quantities of electrolyte when their ζ potential is reduced to a low value near zero; this is expressed approximately in the Hardy—Schulze law. Lyophilic colloids, on the other hand, require very high concentrations of electrolyte before they can be 'salted out'. This difference arises from the origin of the charge and the extent of solvation of the two types. Lyophobic colloidal particles obtain their charge by the adsorption of ions, mainly anions; very small quantities of electrolyte are required to confer the charge necessary for stabilization. When larger amounts of electrolyte are added, the ζ potential at first becomes more negative and then decreases towards zero; at the limiting value of ζ, when the thermal forces exceed the electrical forces of repulsion, precipitation occurs. The coagulation power of ions depends on their sign (ie. of opposite sign to the particle) and not on their chemical nature; thus divalent and trivalent ions are respectively 50—100 and 1000 times as effective as univalent ions. The charge of lyophilic sols is due mainly to the dissociation of ionogenic groups, the value and sign of the charge depending on the pH of the dispersion medium. At an intermediate pH, the value of which depends on the relative strengths of the acidic and basic groups, the sol will be least charged, and hence, least stable. This pH is known as the isoelectric point of the colloid. Solvation plays a large rôle in the stabilization of lyophilic sols and prevention of precipitation by electrolytes.

Experiment 10.8   Prepare colloidal aqueous solutions of arsenic trisulphide, iron(III)hydroxide, silver and gelatine

Theory: The hydrophobic sols are best prepared by chemical methods of condensation in which colloidal particles of sparingly soluble substances are produced. Lyophilic sols, on the other hand, are easily prepared by dissolving the substance in the appropriate dispersion medium; thus starch and gelatine dissolve in hot water, and non-vulcanized rubber in benzene.

Requirements: Dialysis bags, arsenic trioxide, iron(III)chloride, silver nitrate, tannic acid, sodium carbonate, powdered gelatine.

Procedure:

(*a*) Arsenic trisulphide: dissolve 1–2 g of pure powdered arsenic trioxide in boiling distilled water. After prolonged boiling, cool the solution to room temperature and filter if necessary. Dilute this solution with 3–4 volumes of distilled water and bubble in pure hydrogen sulphide gas. The gas must be freed from hydrogen chloride by passing through a soda lime tube and then through three or four wash bottles containing distilled water. The solution will be saturated with gas and arsenic trisulphide formed after 5 min bubbling. Remove the excess hydrogen sulphide by passing a stream of pure hydrogen through the solution for 20–30 min. Remove the small amount of coarse precipitate by filtration. This colloid is quite stable.

(*b*) Iron(III)hydroxide: Pour 12 cm³ of a 32% iron(III)chloride solution into 750 cm³ of boiling distilled water. Hydrolysis of this chloride occurs instantly, forming the deep red sol of iron(III)hydroxide. This colloid is quite stable and may be purified by dialysis. Suspend a cellophane bag filled with the sol in a large beaker full of water; do not tie the top of the dialysis bag. After some hours, test the water for the presence of hydrochloric acid.

(*c*) Silver sol: To 500 cm³ of distilled water, add 20 cm³ 0.1 mol dm⁻³ silver nitrate solution and 5–10 cm³ of a 1% tannic acid solution. Heat the mixture to 70–80 °C and add 10 cm³ of 1% sodium carbonate solution in portions with continual stirring. The silver carbonate formed is instantly reduced by tannic acid giving metallic silver. This remains in the solution as a colloid which has the colour of tea and is quite clear.

(*d*) Gelatine sol: Dissolve 2 g of powdered gelatine in 400 cm³ of water at 80–90 °C. A clear sol is formed on cooling. A 1–2% gelatine solution will set to a jelly on cooling.

Discussion

1.   What purpose is served by the removal of excess hydrogen sulphide and hydrochloric acid during the preparation of arsenic trisulphide and iron(III)-hydroxide sols respectively? Give an account of the procedure of dialysis and explain why the cellophane dialysis bag should not be tied off. Does the water on the outside of the dialysis bag contain any iron(III)hydroxide?

2.   Why are the various sols coloured? How is the colour related to the particle size?

3. Explain why hydrophobic sols cannot be prepared by the dispersion of the macrophase.

Experiment 10.9 Study the coagulation of various sols with electrolytes and alcohol

Requirements: Colloidal solutions of arsenic trisulphide, dialysed iron(III)-hydroxide, silver and gelatine (Expt. 10.8), 0.5 mol dm^{-3} sodium chloride, 0.005 mol dm^{-3} barium chloride, 0.0005 mol dm^{-3} aluminium chloride, 0.005 mol dm^{-3} sodium sulphate, 0.0005 mol dm^{-3} potassium ferricyanide and 0.0005 mol dm^{-3} potassium ferrocyanide solutions and ethanol.

Procedure: To a 10-cm^3 sample of arsenic trisulphide sol, add the sodium chloride solution slowly from a burette, with constant shaking. The solution remains clear for a time, then becomes cloudy, then opaque and finally the arsenic trisulphide separates out in large flakes. It is best to record the last stage, as the precipitation point is easier to judge. Record the amount of electrolyte necessary just to cause precipitation. Repeat for all the colloidal solutions with all the electrolytes. The amount of electrolyte required depends, to some extent, upon the time taken in adding it. Test this by leaving the mixture of sol and electrolyte at the cloudy stage for some hours: separation of a precipitate should be observed. Since, however, the times for the various titrations are roughly the same, the point of precipitation gives good comparative values.

Add 5 cm^3 of each hydrophobic sol dropwise to 5 cm^3 of a 1% gelatine solution with constant shaking. Study the effect of electrolytes on these sols as before, now using 2 mol dm^{-3} solutions.

To a 5 cm^3 sample of each sol, add 10 cm^3 of ethanol and mix thoroughly. Observe the tubes after standing for 5 min.

Study the mutual precipitation of arsenic trisulphide and dialysed iron(III)-hydroxide sols by mixing them together in the ratios 1:9, 2:8, etc. Determine the optimum ratio for precipitation, leaving a colourless supernatant liquid after standing for 2 hours.

Treatment of experimental data and discussion

1. Draw up a table giving the minimum concentration of each electrolyte which is sufficient to cause precipitation of each sol.
2. Use the experimental data to establish the dominant species (ie. anion or cation) of the added electrolyte in causing precipitation of each of the hydrophobic sols. Hence determine the charge carried by the particles in each sol and explain how the charge on each arises.
3. Explain why colloidal particles of like charge do not repel each other under all circumstances, but can aggregate under suitable conditions.
4. Explain the effect of gelatine on the stability of the hydrophobic sols.
5. Why is the stability of the hydrophilic sol markedly affected by the addition of alcohol, while the stability of the hydrophobic sols is unaffected?

Experiment 10.10   Study the effect of colloidal particles on the viscosity of water

Theory: The viscosity of a liquid or a solution is a measure of its resistance to flow. Hydrophobic sols have a viscosity only slightly greater than water, a value almost independent of the concentration. On the other hand, hydrophilic sols have viscosities which, depending on the material, are very much greater than that of water. The viscosity also increases with increase in concentration.

The viscosity of a liquid is determined by measuring the flow time of a definite volume of the liquid through a capillary tube (Ostwald viscometer) and comparing this with the flow time of an equal volume of water.

Requirements: Ostwald or suspended-level viscometer (p. 24), thermostat at 25 °C, stop-watch, colloidal solutions of dialysed iron(III)hydroxide (Expt. 10.8), albumen (0.5 and 1%), glycogen (0.5 and 1%) and gelatine (0.5 and 1%).

Procedure: Using the details of Expt. 2.4, determine the times of flow of equal volumes of water and each of the colloidal solutions. The viscometer should be chosen to have a flow time for water of about 80—100 seconds.

Treatment of experimental data and discussion
1. Tabulate the values of η, η/η_0 and $\eta_{sp} = (\eta/\eta_0 - 1)$ for each sol.
2. Explain the significance of discussing the viscosity of colloidal solutions in terms of the specific viscosity.
3. Why has the hydrophobic sol the lowest viscosity? Explain how the order of increasing viscosity of the hydrophilic sols provides information about the shapes of the different molecules in solution.
4. Outline other methods that are suitable for determining the shape and size of macromolecules in solution.

6   'Colloidal electrolytes'

'Colloidal electrolytes', eg. soaps and salts of long-chain sulphonic acids, in aqueous solution below a certain concentration, 'the critical micelle concentration' (cmc.), behave as typical 1:1 electrolytes. Above this concentration, however, the metal ions are in true solution while the anions exist largely in the form of ionic aggregates, 'micelles', in which all the hydrophobic hydrocarbon groups are oriented towards the centre with the hydrophilic polar groups directed outwards. At the cmc. many properties of these solutions show an abrupt change, eg. electrical conductivity, transport number, surface tension, osmotic coefficient. The variation of conductivity with concentration has been widely used to determine the cmc.

There is, however, another property of these micelles which can be used for the determination of the critical concentration. The phenomenon of 'solubilization' is exhibited by solutions containing micelles; ie. they possess the ability to dissolve appreciable quantities of substances, eg. azobenzene, which are insoluble in water. These substances must dissolve and be incorporated in the hydrophobic interior of the micelle. This solubilization shows a marked increase above the cmc. If dyes with different absorption spectra in water and organic solvents are used, then it is possible to determine the cmc. by direct titration.

Experiment 10.11   Determine the critical micelle concentration of sodium lauryl sulphate from a study of the surface tension of aqueous solutions

Requirements: Capillary rise surface tension apparatus (Expt. 2.7), or du Nouy balance (Expt. 2.11), 0.1 mol dm^{-3} solution of pure sodium lauryl sulphate (sodium dodecyl sulphate).

Procedure: Dilute the stock solution to give solutions of concentrations: 5 and 1×10^{-2}; 9, 8, 7, 6, 5, 4, 3, 2 and 1×10^{-3} and 5×10^{-4} mol dm^{-3}. All glassware must be scrupulously clean and should be steamed out before use.

Using the appropriate technique, determine the surface tension of water and each of the solutions of sodium lauryl sulphate; record the temperature of measurement. Use water to calibrate the apparatus. Determine the density of the solutions of higher concentration (Expt. 2.2).

Treatment of experimental data and discussion

1.   Plot the graph of surface tension against concentration and hence estimate the critical micelle concentration from the break in the curve.
2.   Explain why there is a sharp break in the γ-c curve.
3.   Explain the effect on the critical micelle concentration of (a) adding a neutral salt, (b) adding alcohol, (c) increasing the temperature and (d) increasing the chain length.
4.   Why is it essential that all the apparatus be scrupulously clean and the sodium lauryl sulphate as pure as possible?

Experiment 10.12   Determine the critical micelle concentration of sodium lauryl sulphate from a study of the conductivity of aqueous solutions

Requirements: Conductivity bridge and bottle-type cell, thermostat at 25.00 ± 0.01 °C. Pure sodium lauryl sulphate.

Procedure: Following the operational instructions for the bridge determine the cell constant, J, of the conductivity cell (Expt. 9.16).

Prepare solutions of sodium lauryl sulphate, in conductivity water, of concentrations 2 and 1×10^{-2}; 9, 8, 7, 6, 5, 4, 3, 2 and 1×10^{-3} mol dm^{-3}. Taking the normal precautions measure the conductance of duplicate samples of water and each of the solutions.

Treatment of experimental data and discussion

1.   From the tabulated values of the conductance, calculate the conductivity, κ, and the molar conductivity, Λ (equation (9.39)). Plot graphs of κ against c and Λ against $\sqrt{c}$.
2.   From the graphs estimate the critical micelle concentration.
3.   Explain why the sodium lauryl sulphate solution approximately obeys the

limiting Debye–Hückel relationship at low concentrations but changes abruptly at the micelle concentration.
4. Give an account of the accepted structure of a micelle.
5. How is the critical micelle concentration affected by (a) the presence of a neutral salt, (b) the presence of acetone and (c) an increase in temperature?

Experiment 10.13   Determine the critical concentration for the formation of micelles from the spectral behaviour of a dye

Theory: In a solution of a surfactant at a concentration above its cmc. a dyestuff has a visible absorption spectrum similar to that in an organic solvent; below the cmc. the absorption spectrum is markedly different. This change indicates that the dye is solubilized by the micelle. Pinacyanol chloride (M_r = 388.5) has an absorption spectrum in water with bands at 550 and 600 nm. These disappear with the addition of minute amounts of sodium lauryl sulphate with the formation of a new band at 485 nm. At higher concentrations this short wavelength band begins to diminish, and simultaneously, the intensity of bands at 560–570 nm and at 600–610 nm increases. These latter bands are characteristic of the spectrum of the dye in organic solvents.

Requirements: Recording spectrophotometer (thermostatted cells are an advantage), 1-cm glass cells, solid pinacyanol chloride (1,1'-diethyl-2,2'-carbocyanine chloride), pure sodium lauryl sulphate.

Procedure: Prepare a stock solution of pinacyanol chloride in water of concentration 1×10^{-4} mol dm^{-3}, and stock solutions of 1×10^{-1} and 1×10^{-2} mol dm^{-3} of sodium lauryl sulphate.

Make up the following solutions each containing 2×10^{-5} mol dm^{-3} pinacyanol chloride (2 cm^3 of stock dye solution in 10 cm^3 of solution): 9, 8, 7, 6.5, 6, 5.5, 5, 4, 3, 2 and 1×10^{-3} and 5×10^{-4} mol dm^{-3} of sodium lauryl sulphate. Note the colour of the highest and lowest concentration of sodium lauryl sulphate solution.

Record the absorption spectrum for each of these solutions over the wavelength range 450–670 nm, preferably controlling the temperature of the solution during the scanning.

Plot the absorption spectra of (i) a solution of the dye in acetone, and (ii) a solution of the dye in 10^{-4} mol dm^{-3} KCl.

Repeat the experiment at another temperature.

Repeat the experiment this time in the presence of a fixed concentration of potassium chloride (0.04 mol dm^{-3}).

Treatment of experimental data and discussion
1. From the absorption spectra plot the maximum absorbance values (at 562 and 608 nm) against the concentration of sodium lauryl sulphate. Deduce the critical micelle concentration from the point at which the graphs show an abrupt change.

2. Explain the results of the determination at a different temperature and in the presence of potassium chloride.

3. Account for the different absorption spectra of the dye in water, in the surfactant solutions below and above the cmc., in acetone and in KCl solution.

11

Chromatography and paper electrophoresis

Chromatography is a technique for the separation of mixtures and the purification of compounds. While it is not strictly an analytical method, it is a device whereby mixtures of molecular species can be separated into fractions, that are more suitable than the original mixture for the application of analytical methods.

1 Adsorption chromatography

Adsorption chromatography depends on the selective adsorption of material from solution by powdered solids, eg. alumina. The components first adsorbed will be those with the greatest affinity for the surface of the solid particle; substances less strongly adsorbed will accumulate below this first 'zone', etc. The general procedure consists first of the preparation of a packed column of suitable adsorbent in the solvent to be used for dissolving the mixture. An aliquot of the sample is then added to the top of the column, and the column developed by washing through with the solvent. As the solvent passes through, the various components will pass through at different rates, and can, under suitable conditions, be washed out, or eluted, and collected as separate fractions.

Various materials can be used as adsorbents, eg. (in order of decreasing adsorbent power) activated charcoal, magnesium silicate, aluminium oxide, silica gel, calcium carbonate, calcium sulphate, Fuller's earth, sugar, starch and powdered cellulose. The solvent selected is generally limited by the solubility characteristics of the sample, since it is advisable to apply only small volumes (5–25 cm^3) of the sample to small columns. In general, the adsorptive power of the adsorbent is reduced when very polar solvents are used. The order of solvents, from the most elutive is: water, alcohols, acetone, ether, chloroform, benzene, toluene, carbon tetrachloride, cyclohexane, hexane (Trappe's Series). In consequence of this, the weight of sample, type of adsorbent, length of column, etc., may have to be adjusted to give suitable separation of the mixture. The rate of flow of the solvent should be moderately slow, 50–100 cm^3 per hour for a 1.5-cm diameter column.

A standard column arrangement is shown in Fig. 11.1.

Thin layer chromatography (TLC) is a special example of adsorption chromatography, in which thin layers of adsorbents such as silica and alumina are

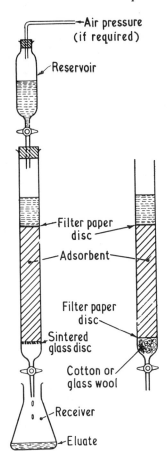

Fig. 11.1 Apparatus for column chromatography

spread on glass plates. The adsorbent layer, 0.25 to 1 mm thick, is held firmly on the glass surface with binders, such as plaster of Paris. Spots of the materials to be separated are made at one end of the plate and the chromatogram developed by standing the plate in a dish of solvent in an enclosed vessel. The eluent rises through capillary action and a rising chromatogram is obtained. After the liquid has risen through 8–10 cm (20–40 min) the plate is taken out of the developer and dried. If necessary the separated spots (components) can be made visible with a suitable reagent. The spots can also be scraped off, extracted and subjected to further analysis.

Since the basic principles of TLC and column adsorption chromatography are the same, TLC provides a rapid means of establishing the conditions (ie. correct solvent and adsorbent) for the separation of a mixture. Using large plates (40 × 40 cm or 40 × 100 cm) it can be used as an alternative to column chromatography for larger scale separations.

Applicators, for applying the layers of adsorbent to the glass plates are available commercially. A slurry of the adsorbent, in water, is poured over the plates, spread and smoothed by a glass straight edge and the whole dried at 50–70 °C. The covered glasses are then removed from the block and stored in a desiccator until required. The consistency of slurry which will allow uniform spreading must be found by experiment.

Another technique frequently used in adsorption chromatography is displacement development in which the column is washed with a solution containing a solute which is more tightly held by the adsorbent than the components of the sample under test. This type is best illustrated by *ion-exchange* columns. The organic ion-exchangers are resins usually made by the copolymerization of styrene and divinyl benzene. Suitable groups, on which the ion-exchange principle depends, can be introduced before or after condensation. Various degrees of cross linking are available for different purposes. Modified celluloses with ion-exchange properties are also available.

The *cation exchange* resins, containing $-SO_3H$ or $-COOH$ groups designated strong and weak acid cation resins respectively, can be regarded as insoluble acids. *Anion exchange resins* contain quaternary ammonium and amino groups; and are designated strong and weak base anion exchange resins respectively.

The reaction between a cation resin and the sodium salt of a carboxylic acid can be represented:

$$R . SO_3H + CH_3COONa \longrightarrow R . SO_3Na + CH_3COOH$$

resulting in the formation of the sodium salt of the exchanger and the free acid. From the principle of the ion-exchange process, it is obvious that the adsorbed component cannot be removed merely by washing with the solvent. In this example a 10% hydrochloric acid solution is required; this 'strips' the column regenerating the resin in the H^+ form. When two or more components are adsorbed, development of the column results in the removal of the components one by one, with a clear solute free zone between the components.

Experiment 11.1   Study the adsorption of various dyes on alumina

Theory: The dyes chosen are those given in Brockmann's standard test for alumina, in which, according to the grade, two dyes are separated, either both remaining on the column as separate zones or one remaining on the column and the other passing through in the eluate.

Requirements: Five chromatography tubes (20 cm × 1.5 cm diameter), alumina, benzene, petroleum ether (bp. 60–80 °C); the following dyes: (*a*) azobenzene, (*b*) *p*-methoxyazobenzene, (*c*) Sudan yellow, (*d*) Sudan red, (*e*) *p*-aminoazobenzene, and (*f*) *p*-hydroxyazobenzene.

Procedure: Prepare five columns (15 cm length) of alumina in 1:4 benzene-petroleum ether mixture and cover each with a filter paper disc. Prepare the following solutions containing 20 mg of each dye dissolved in 10 cm³ benzene: a + b, b + c, c + d, d + e and e + f; dilute each solution to 50 cm³ with petroleum ether.

To the first column, when the surface is about to become dry, add 10 cm^3 of solution a + b and, when the last drop of this is passing into the column, add 20 cm^3 of 1:4 benzene-petroleum ether mixture. Collect the eluate from the bottom of the tube. Repeat for the remaining solutions, each on a fresh column.

Treatment of experimental data and discussion

1. Record the positions of the bands and colour of the eluates.
2. Compare your observations with those of the standard test and hence classify the sample of alumina.

Standard results:

Grade	1 (highest activity)	2		3		4		5 (lowest activity)
Top band	b	b	c	c	d	d	e	f
Lower band	a	—	b	—	c	—	d	e
In eluate	—	a	—	b	—	c	—	—

3. Explain why it is important (*a*) that the column of adsorbent should be tightly packed and (*b*) that the column should never be allowed to dry.

Experiment 11.2   Separate leaf pigments on a column of several adsorbents and plot their absorption spectra

Theory: The separation of leaf pigments was originally accomplished by Tswett using a calcium carbonate column; better separation can be achieved using three adsorbents: sucrose, calcium carbonate and alumina.

Requirements: Chromatography tube (20 cm × 1.5 cm diameter) with no sintered disc or tap, recording spectrophotometer and 1 cm glass cells, fresh spinach leaves, powdered sucrose (icing sugar), calcium carbonate, alumina, benzene, petroleum ether (bp. 60–80 °C), methanol.

Procedure: Immerse three or four leaves in a mixture of 45 cm^3 petroleum ether, 5 cm^3 benzene and 15 cm^3 methanol in a conical flask for 1 hour. Remove the almost white residue by filtration at the pump. Wash the filtrate four times with 50-cm^3 aliquots of water to remove the methanol, separating the layers in a funnel. Dry the solution over the minimum amount of anhydrous sodium sulphate; filter and concentrate it, under reduced pressure, to 5 cm^3.

Pack the column dry on a cotton-wool support, with alumina (4 cm), calcium carbonate (6 cm) and sucrose (6 cm), inserting a filter paper disc between the adsorbents. Each powder (sieved in a 80–100 mesh) must be gently tamped down using a tight-fitting glass rod. Run in petroleum ether from a dropping funnel, with gentle suction at the bottom of the tube, if necessary.

When the top of the column is almost dry, add the solution of leaf extract and develop the chromatogram with a 1:4 mixture of benzene and petroleum ether. Zones are soon formed: upper yellowish layer of chlorophyll *b* (on sucrose), followed by bluish-green chlorophyll *a* (on sucrose or calcium carbonate), yellow xanthophylls (on calcium carbonate) and orange carotene (on alumina). Lyocopene is held at the top of the alumina layer. If the flow rate is unduly slow, apply air pressure (in preference to suction) to the top of the funnel. When fully developed, wash the column with petroleum ether and dry by suction in a stream of carbon dioxide. Using a glass rod, push out the column on to a sheet of paper and cut out the coloured zones. Elute the pigment from each of these with 10 cm³ ether containing 0.2 cm³ methanol. Filter the solutions and plot the absorption spectrum of each.

Treatment of experimental data and discussion
1.   Identify the various components from the absorption spectra.
2.   Why is it necessary to dry the column in the absence of oxygen?
3.   Explain why it is preferable to apply pressure to the top of the column rather than to apply suction.
4.   Discuss the importance of this method for the separation of naturally occurring compounds.

Experiment 11.3   Separate a mixture of *o*- and *p*-nitroanilines on an alumina column

Requirements: Chromatography tube (30 cm × 3 cm diameter) fitted with sintered glass disc and tap, alumina, benzene, *o*- and *p*-nitroanilines.

Procedure: Fill the tube with a slurry of alumina in benzene to give a column about 20 cm long; cover with a filter paper disc. Weigh out approximately 0.25 g of both *o*- and *p*-nitroanilines and dissolve in 30 cm³ benzene. Add this solution to the top of the column, and, when it is all on the adsorbent, develop the chromatogram by running in benzene from the funnel. Two yellow bands form and travel down the column; continue washing until the eluates from both bands have been collected separately. If available, a fraction collector (5-cm³ fractions), can be used; this will give a series of tubes for each band, showing the distribution of the amount of component collecting. Concentrate the two solutions, crystallize, weigh and identify the pure compounds.

Experiment 11.4   Demonstrate the application of thin layer chromatography to the separation of azobenzenes

Requirements: Glass plates and commercial applicator, silica gel for TLC, covered tanks to take the microscope slides vertically, azobenzene, *p*-methoxyazobenzene, Sudan yellow, Sudan red, *p*-aminoazobenzene and *p*-hydroxyazobenzene, benzene.

Procedure: Prepare several plates coated with silica gel (10 g in 10 cm^3 water), and dry. Prepare solutions of each of the dyes in benzene (20 mg in 10 cm^3) and suitable mixtures of two or three of the dyes. About 1 cm from the end of a plate, make 3 equally spaced marks on the adsorbent. On the centre mark, apply by means of a drawn-out melting point tube, a small sample of one of the mixtures. (*NB.* the spot must be kept as small as possible, drying by a current of hot air, if necessary.) On the marks on either side apply similar spots of the reference solutions. Similarly prepare the remaining slides with the other solutions and mixtures. Place each plate (spots at the bottom) in a tank, containing benzene to a depth of 0.5 cm, replace the cover to prevent evaporation of the solvent. Allow the benzene to flow up the plates until the solvent front nearly reaches the top. Remove the plates, mark the positions of the fronts and calculate the R_F values (p. 288) of the various components in the single solutions and in the mixtures.

Discussion
1. Discuss the relative advantages of TLC and paper chromatography.
2. What methods are available for the location of spots of colourless substances.
3. Why is it necessary to keep the tank sealed during the development of the chromatogram?
4. Explain why the applied spots must be kept as small as possible.

Experiment 11.5   Convert the salt of an organic acid or base into the free acid or base respectively using an ion-exchange resin

Theory: Organic acids can easily be liberated from their salts by displacement with dilute hydrochloric acid when the free acid is insoluble in water, or volatile. Citric acid is very soluble in water and so is not easily obtained from its salts using the normal chemical means. Using an ion-exchange column, this conversion is readily achieved.

Requirements: Chromatography tube with sintered glass disc and tap (20 cm × 1.5 cm diameter), cation exchange resin (eg. Zeo-Karb 225), anion exchange resin (eg. De-Acidite FF), 3 mol dm^{-3} hydrochloric acid and 0.5 mol dm^{-3} sodium carbonate solutions, sodium citrate, and aniline hydrochloride.

Procedure: Fill the column with a slurry of the cation exchange resin in distilled water and allow it to settle to give a column 15 cm long. Allow most of the water to drain off, and pass 100 cm^3 of hydrochloric acid through the column followed by excess distilled water, until the washings are neutral to methyl red. (The resin is now in the H$^+$ form.) Turn off the tap, making sure that the resin is covered with water; *never* allow a column to become dry. Now add 100 cm^3 of a solution of sodium citrate containing 0.05 g of the salt and allow it to pass down the column at a rate of 5 cm^3 a min (adjust the tap to give this rate). Collect the eluate, wash the column with a further 100 cm^3 of water and add the washings to the eluate. Evaporate the solution of citric acid so formed to small bulk and allow it to crystallize. Confirm its purity.

Regenerate the column, ie. displace the sodium ions by washing off with hydrochloric acid followed by excess water.

Convert the aniline hydrochloride to free aniline in solution by passing a solution of the salt (0.5 g in 100 cm^3) through a column of anion exchange resin, which has been regenerated with sodium carbonate solution and well washed.

Discussion

1. Explain how the method could be made quantitative to estimate the concentration of sodium chloride in solution.

2. Explain how a mixture of anionic and cationic exchange resins can be used to produced de-ionized water.

Experiment 11.6   Study the separation of the chromium(III)thiocyanate cationic complexes by the ion-exchange technique

Theory: Bjerrum has demonstrated the existence of the complexes $Cr(CNS)_n(H_2O)_{6-n}^{+3-n}$ where n can have any value from 0—6. Generally the equilibrium constants for the formation of the complexes, in metal-anion complex ion systems, are such that one of the intermediate complexes cannot predominate in a solution at equilibrium. In systems, such as the one under study, in which equilibrium is only slowly established, it is possible, in principle, to prepare solutions which contain a single intermediate complex. Since the charges on the various complexes differ, this separation may be achieved using a cationic ion-exchange resin. The ions to be separated are those in which $n = 0$, 1 and 2; ie. the hydrated ion $Cr(H_2O)_6^{+++}$, $Cr(CNS)(H_2O)_5^{++}$ and $Cr(CNS)_2(H_2O)_4^{+}$ respectively; these are eluted from the ion-exchange resin with increasing concentrations of perchloric acid.

Requirements: Spectrophotometer, with 1-cm glass cells, chromatography column, 30 × 1 cm diam. (Fig. 11.2), cationic exchange resin, eg. Dowex 50 or Zeo-Karb 225, chromium (III) nitrate, $Cr(NO_3)_3 \cdot 9H_2O$, 0.15, 1.0 and 5.0 mol dm^{-3} solutions of perchloric acid, 2 mol dm^3 sodium hydroxide solution, 100 volume hydrogen peroxide, 0.1 mol dm^{-3} iron(III)ammonium sulphate, 0.1 mol dm^{-3} iron (III)-nitrate; 10% potassium persulphate solution and standard solutions: 0.1 mol dm^{-3} silver nitrate, 0.3 mol dm^{-3} potassium thiocyanate, and 0.017 mol dm^{-3} potassium dichromate.

Procedure: Owing to the varying extent of hydration and deliquescence of samples of chromium(III)nitrate, solutions of this compound should be standardized. To 25 cm^3 of approximately 0.1 mol dm^{-3} chromium nitrate solution, add 20 cm^3 of approximately 0.1 mol dm^{-3} silver nitrate solution (catalyst) and 50 cm^3 of 10% potassium persulphate solution. Boil gently for 20 min, cool and dilute to 250 cm^3. To a 50-cm^3 aliquot of this solution add 50 cm^3 0.1 mol dm^{-3} iron(II)ammonium sulphate solution, 200 cm^3 of 1 mol dm^{-3} sulphuric acid and 0.5 cm^3 of N-phenyl-anthranilic acid as indicator. Titrate the excess iron(II) salt with standard 0.017 mol dm^{-3} potassium dichromate solution (colour change green to violet-red).

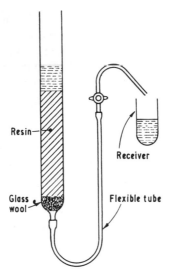

Fig. 11.2   Ion exchange column arrangement

Prepare a 0.3 mol dm^{-3} solution of potassium thiocyanate and standardize with standard silver nitrate solution.

Prepare the mixture of complex ions by heating, under reflux at 95 °C, a 1:1 mixture of these two standardized solutions (final stoichiometric composition 0.05 mol dm^{-3} chromium nitrate and 0.15 mol dm^{-3} potassium thiocyanate), for at least 3 hours.

Prepare the column by pouring in a slurry of resin in water to give a bed 20—25 cm high. Convert the resin to the H$^+$ form by washing first with 2 mol dm^{-3} hydrochloric acid (200 cm^3) and then with distilled water until no chloride ion is detectable in the eluate. Without allowing the surface to become dry, add an aliquot (50 cm^3) of the solution of the complex to the column, and wash thoroughly with water (flow rate 2—3 cm^3 per min) until the eluate is colourless. Reject this solution which contains neutral and anionic complexes and free acid. Elute the cationic complexes, retained on the column as a darker band near the top, with perchloric acid of increasing concentration (flow rate 2—3 cm^3 per min) and collect the eluate in 25-cm^3 fractions. Suitable volumes and concentrations are: 200 cm^3 of 0.15 mol dm^{-3}; 300 cm^3 of 1.0 mol dm^{-3} and 200 cm^3 of 5.0 mol dm^{-3} acid.

Follow the elution spectrophotometrically at 410 nm, where all the solutions show maximum absorption. Construct a graph of absorbance of eluate against the total volume and demonstrate the existence of three distinct peaks corresponding to the three ionic species. Plot the complete absorption spectra of one sample selected from each of the three series over the range 300—700 nm.

The separate fractions must now be analysed to establish the nature of the various complexes (ie. the ratio thiocyanate/chromium).

Chromium: The complex ion in solution or the standard chromium nitrate solution, used to construct the calibration curve, is converted to chromate which is estimated spectrophotometrically. To 10, 8, 6, 4, 2 and 0 cm³ of 0.005 mol dm⁻³ chromium nitrate solution add 10 cm³ of 2 mol dm⁻³ sodium hydroxide solution and 10 drops of 100 volume hydrogen peroxide solution. When the excess peroxide has decomposed, make each solution up to 100 cm³ with distilled water and determine the absorbance in the region of maximum absorption of the chromate ion (400–500 nm) against the blank. Treat 5 or 10-cm³ samples of the separate fractions similarly, and, measure the absorbance.

Thiocyanate is converted to the red iron(III) thiocyanate complex for spectrophotometric analysis. To 10, 8, 6, 4, 2 cm³ of 0.015 mol dm⁻³ potassium thiocyanate solution add 10 cm³ of 2 mol dm⁻³ sodium hydroxide and acidify with perchloric acid. Add 10 cm³ of a solution of iron(III)nitrate and dilute to 50 cm³ with distilled water. Determine the absorbance of each solution against a blank, in which the iron(III) salt is omitted, at 402, 460 and 490 nm. Treat 10-cm³ samples of the separate fractions similarly.

Treatment of experimental data and discussion

1. Plot the calibration (Beer–Lambert) curve of absorbance against the concentration of Cr^{+++} and hence determine the concentration of Cr^{+++} in the various fractions from the column.
2. Plot calibration curves of the absorbance against the concentration of thiocyanate at the three wavelengths and hence determine the concentration of thiocyanate in each fraction.
3. Calculate the ratio thiocyanate/chromium for each fraction and discuss the order of elution of the complexes in terms of the charges carried.

2   Partition chromatography

(a) Counter current distribution

When a solute soluble in two immiscible solvents is shaken with a mixture of the two, partition occurs, and the solute is distributed between the two solvents. The extent of partition depends on the properties of the solute and solvents and is expressed as the partition coefficient (p. 222):

$$D = \frac{\text{Concentration in solvent A}}{\text{Concentration in solvent B}}$$

The technique of counter current distribution is based on the different partitioning of the different components in a mixture in the two phases of a system of two immiscible liquids. One of these phases, A, remains stationary and is divided over a series of extraction tubes or cells (separating funnels in Expt. 11.7). The other phase, B, in each cell is moved after equilibration with phase A by simple transfer to the adjoining cell while the first cell is recharged with a fresh sample of phase B. This procedure of transfer, equilibration, and transfer of one of the phases, is repeated many times. In the large automated machines, the heavier phase remains stationary and the lighter phase is transferred; using funnels for equilibration, separation and transfer of the heavier phase, is more convenient.

If the volumes of the two phases are equal and if the fractions of the solute in each solvent at equilibrium are X and Y then $X + Y = 1$. The partition coefficient, D, defined as the concentration of solute in the lower phase divided by the concentration in the upper phase (since it is the lower phase that is transferred) is given by $D = X/Y$. The distribution of the solute between the two phases, for a series of n transfers, represents the simplest of all counter current procedures, and requires only the direct application of the binomial expansion $(X + Y)^n$ where X is the fraction transferred in the lower layer [ie. $D/(D + 1)$] and Y the amount remaining [ie. $1/(D + 1)$]. The expansion is thus:

$$\left[\frac{D}{D+1} + \frac{D}{D+1} \right]^n .$$

for which the rth term is:

$$T_{n,\, r} = \frac{n!}{r!(n-r)!} \left(\frac{1}{D+1} \right)^n D^r \tag{11.1}$$

$T_{n,\, r}$ is the fraction of the original solute present in the rth tube in a distribution of n transfers ($r = 0, 1, 2 \ldots n$; ie. $(n + 1)$ values). From equation (11.1), for given values of D, the total amount of material in each tube can be calculated and plotted against the tube number to give a distribution curve. As an extension of this, it is possible to calculate the distribution of two different substances, with different values of D in the two solvents, and hence predict the conditions for their separation in a mixture.

The position of the maximum after n transfers can be estimated approximately by the formula:

$$N = n \left(\frac{D}{D+1} \right) \tag{11.2}$$

this equation is only exact if $D = 1$ or if there is an infinite number of transfers. For values of $n > 20$ it gives a reasonable estimate of the position of the maximum. The fraction of solute (y) present in each funnel, near the maximum, can be calculated from the equation:

$$y = \left\{ \frac{(D + 1)^2}{2\pi n D} \right\}^{\frac{1}{2}} \exp \left(\frac{-x^2(D + 1)^2}{2Dn} \right)$$

or

$$\log y = \log \left\{ \frac{(D + 1)^2}{2\pi n D} \right\}^{\frac{1}{2}} - \frac{(D + 1)^2 x^2}{2.303 \times 2Dn} \tag{11.3}$$

where x is the distance, in units, from the maximum tube (ie. N in equation (11.2)). Thus by substituting values of x from 0 to 7 or 8, the fraction, and hence total amount of solute, present in each tube around the maximum, can be determined.

Experiment 11.7   Study the separation of bromocresol green (BCG) and phenol red (PR) by a counter current distribution method

Theory: The principles of counter current distribution described above are illustrated by the separation of BCG and PR between butan-1-ol and sodium carbonate solution. While the full experiment requires 29 separating funnels, the basic principles of separation can be readily demonstrated with 6. The equations given previously cannot be applied to such a small number of transfers.

Requirements: Twenty-nine stoppered separating funnels (125 cm^3), spectrophotometer and 1 cm-glass cells. Solid bromocresol green, solid phenol red, butan-1-ol (butanol), 0.05 mol dm^{-3} sodium carbonate solution.

Procedure: Shake together equal volumes of butanol and 0.05 mol dm^{-3} sodium carbonate solution in a large separating funnel to produce saturation of each solvent in the other. Separate the two phases and use as stock solutions (about 1 dm^3 of each is required).

   (*a*) *Preparation of calibration curves for the dyes:* Two calibration curves are required for each dye, one in butanol and the other in the carbonate solution. Dissolve about 2 mg of each dye in 100 cm^3 of each solvent (20 μg cm^{-3}) and from these prepare by dilution standard solutions of known concentration in the range 0.5–15 μg cm^{-3}. Plot the absorption spectrum for an intermediate concentration of each dye in each solvent, BCG from 450–700 nm and PR from 450–600 nm. Select the wavelengths of maximum absorption for each dye solution (approx. values BCG in carbonate 620 nm, in butanol 630 nm; PR in carbonate 560 nm, in butanol 570 nm) and determine the absorbances of the remaining solutions in the series.

   (*b*) *Determination of the partition coefficients:* Prepare stock solutions of PR; 3–4 mg PR in 100 cm^3 sodium carbonate (saturated with butanol); and BCG 4–5 mg BCG in 100 cm^3 butanol (saturated with carbonate).

   (i) Phenol Red: Transfer 20 cm^3 of stock PR solution to a separating funnel containing 20 cm^3 of butanol. Equilibrate for 3 min and carefully separate the two layers. Measure the absorbance of each solution at the appropriate wavelength of maximum absorption (dilute the carbonate solution 10 times).

   (ii) Bromocresol green: Determine the partition coefficient as for PR, using 20 cm^3 of stock solution of BCG and 20 cm^3 of carbonate for partition.

   (*c*) *Counter current distribution:* Set up the funnels (numbered 0–28) on a suitable frame and add 20 cm^3 of butanol (saturated with carbonate) to funnels 1–28. Pipette 20 cm^3 of each stock solution, PR in carbonate, and BCG in butanol, into funnel 0 and shake vigorously for 30 s. After complete separation of the layers, transfer the lower aqueous layer, as carefully as possible, from funnel 0 to funnel 1, and add 20 cm^3 of fresh carbonate to funnel 0. Shake both funnels for 30 s and transfer the lower layer of 1 to 2 and of 0 to 1 and add 20 cm^3 of fresh carbonate to 0. Repeat this operation until all 29 funnels are in operation. At this stage the dyes are well separated, the PR has migrated to the lower phases of the funnels 20–29 while the BCG has remained in the upper phases of the first ones. Measure the absorbance of a sample of each phase (diluted if necessary), of the funnels containing the dye, at the appropriate wavelength maximum.

Treatment of experimental data and discussion

1. Construct a calibration curve for each dye in each solvent (four in all) and show that the Beer–Lambert law is valid.
2. From the calibration curves, calculate the concentration and amount of dye in each phase; check that the total amount of dye in both solutions is the same as that added originally. Since the lower phase is to be transferred later

   $$D = \frac{c_{aq}}{c_{but}} ; \text{hence calculate } D(D_{PR} \sim 9\text{--}10, D_{BCG} \sim 2).$$

3. From the measured absorbances and the calibration curves, determine the concentration of each dye in each phase and hence, the total amount of dye in each funnel. Plot the experimentally determined diagram showing the distribution of the two dyes in the various funnels.
4. From the experimentally determined values of D for the two dyes, calculate the approximate position of the maximum for each (equation (11.2)), and, from equation (11.3) calculate the fraction, and hence the total amount, of each dye in the funnels on either side of the maximum. Plot the theoretical distribution curves on the same diagram as the experimental curves and comment on any difference.
5. Discuss the advantages of this method of separation of two or more components and list the factors that govern the choice of the appropriate solvents.

(b) Paper partition chromatography

Partition chromatography is based on the different partition coefficients of two or more solutes between a stationary aqueous phase supported on a porous material and a mobile, flowing, solvent phase which is immiscible or only slightly miscible with water. The water is held adsorbed in a solid, eg. silica gel, and the other liquid in which the mixture is dissolved is slowly passed through the column. In this way, the solutes are subjected to an infinite number of partitions by flowing through the column and can be collected separately in the eluate.

In *filter-paper partition chromatography*, the stationary phase is the adsorbed water always present in the filter paper itself. Essentially the method (a micromethod) consists of applying a spot of the solution to the filter paper and drying off the solvent. The paper is then irrigated, with an organic solvent, the mobile phase, in an airtight container to prevent losses by evaporation. This is known as the development of the chromatogram. After flowing for the appropriate time, the paper is dried and the separated spots located by a suitable colour-producing reagent. The fundamental measurement in paper chromatography is the R_F value defined as:

$$R_F = \frac{\text{Distance travelled by component}}{\text{Distance travelled by solvent front}}$$

$$= \frac{\text{Distance from origin to centre of spot}}{\text{Distance from origin to solvent front}}$$

These ratios are useful but do not have an absolute value, the value depends on such factors as the dimensions and quality of the paper, the direction of flow (descending, ascending or radial), the nature of the solvent, the temperature, equilibration time of solvent and paper (if any) and the nature of the mixture to be chromatographed. R_F values of various compounds in a wide range of solvents are available in the literature.

Essentially, three different experimental procedures for one dimensional chromatography are available, depending on the direction of flow of the mobile phase.

(*a*) *Descending method:* This method is particularly useful for slow moving components (ie. with low R_F values) since the solvent can be allowed to drip off the end of the paper, thereby increasing its effective length.

(*b*) *Ascending method:* In this method, the solvent, contained in a tray in the bottom of the tank, flows up the paper by capillarity. In this technique, the volume of solvent is not important, since when the front reaches the top of the paper, chromatography ceases. It is, however, not advisable to leave a paper in this state for any length of time as diffusion of the components on the paper gives rise to diffuse spots on the developed chromatogram.

(*c*) *Radial or horizontal method:* This is a convenient method for the rapid separation of mixtures. In this, a filter paper held in the horizontal plane is fed with solvent from a wick in the centre. A spot of mixture applied at the centre will move out and produce concentric arcs which can be located.

Since a given substance has, in general, different R_F values in different solvents, confusion in the identification of substances with similar R_F values in a single solvent can be avoided using two-dimensional chromatography. In this

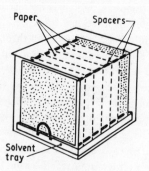

Fig. 11.3   Typical multisheet frame for ascending chromatography

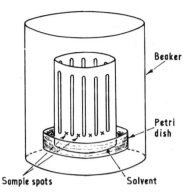

Fig. 11.4 Whatman CRL/1 paper and assembly

method the spot is applied near one corner of a square sheet of filter paper and chromatography is performed in one direction with a solvent. The paper is then dried to remove all the solvent, turned through a right angle and chromatography carried out in the second direction with a different solvent. The line of spots produced in the first solvent acts as a number of new origins for the second solvent. When colour location is complete, a series of spots is produced across the paper.

Application of the substance to paper. As in other chromatographic techniques, it is important that the paper should not be overloaded. The applied spots must not be too large (1 cm diameter is sufficient), so, if the solution is very dilute, several applications must be made, ensuring that the paper is well dried with a hot air blower between applications. The spots must be 2—3 cm apart on the paper, to prevent interference by overlapping.

The spots can be applied (i) with a platinum loop: a loop 2—3 mm diameter will hold 3—4 mm^3 and will deliver constant drops; it has the advantage that it can be easily cleaned by heating to redness in a flame; (ii) with a capillary tube; melting point tubes are satisfactory, using a fresh tube for each solution; (iii) with a micro-pipette calibrated to deliver various amounts from 5—50 mm^3.

Solvents. The solvent, ie. the mobile phase, for a particular separation is best selected from published R_F values. Two of the most useful solvent systems are: butan-1-ol/glacial acetic acid/water in the ratio 120/30/50, and 80% phenol/ethanol/water/0.880 ammonia solution 150/40/10/1.

Location of components. Unless the substances applied to the paper are coloured, their final position on the paper must be revealed by some suitable technique. Observation of the paper under ultraviolet light may reveal the position of certain substances (eg. purines); the incorporation of a radioactive isotope in the original substance aids location of the spots by scanning with counting equipment. The most widely used method is the production of a colour by reaction with a specific reagent. It is recommended that the paper should be dipped in the locating reagent, contained in a shallow tray, rather than the older technique of spraying the reagent on to the paper. Location reagents:

For *amino acids* (general and specific reagents):

Ninhydrin: 0.2% (w/v) ninhydrin (indane-trione hydrate) in acetone; after dipping, allow the acetone to evaporate and heat at 105 °C for 1–2 min. Most amino-acids give blue or purple colours; proline and hydroxyproline, yellow colours. The spots fade and their outlines should be marked in pencil.

Isatin: 0.2% (w/v) isatin in acetone; after dipping, allow the acetone to evaporate and heat at 105 °C for 2–3 min. Prolines give a blue colour, derivatives of phenylalanine and some S-containing amino-acids give a light blue colour. Reagent less sensitive than ninhydrin.

Ehrlich's reagent. Immediately before use, dissolve 1 g *p*-dimethyl-amino-benzaldehyde in 90 cm^3 acetone and 10 cm^3 concentrated hydrochloric acid. As the acetone dries, the spots given by tryptophan (purple), and citrulline and urea (yellow) appear. The colours are not permanent.

Sakaguchi reagent. Dip the paper through a 0.1% (w/v) solution of 8-hydroxy-quinoline solution in acetone. When dry, dip through a solution of 0.2 cm^3 bromine in 100 cm^3 of 0.5 mol dm^{-3} sodium hydroxide. Arginine and other guanidines give orange-red colours.

Diazo reagent. Dip the paper through a 1:1 mixture, prepared 5 min previously, of 1% (w/v) *p*-anisidine in ethanolic 0.11 mol dm^{-3} hydrochloric acid and 10% (v/v) amyl nitrite in ethanol. Develop the colour by holding the paper in ammonia vapour. Histidine and other iminazoles give dark red colours.

For *Sugars:*

Aniline phthalate. Dip the paper through a freshly prepared solution of 1.5 cm^3 aniline and 1.6 g phthalic acid in 95 cm^3 of acetone and 5 cm^3 water. When dry, warm the paper at 105 °C for 5–15 min to give maximum colour. Hexoses and hexosamines give brown, pentoses red, and rhamnose, salmon pink colours.

Phloroglucinol. Dip the paper in a solution of 0.2 g phloroglucinol in 80 cm^3 of 90% ethanol and 20 cm^3 of 25% (w/v) trichloracetic acid. When dry, warm the paper to 105 °C for maximum colour. Fructose gives brown, galactose faint brown, and pentoses, grey-green colours.

Experiment 11.8   Demonstrate the principles of filter paper partition chromatography

Requirements: Apparatus for paper chromatography; four samples of commercial writing inks, red, brown, green and blue: solutions of Congo red, phenol red and bromophenol blue (80 mg in 100 cm^3 ethanol).

Solvents: butan-1-ol/ethanol/2 mol dm^{-3} ammonia solution, ratio 60/20/20.

Ethanol/water/saturated aqueous solution of ammonium sulphate, ratio 15/75/10.

Procedure: Using the apparatus and techniques previously described, apply spots, individually and as mixtures, of the inks and the indicators to the filter paper. To ensure that the indicators are in the correct colour form for chromatography, hold the papers over a bottle of ammonia until the colours are characteristic of the basic form of the indicator. Flow, ie. develop, the papers with each solvent and periodically observe the separation.

Discussion
1.   Compare the separation achieved with the two different solvent mixtures.
2.   Determine the R_F values of the indicators.
3.   Discuss the part played by adsorption during the development of a chromatogram.
4.   Comment on the various components present in the inks.

Experiment 11.9   Study the separation of dicarboxylic acids by one-dimensional paper partition chromatography

Requirements: Multisheet frame and tank (Fig. 11.3), suitable sheets of Whatman No. 1 filter paper. Solution of the following acids (1 g acid in 100 cm^3 water): oxalic, malonic, tartaric, malic, succinic, fumaric, adipic, azelaic and sebacic. Solutions of bromocresol green (50 mg in 50 cm^3 ethanol) and lead acetate (2.5% w/v in water). Solvent: 25% aqueous ammonia/water/ethanol, ratio 5/15/80.

Procedure: On a pencil line 3 cm from the bottom of the paper apply spots of each acid (not exceeding 1 cm diameter) at 2–3 cm intervals. Prepare mixed spots by applying several solutions to the same origin, always ensuring that the previous application is dry. Stand the frame in a dish of solvent in an enclosed tank and develop for 16 hours. Mark the position of the solvent front on each sheet and dry the papers for 3 hours in air. Spray each sheet with the bromocresol green solution and dry in a stream of hot air. Finally spray each sheet with lead acetate solution; the acids will be revealed as yellow spots on a purple background.

Discussion
1.   Determine the R_F value for each acid when run singly and as a component in a mixture. Is the R_F value affected by the other components in a mixture?
2.   Is the separation affected by the presence of neutral salts in the mixture?

Experiment 11.10   Study the separation of amino-acids in mixtures by one- and two-dimensional filter paper chromatography

Requirements: Multisheet frame and tank (Fig. 11.3), suitable sheets of Whatman No. 1 filter paper, 0.01 mol dm^{-3} solutions of: glycine, arginine, histidine, proline, hydroxyproline, tyrosine, tryptophan, in 10% aqueous propan-2-ol with the addition of one drop of concentrated hydrochloric acid to aid solution. Mixture of amino-acids, egg albumen. Solvents: butan-1-ol, glacial acetic acid, water; and phenol, ethanol, water, ammonia (p. 289). All colour reagents for amino-acids (p. 289).

Procedure: Hydrolyse the egg albumen, or other protein, by refluxing 1 g of the protein for 24 hours with 10 cm^3 of constant boiling hydrochloric acid. Vacuum distil at 40 °C until dry, add water and repeat the distillation two or three times to remove excess acid. If the hydrolysate is dark brown, decolorize rapidly with the minimum amount of charcoal. It may be necessary to dilute this solution for chromatography.

Prepare the papers for chromatography, taking care that the papers are only handled at the edges, as follows:

1. One-dimensional chromatography: draw a pencil line 3 cm from the bottom of the paper and mark the position of the spots at 2-cm intervals. Prepare six sheets, each containing a spot of the pure amino-acids, the synthetic mixture of amino-acids and the protein hydrolysate.

2. Two-dimensional chromatography: mark the position of the spot in the right-hand corner, 3 cm from each edge. Prepare two papers, one containing a spot of the mixture and the other a spot of the hydrolysate.

Now mount three one-dimensional and the two two-dimensional papers in the frame and develop in the butanol-acetic acid mixture by ascending chromatography for 15−17 hours. Make sure that the tank is airtight, and that it is shielded from draughts. Remove the frame from the tray; mark the position of the solvent front on both edges of each paper; allow excess solvent to drip off and dry the papers (in the frame) by a jet of warm compressed air.

Remove the three one-dimensional papers, turn the frame through a right angle and insert the three remaining, unused, one-dimensional papers. As before, flow the chromatograms by the ascending method, using the phenol solvent for 17 hours. When complete, take the frame out and dry the papers, making sure that all the phenol has been removed; this takes several hours. The paper must, on no account, be heated above 40 °C, as phenol attacks some amino-acids which, in consequence, will not be revealed on the chromatogram. Roll each paper into a small cylinder and dip in a measuring cylinder containing acetone. Hang the paper in air to dry. This treatment removes any residual phenol.

The chromatograms now have to be treated with suitable colour-producing reagents and it is suggested that the multiple dip technique be applied. In this, the papers are dipped first in ninhydrin to reveal all amino-acids and then in a second and/or third reagent to reveal the location of specific acids. The order of dipping in these several reagents is important. The spots must be outlined with pencil as they are formed; for, on subsequent treatment, only certain spots will remain, the other disappearing.

Dip the two-dimensional papers only in ninhydrin and the one-dimensional ones (1, 2, 3, from first solvent, 1′, 2′, 3′, from second solvent) as follows:

1 1′   Ninhydrin, followed by Ehrlich, followed by Diazo reagent.
2 2′   Ninhydrin, followed by Sakaguchi reagent.
3 3′   Isatin, followed by Ehrlich, followed by Diazo reagent.

Discussion

1. Mark and identify as many spots as possible in the protein hydrolysate. Record all the R_F values of the amino-acids in the different solvents.
2. Explain the advantages of two-dimensional chromatograms.
3. Why is it necessary to remove the excess acid from the protein hydrolysate?
4. Explain why the tank must be airtight and shielded from draughts.
5. Give an account of the advantages of partition chromatography in the rapid analysis of biological fluids, eg. urine.

6.   Explain how the method could be adapted for the quantitative determination of the amino-acids in a mixture and indicate the accuracy that might be expected.

Experiment 11.11   Demonstrate the separation of inorganic ions by filter paper chromatography

Theory: The solubility of the inorganic compound in the organic solvent will largely determine how far the particular compound will move from the origin. Very soluble compounds will move with, or closely behind, the front; less soluble compounds will move to intermediate positions, while the insoluble compounds will not move at all. The solubility of some inorganic compounds in organic solvents is a result of the formation of a soluble complex with the solvent itself. This will have a marked effect on the R_F value. In some solvents, a complex forming reagent is deliberately added, thereby changing the extent of separation.

Requirements: Ascending chromatography tank (technique using Whatman CRL/1 paper, Fig. 11.4, is suitable). Aqueous solutions of nitrates of metals, sodium or potassium salts of anions; concentrations not exceeding 5% (w/v). Solvents and reagents given under procedure.

Procedure: Study the separation of the following ions:

(*a*) Iron, copper, cobalt and nickel (Whatman No. 41 paper), solvent: acetone/concentrated hydrochloric acid/water (87/8/5). Dry the paper in warm air, spray with rubeanic acid (0.1 g dissolved in 60 cm³ ethanol and 40 cm³ water), dry and expose to ammonia vapour. Colours: Iron—grey, copper—deep grey, cobalt—orange, nickel—blue.

(*b*) Sodium, potassium and lithium (Whatman No. 1 paper), solvent: absolute methanol. Dry the paper and dip in 2% (w/v) alcoholic solution of violuric acid. Colours: sodium and potassium—yellow brown, lithium—violet.

(*c*) Aluminium and zinc (Whatman No. 1 paper), solvent: tert.-butanol/ acetone/3 mol dm⁻³ hydrochloric acid (40/40/20). Dry the paper and dip in 2% (w/v) alcoholic solution of alizarn. Colours: aluminium—red, zinc—violet.

(*d*) Nickel, manganese, cobalt and iron(III) (Whatman No. 1 paper), solvent: acetone 6 mol dm⁻³ hydrochloric acid (90/10). Dry the paper and dip in ammoniacal solution of dimethylglyoxime. Colours: nickel—red, manganese—red brown, cobalt—dark brown and iron—brown.

(*e*) Chloride, bromide and iodide (Whatman No. 1 paper), solvent: butanol/ pyridine/1.5 mol dm⁻³ ammonia solution (40/20/40). Dry the paper and dip in ammoniacal silver nitrate solution.

Record the R_F values of the ions and comment on the usefulness of this method in qualitative analysis.

(c) Gas-liquid partition chromatography

Gas chromatography (GLC) is essentially the separation of a mixture into its component substances according to the variation of solubility and volatility of the

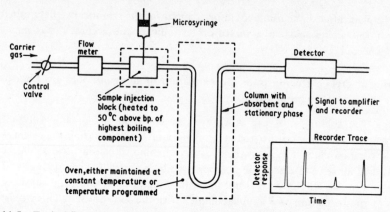

Fig. 11.5   Typical flow sheet of an analytical gas chromatograph

components. The technique is used on a small scale for the analysis of mixtures, and, on a large scale, for the separation and collection of the various components of a mixture. The chromatographic column (Fig. 11.5) consists of a tube (often in the form of a coil) packed with a powder (eg. Celite) of relatively uniform particle size coated with a non-volatile liquid — the stationary phase. The stationary phase should, if possible, chemically resemble the sample, eg. squalene for paraffins, a polyglycol for alcohols, a polyester for methyl esters. In capillary columns the stationary phase is coated on the walls of a long capillary tube. The moving phase of paper-partition chromatography is replaced by a mobile gas phase — the carrier gas. The temperature of the column can be kept constant over a range of temperatures (isothermal) or varied (temperature programming) according to the material under investigation. Suppose a small sample ($0.1-10$ mm^3) of a single substance is introduced or injected through a replaceable rubber septum on to the column and an inert gas passed through the column. In its initial contact with the column material, equilibrium is immediately set up between the substance dissolved in the stationary phase and the substance in the vapour phase in the interstices of the particles. Under the applied pressure of inert gas, the vapour fraction moves down the column and the initial equilibrium is disturbed. So, material passes out of solution and material in the vapour phase passes into solution to give new positions for equilibrium. The process of distribution between the two phases continues repeatedly down the column and in this manner a chromatographic zone develops. The speed at which the zone passes through the column depends on the gas pressure, the volatility of the substance and its solubility in the stationary phase and the temperature of the oven. When more than one substance is injected on to the column the most volatile component moves fastest, followed by the next most volatile, etc. In consequence, separation occurs into single components as the mixture traverses the column.

The various zones must be detected as they are eluted from the column. The three most common detectors are the katharometer, argon ionization detector and the hydrogen ionization detector.

The peaks which appear on the recorder trace (Fig. 11.5) correspond to the various components in the mixture. The 'retention volume', ie. the volume of carrier gas which has passed through the column before a given component is eluted from the column, is characteristic of that particular component. Since the flow rate of the carrier gas is usually kept constant the 'retention time' is generally used for characterization purposes.

Many samples for analysis by gas chromatography have a large number of components with a wide range of boiling points. If such a sample is analysed isothermally at a relatively low temperature the early peaks will be sharp and well resolved, but the less volatile components will be eluted more slowly and the peaks will be broader and not so sharp. Isothermal analysis of the same mixture at higher temperatures will give good separation and resolution of the less volatile components, but the more volatile components will be eluted so rapidly that they may not be separated. Isothermal analysis is, therefore, inadequate where good separation is required for components which have widely differing volatility. Temperature programming is an attempt to get each component to pass through the column at its optimum temperature; it can be considered as a sequence of isothermal stages where each component travelling through the column meets its 'optimum temperature' as the column temperature is gradually increased. Normally a linear rise in temperature is used, and, if the rate of increase of temperature is correct, the members of a homologous series will emerge linearly with time.

Experiment 11.12   Identify the members of a homologous series of methyl esters in solution and estimate their concentration by gas–liquid chromatography

Requirements: Gas chromatograph with polyester-type or general purpose column and, if possible, temperature programmer. Microsyringe ($1-10$ mm^3 capacity). Unknown solution in ethanol of the following straight chain fatty acid methyl esters: C_8, C_{10}, C_{12}, C_{14}, C_{16} and C_{18} at different concentrations in the range 0.1 to 1.0 g in 25 cm^3 ethanol (total quantity of esters should not exceed 2.5 g). Two standard solutions of different esters each containing $0.1-0.2$ g of one of the above esters, dissolved in 10 cm^3 of ethanol.

Procedure: Carefully read the instruction manual of the gas chromatograph and set the carrier gas pressure and other operating conditions accordingly. Set the injection block temperature to about 300 °C and the oven temperature to 150 °C. Inject 1 mm^3 of the unknown solution on to the column, taking care to exclude air bubbles from the syringe. Adjust the sensitivity and, if necessary, the sample volume so that all the 'peaks' except that due to the solvent are on the recorder scale and trace. The first peaks in this isothermal run are well-shaped but subsequent ones are wide and flat and take a long time to appear. The last components will probably not be eluted until the oven temperature is increased to about 200 °C; all the components must be removed as they will be eluted eventually and be detected and recorded on subsequent chromatograms.

Now increase the temperature to 250 °C and repeat the injection. The early peaks will be crowded together, possibly not separated at all, only the later peaks will be acceptable.

Reduce the oven temperature to 150 °C again and repeat the injection, this time using temperature-programming. An initial isothermal period, approximately the retention time of the solvent, is usually required and then the temperature is smoothly increased to 250 °C, by which time the last component should have been eluted. Repeat this procedure, varying the initial temperature and rate of increase of temperature until the peaks are equally spaced. If no temperature programmer is available the increase of temperature can be achieved manually, using the temperature control and a stop-watch. Never allow the temperature to rise above the stated maximum for the column, since this will permanently damage it.

Working with the conditions previously established, obtain chromatograms of the two standard ester solutions.

Treatment of experimental data and discussion

1. Explain the significance of the retention time of a compound and list the factors which determine it. Measure the retention time of the two standard esters and locate the corresponding peaks in the chromatogram of the unknown solution and hence identify all the peaks.

2. Measure the area (height × width at half-peak height) of each peak of the mixture and the known solutions. Compare the peak areas and, from the known amounts of single esters injected, estimate the concentration of each component in the mixture. The peak area can be assumed to be proportional to the weight of each component present.

3. From two separate chromatograms of the unknown solution estimate the reproducibility of the method.

4. Explain the principles of operation of the detector in the gas-chromatograph used.

5. Discuss the advantages and limitations of the method for the analysis of mixtures.

3   Paper electrophoresis

The earliest studies of the transport of dissolved substances in an electric field were those of Hittorf and Lodge in their investigations of the migration of ions. Later, the migration of colloidal particles (eg. proteins) under similar conditions was observed. The migration is due to the charge carried by the ion or particle in solution.

The moving boundary method of electrophoresis, developed extensively by Tiselius, suffers from the disadvantage that complete separation of a mixture into its components is never achieved. If electrophoresis is carried out in a supporting medium; eg. in columns packed with gels or powders, or in a strip of filter paper, complete separation of the various components leads to the formation of zones; hence, the correct name 'zone electrophoresis'.

In principle, a potential difference is applied to the ends of a strip of filter paper (or column of inert supporting material) impregnated with a suitable buffer solution, near the centre of which is a drop of the solution to be studied. The differently charged components move with different velocities and form discrete zones which can be located by suitable means; eg. staining. The most widely used supporting medium is filter paper. The migration is stabilized and cannot be influenced by convection currents in the same way as in the free boundary method. Migration in paper depends on (a) the properties of the ion, its sign and net charge, whether it is dispersed as a colloid (proteins) or as molecules (amino-acids); (b) the properties of the buffer solution, pH, ionic strength, temperature and viscosity; (c) the texture of the paper; and, (d) the current density and field strength. The filter paper acting as a series of capillaries offers a certain resistance to electromigration, and, in consequence, the electrophoretic mobility (velocity/applied field strength) calculated from such studies is always lower than that determined by the free boundary method, under the same conditions. The ratio of the mobility in paper to that in free solution, for a range of ions, is remarkably constant for a given batch of filter paper.

The migration velocity of a component, and hence the time necessary for complete separation of a mixture, depends on the applied field strength and the current density. There is an upper limit to the field strength, determined by the electrical heat applied to the paper. This, for a filter paper suspended in air between two electrode vessels, produces evaporation of water from the paper and the enrichment of buffer ions in the paper, thus affecting the conductance along the paper. Further, heating produces a liquid flow to replace these losses with the concomitant and undesirable movement of the substances under study. Experimentally, these difficulties can be overcome by cooling the apparatus and by maintaining a high humidity within the chamber. The variation of mobility, as reflected by the direction and distance of migration in a given time, with pH and ionic strength of the buffer solution, depends mainly on the nature of the migrating substances. Most amino-acids in buffer solutions of low pH value have a positive mobility, decreasing with increasing pH to zero at the isoelectric point, thereafter increasing in the negative direction. Buffer solutions based on diethylbarbituric, phosphoric and acetic acids are recommended for proteins and amino-acids, and on boric acid for sugars; boric acid forms anionic complexes with cis-OH groups.

The migration of materials in paper is also affected by the electro-osmotic flow of liquid through the paper. The paper is, in general, negatively charged with respect to the buffer solution; in consequence, in an electric field, liquid will flow towards the negative electrode, thereby increasing the movement of positively charged molecules and decreasing that of negatively charged ones. The liquid flow is approximately proportional to the potential gradient and can be estimated by incorporating uncharged particles, which are not absorbed by the paper, eg. xylose, glucose, urea, in the applied mixture. Such molecules will move due to electro-osmosis and an apparent mobility can be calculated. This value when added to (subtracted from) the apparent mobility of anions (cations) gives mobility values for these ions approximately corrected for electro-osmosis.

The apparatus for paper electrophoresis consists of an airtight chamber in which the wet paper strips can be supported under tension between the electrode compartments, electrodes and a d.c. supply. Various types of apparatus are described in the literature; Fig. 11.6 shows one type available commercially in which electrophoresis is conducted horizontally. In this the compartments containing the electrodes, platinum wires or carbon rods, are separated from the compartments into which the paper dips. On both sides, the two adjacent compartments are connected by wicks or glass tubes packed with cotton-, glass- or asbestos-wool. In this way, electrical connection to the paper is made and at the same time the diffusion of the products of electrolysis on to the paper is considerably reduced. The pH of the inner compartments (into which the papers dip) is thus kept constant thereby avoiding a pH gradient along the paper. The level of the buffer solution in the four compartments must be the same to prevent siphoning during electrophoresis. The adjacent compartments readily come to the same level through the wicks; levelling between the positive and negative compartments is best achieved by a Y-piece siphon and suction tube.

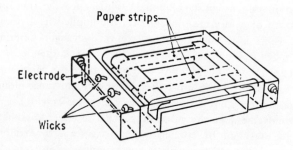

Fig. 11.6   Apparatus for horizontal paper electrophoresis

Experimental procedure: Load the dry filter paper strips (3 cm width and the appropriate length), with the origin marked by a pencil line, on the horizontal frame, Fig. 11.6. Pour the buffer solution into the four chambers so that the liquid level is constant; the solution immediately rises on the paper which soon becomes completely wetted. Slightly stretch the paper on the frame so that there is no sag. Apply 5–10 mm^3 of the sample across the origin of the wet paper, using a micropipette, or capillary tube, with a blunt tip to avoid roughening the paper; take care that diffusion is kept to a minimum. The application should be made only across the centre 2 cm thereby preventing subsequent marginal distortion. Immediately cover the tank with a heavy glass plate and apply a measured potential to the strips, recording the polarity of the electrodes. A strip of plastic foam attached to the inside of the glass cover prevents the dripping of condensed water back on to the strips. After electrophoresis, switch off the current, immediately remove the strips and dry. If the components are colourless, locate them by dipping the paper in a suitable location reagent.

Experiment 11.13   Demonstrate the principle of paper electrophoresis using dyestuffs

Requirements: Paper electrophoresis apparatus (Fig. 11.6) and power pack; 3-cm strips of Whatman No. 1 filter paper; dilute solutions of bromophenol blue, bromo-thymol blue, bromocresol green, xylene cyanol FF, acid fuchsin (wet the solids with a drop of ethanol before adding water); Michaelis veronal buffer solution pH = 8.6, 1 = 0.2 mol dm^{-3} (p. 313).

Procedure: Using the technique previously described, set up strips of paper and apply a 5 mm^3 sample of each dye solution to separate strips. The origin should be nearer the negative electrode. Apply a potential of 300–350 V for 2–3 hours; record the time and potential. Switch off the current, remove and dry the papers. To reveal the acid fuchsin zone, dip the dry paper in dilute acid solution.

Treatment of experimental data and discussion

1. Comment on the usefulness of this method for the separation of a mixture of dyestuffs.
2. Explain why the origin should be nearer the negative electrode.
3. Calculate the apparent mobility of each dyestuff and explain why this differs from the mobility of the dyestuff in free solution.

Experiment 11.14   Study the separation of amino-acids in a mixture by paper electrophoresis

Requirements: Paper electrophoresis apparatus (Fig. 11.6) and power pack, 3-cm strips of Whatman No. 1 filter paper; 0.01 mol dm^{-3} solutions of lysine, valine and glutamic acid, and a mixture of these acids at the same concentration; 1% (w/v) glucose solution, 0.1 mol dm^{-3} phosphate buffer solution pH = 7.0, 0.2% ninhydrin in acetone (p. 289), aniline phthalate solution (p. 290).

Procedure: Using the techniques previously described, set up strips and apply 5 mm^{-3} of the solutions of (*a*) lysine, (*b*) valine, (*c*) glutamic acid, (*d*) mixture, and (*e*) glucose. Apply a potential of 150 V for 2–3 hours; record the time and applied potential. Switch off the current, dry the papers at 100 °C and locate the amino-acids with ninhydrin and the glucose with aniline phthalate. Record the direction and distance of migration of the various components.

Treatment of experimental data and discussion

1. Determine the apparent mobility of each amino-acid from the distance migrated in the known time and the applied field strength.
2. Explain why the glucose moves under the applied field strength. Hence obtain a corrected mobility value for each amino-acid.
3. Compare these values of the mobility with the values in free solution and offer reasons for any differences.
4. Compare and contrast the advantages of the electrophoretic and partition chromatographic techniques for the separation of a mixture of amino-acids.

Experiment 11.15   Determine the isoelectric point of glutamic acid by paper electrophoresis

Requirements: Paper electrophoresis apparatus (Fig. 11.6) and power pack, 3-cm strips of Whatman No. 1 filter paper; 0.01 mol dm^{-3} solution of glutamic acid, buffer solutions in range pH 2–6 (p. 311), 0.2% ninhydrin in acetone (p. 289), alkaline ninhydrin for very acidic buffer solutions; ie. pH 2.0, 0.2 g ninhydrin in 100 cm^3 ethanol plus 0.5 cm^3 of 1.0 mol dm^{-3} potassium hydroxide solution.

Procedure: Using the technique described, set up three papers in one of the buffer solutions and apply 5 mm^{-3} of the glutamic acid solution to the marked origin of each. Apply a constant potential of 120 V; after 3 hours, switch off the current and remove one strip; continue electrophoresis for a further 3 hours and remove a second strip. Leave the third paper on for a total of 17 hours. Dry the papers and locate the position of the amino-acid with ninhydrin.

Repeat the experiment with the different buffer solutions, always keeping the applied potential constant and removing the papers at exactly the same times.

Treatment of experimental data and discussion

1. For each strip, measure the distance moved from the origin and record the direction. Assuming that the field strength and the electro-osmotic migration is constant for all the buffer solutions at a fixed time, the distance moved is proportional to the mobility. Plot a graph of distance moved (positive or negative) against pH, and hence, determine the isoelectric point. Alternatively, calculate the mobility in each buffer solution, from the velocity of migration and the known applied field strength; for this, strips containing glucose should be run in parallel so that a correction for the electro-osmotic flow can be made.

2. Explain the significance of the iso-electric point. How is this quantity related to the dissociation constants of the acid?

Appendix

Table 1   Fundamental physical constants

Physical constant	Symbol	Value
acceleration due to gravity	g	9.81 m s^{-2}
atomic mass unit	amu	$1.660\ 53 \times 10^{-27} \text{ kg}$
Avogadro constant	N_A	$6.022\ 17 \times 10^{23} \text{ mol}^{-1}$
Boltzmann constant	k	$1.380\ 62 \times 10^{-23} \text{ J K}^{-1}$
electronic charge	e	$1.602\ 192 \times 10^{-19} \text{ C}$
Faraday constant	$\mathscr{F}$	$9.648\ 67 \times 10^{4} \text{ C mol}^{-1}$
gas constant	R	$8.314 \text{ J K}^{-1} \text{ mol}^{-1}$
'ice-point' temperature	T_{ice}	273.150 K
molar volume of ideal gas at stp.	V_m	$2.241\ 36 \times 10^{-2} \text{ m}^3 \text{ mol}^{-1}$
permittivity of a vacuum	ϵ_0	$8.854\ 185 \times 10^{-12} \text{ kg}^{-1} \text{ m}^{-3} \text{ s}^4 \text{ A}^2$
Planck constant	h	$6.626\ 20 \times 10^{-34} \text{ J s}$
Rydberg constant	R_∞	$1.097\ 373\ 1 \times 10^{7} \text{ m}^{-1}$
standard pressure (atmosphere)	P	$101\ 325 \text{ N m}^{-2}$
triple point of water		273.16 K
velocity of light in a vacuum	c	$2.997\ 925 \times 10^{8} \text{ m s}^{-1}$

Table 2   General properties of some organic compounds

Compound	mp./ °C	bp./ °C	Density/ kg m^{-3} (298 K)	Refractive index (n_D) (293 K)	10^4 x Viscosity/ N s m^{-2} (298 K)	10^3 x Surface Tension/ N m^{-1} (293 K)
Acetic acid	16.7	117.9	1 044.0	1.3716	11.55	27.8
Acetone	−94.7	56.1	785.0	1.3588	3.16	23.7
Aniline	−6.3	184.1	1 022.0 (293)	1.5863	3.71	42.9
Benzene	5.51	80.1	874.0	1.5011	6.01	28.89
Benzoic acid	122.4	249.0	1 266.0 (288)	1.504 (405)	−	−
Carbon tetra-chloride	−22.9	76.5	1 584.0	1.4601	8.8	26.95
Chloro-benzene	−45.2	132.0	1 106.0	1.5241	7.97	33.56
Chloroform	−63.5	61.7	1 480.0	1.4459	5.42	27.14
Cyclohexane	6.6	80.7	774.0	1.42662	9.8	25.5
Ethyl acetate	−82.4	77.1	900.0 (293)	1.3723	4.41	23.9
Ethanol	−114.1	78.3	785.0	1.3611	10.6	22.75
Ethyl ether	−116.2	34.51	714.0	1.3526	2.22	17.01
Glycerol	18.07	290.0	1 264.4	1.4746	942	63.4
Hexane	−95.3	68.7	655.0	1.37506	2.94	18.43
Methanol	−97.7	64.5	787.0	1.3288	5.47	22.61
Naphthalene	80.3	218.0	1 180.0	1.4003 (297)	−	−
Phenol	40.9	181.8	1 132.0	1.5509	−	−
Toluene	−95.1	110.6	862.0	1.4961	5.50	28.5

Table 3   Buoyancy correction for bodies of different densities (ρ_b/g cm^{-3}) weighed in air with brass (or steel) weights

ρ_b	k	ρ_b	k	ρ_b	k
0.5	+2.26	1.9	+0.49	11.0	−0.03
0.6	+1.86	2.0	+0.46	12.0	−0.04
0.7	+1.57	2.5	+0.34	13.0	−0.05
0.8	+1.36	3.0	+0.26	14.0	−0.06
0.9	+1.19	3.5	+0.20	15.0	−0.06
1.0	+1.06	4.0	+0.16	16.0	−0.07
1.1	+0.95	4.5	+0.13	17.0	−0.07
1.2	+0.86	5.0	+0.10	18.0	−0.08
1.3	+0.78	5.5	+0.08	19.0	−0.08
1.4	+0.72	6.0	+0.06	20.0	−0.08
1.5	+0.66	7.0	+0.03	21.0	−0.09
1.6	+0.61	8.0	+0.01	22.0	−0.09
1.7	+0.56	9.0	−0.01	23.0	−0.09
1.8	+0.52	10.0	−0.02	24.0	−0.09

Correction for buoyancy of air

The weight of a body in air is less than its true weight by an amount equal to the weight of air displaced. When the object to be weighed and the weights have the same density, and in consequence the same volume, the true and observed weights are identical. The correction to be applied to the observed weights increases as the difference between the density of the body and that of the weights increases; it is positive (negative) if the density of the body is less (greater) than the density of the weights. Let W_v and W_b be the true (*in vacuo*) and apparent (in air) weights of the body respectively and ρ_b, ρ_a, and ρ_w be the densities of the body, air and the weights respectively.

Then by Archimedes' principle:

$$\begin{array}{ccc} \text{Wt. of body} \\ \textit{in vacuo} \end{array} = \begin{array}{c} \text{Wt. of body} \\ \text{in air} \end{array} + \begin{array}{c} \text{Wt. of air} \\ \text{displaced} \\ \text{by body} \end{array} - \begin{array}{c} \text{Wt. of air} \\ \text{displaced} \\ \text{by weights} \end{array}$$

$$\therefore W_v = W_b + \frac{W_b}{\rho_b}\rho_a - \frac{W_b}{\rho_w}\rho_a$$

$$= W_b + W_b\rho_a\left(\frac{1}{\rho_b} - \frac{1}{\rho_w}\right) \tag{1}$$

ρ_a lies in the range $0.0011{-}0.0013$ g cm^{-3} under normal conditions of temperature and pressure and for a humidity of 50%; it is therefore permissible for most purposes to take $\rho_a = 0.0012$ g cm^{-3}. For brass weights $\rho_w = 8.4$ g cm^{-3}, and for chrome steel weights, as used in most single pan balances $\rho_w = 8.0$ g cm^{-3}.

Thus equation (1) for brass weights can be written:

$$W_v = W_b + 0.0012W_b\left(\frac{1}{\rho_b} - \frac{1}{8.4}\right) \tag{2a}$$

$$= W_b + \frac{W_b k}{1000} \tag{2b}$$

where

$$k = 1.20 \times \left(\frac{1}{\rho_b} - \frac{1}{8.4}\right)$$

Values of k for bodies of different density are given in Table 3. Thus for a body of weight W_b an amount $W_b \times k$ mg must be added to give the weight *in vacuo*.

Table 4   Weight *in vacuo* of 1 cm^3 of water

$t/°C$	Wt./g	$t/°C$	Wt./g
4	1.00000	21	0.99802
10	0.99973	22	0.99780
11	0.99963	23	0.99756
12	0.99952	24	0.99732
13	0.99940	25	0.99707
14	0.99927	26	0.99681
15	0.99913	27	0.99654
16	0.99897	28	0.99626
17	0.99880	29	0.99597
18	0.99862	30	0.99567
19	0.99843	40	0.99224
20	0.99823	50	0.98807

(From Vogel, *A Textbook of Quantitative Inorganic Analysis*. Longman, 1961 (3rd edn.), p. 155, Table II.)

For the accurate weighing of large objects it is customary to use a vessel of approximately the same size as a counterpoise; this reduces the buoyancy correction considerably.

For the calibration of volumetric glassware the weight of 1 cm^3 of water *in vacuo* must be known. Values of this, at various temperatures, are given in Table 4.

Table 5   Apparent weight of water using brass or steel weights to give 1 dm^3 at 20 °C

$t/°C$	Weight/g	Volume of 1 g of water/cm^3	$t/°C$	Weight/g	Volume of 1 g of water/cm^3
10	998.39	1.0016	23	996.60	1.0034
11	998.32	1.0017	24	996.38	1.0036
12	998.23	1.0018	25	996.17	1.0038$_5$
13	998.14	1.0018$_5$	26	995.93	1.0041
14	998.04	1.0019	27	995.69	1.0043
15	997.93	1.0021	28	995.44	1.0046
16	997.80	1.0022	29	995.18	1.0048
17	997.66	1.0023	30	994.91	1.0051
18	997.51	1.0025	31	994.64	1.0054
19	997.35	1.0026	32	994.35	1.0057
20	997.18	1.0028	33	994.06	1.0060
21	997.00	1.0030	34	993.75	1.0063
22	996.80	1.0032	35	993.45	1.0066

Note. – For the calibration of flasks of capacity other than 1 dm^3, the corresponding multiple or sub-multiple of the above values is taken.

(From Vogel, *A Textbook of Quantitative Inorganic Analysis*. Longman, 1961 (3rd edn.), p. 195, Table IX. See also BS 1797, 1952.)

Table 6   Saturation vapour pressure of water (mmHg) at various temperatures

$t/°C$	0	2	4	6	8
0	4.579	5.294	6.101	7.013	8.045
10	9.209	10.518	11.987	13.634	15.477
20	17.535	19.827	22.377	25.209	28.349
30	31.824	35.663	39.898	44.563	49.692
40	55.324	61.50	68.26	75.65	83.71
50	92.51	102.09	112.51	123.80	136.08

Table 7   Boiling point of water at various pressures

Pressure/mmHg	700	710	720	730	740	
$t/°C$	97.71	98.11	98.49	98.88	99.26	
Pressure/mmHg	750	760	770	780	790	800
$t/°C$	99.63	100.00	100.37	100.73	101.09	101.44

Table 8   Limiting molar conductivities at 298 K

	$10^4 \Lambda_0/\Omega^{-1} m^2 mol^{-1}$		$10^4 \Lambda_0/\Omega^{-1} m^2 mol^{-1}$
Ag^+	61.92	Br^-	78.4
Ba^{++}	127.3	CH_3COO^-	40.9
Ca^{++}	119.0	$C_6H_5COO^-$	32.3
H_3O^+	349.82	Cl^-	76.34
K^+	73.52	I^-	76.8
Li^+	38.69	NO_3^-	71.44
Mg^{++}	106.1	OH^-	198.0
NH_4^+	73.4	SO_4^{--}	159.6
Na^+	50.11		

$$\text{Mobility of an ion/m}^2 \text{ s}^{-1} \text{ V}^{-1} = \frac{(\Lambda_0/\Omega^{-1} m^2 mol^{-1})}{96500 \text{ C mol}^{-1}}$$

Table 9   Decomposition potentials of 1 mol dm^{-3} solutions between smooth platinum electrodes

Substance	Decomposition potential/V	Substance	Decomposition potential/V
$ZnSO_4$	2.55	H_3PO_4	1.70
$NiSO_4$	2.09	HNO_3	1.69
$CdSO_4$	2.03	H_2SO_4	1.67
$CoSO_4$	1.92	HCl	1.31
$ZnBr_2$	1.80	HBr	0.94
$Pb(NO_3)_2$	1.52	HI	0.52
$CuSO_4$	1.49	NaOH	1.69
$AgNO_3$	0.70	KOH	1.67

Table 10   Cathodic and anodic overvoltages

	Overvoltage/V	
	Cathodic (H_2) (in 0.5 mol dm^{-3} H_2SO_4)	Anodic (O_2)
Platinized platinum	−0.005	0.24
Bright platinum	−0.09	0.44
Silver	−0.15	0.41
Nickel	−0.21	
Copper	−0.23	
Tin	−0.53	
Lead	−0.64	0.31
Zinc	−0.70	
Mercury	−0.78	

Preparation of electrodes

(a) *Silver electrode:* For accurate work the silver electrode is coated with a fresh film of silver as follows: treat a solution of silver nitrate (3.5 g dm^{-3}) with dilute potassium cyanide solution (care) with constant stirring until the first formed precipitate of silver cyanide is redissolved. Clean the electrode, with fine emery paper, and dip in 1 : 1 nitric acid until effervescence occurs, finally wash with water and support with a silver anode in the cyanide solution. Pass a current of 0.2 mA until the surface is covered with a thin electrolytic deposit. Wash thoroughly with water and store in water. An electrode prepared from 22 SWG. silver wire is adequate for most purposes.

(b) *Platinum and gold electrodes:* Platinum may be used either in the shiny form (oxidation-reduction electrodes), black form (hydrogen electrode, and conductance measurements) or grey form (conductance measurements).

The platinum black, ie. platinized platinum is deposited as follows: clean two platinum electrodes in acid cleaning mixture and thoroughly wash with water. The platinizing solution consists of 3 g chloroplatinic acid and 0.02−0.03 g lead acetate dissolved in 100 cm^3 of water. Support the electrodes in this solution and connect through a reversing switch to a rheostat and 4 V accumulator. Pass the current, adjusted so that there is only a moderate evolution of gas, for 10−15 min reversing the polarity every 30 s. A moderately thick coat of platinum black is preferable to a thin one. The deposit now contains occluded gas and liquid. Remove this by immersing the electrodes in 0.3 mol dm^{-3} sulphuric acid and connecting them through a reversal switch to a 4 V accumulator. Electrolyse the solution for 30 min with reversal of polarity every 30 s, so that gas bubbles vigorously from both electrodes. Thoroughly wash with distilled water and store in water.

The coating may be stripped off by electrolysing either *aqua regia* or 5 mol dm^{-3} hydrochloric acid with the electrode connected to the positive pole of the accumulator. For a pair of electrodes connect to a 4 V accumulator through a reversing switch and pass the current alternately in both directions.

For conductance measurements it is sometimes advisable to use grey platinum electrodes; these are prepared by heating blacked electrodes to dull redness in a flame.

(*c*) *Silver–silver chloride electrode:* Immerse a freshly plated silver electrode (above) in 0.1 mol dm^{-3} hydrochloric acid and connect to the positive pole of an accumulator. Using a platinum electrode as cathode, pass a current of 2.5 mA cm^{-2} over the whole surface area for 30 min. Current densities below 0.6 mA cm^{-2} produce grey patches and the electrode behaves erratically. An even purple deposit of silver chloride will be formed, this is unaffected by sunlight. The electrode potential is reproducible to ±0.02 mV. Store the electrode in distilled water.

(*d*) *Quinhydrone electrode:* The electrode is simply constructed by immersing bright platinum wire or foil into the test solution containing excess solid quinhydrone (0.5–1 g in 100 cm^3 of solution). The platinum, which must be cleaned with chromic acid, washed with water and ignited in an alcohol flame, and the reference electrode are connected to the potentiometer.

Agar salt bridges

Add 3–5 g of powdered agar, in small portions at a time so that the solution does not froth and boil over, to 100 cm^3 of a saturated solution of potassium chloride at 100 °C on a steam bath. Keep the solution at 100 °C until all the agar has passed into solution and then add 10–15 g solid potassium chloride to produce excess of the solid. The simplest method to fill a bridge is to use an inverted U-tube with a side arm connected to a rubber tube. The hot solution is sucked up and the rubber tubing clipped off. On cooling the agar solidifies.

Other agar bridges can be made by dissolving 3–5 g powdered agar in 100 cm^3 of 2 mol dm^{-3} potassium or ammonium nitrate solution.

Silver coulometer

A silver coulometer (Fig. A.1) consists of a sintered glass crucible (A) of No. 2 or No. 3 porosity, supported on a glass frame (B) so that there can be free movement of the solution through the sintered disc. The cathode consists of a piece of platinum foil (1 cm^2) welded on to platinum wire, while the anode is a coil of stout silver wire.

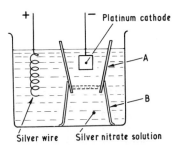

Fig. A.1   Silver coulometer

Clean and wash the electrodes and crucible. Dry the cathode and crucible together in an oven at 150 °C to constant weight, taking the usual precautions of cooling in a desiccator, etc. Connect the electrodes in the circuit and fill the beaker with 15% silver nitrate solution, this must be of high quality and free from organic material. Pass the current, which should not exceed 10 mA cm^{-2} of cathode surface, through the solution; silver will be deposited on the cathode, any small fragments which break away will be collected in the crucible. After switching off the current, disconnect the platinum cathode, rest it in the crucible and carefully remove crucible from beaker. Wash the outside of the crucible with a jet of water and then the inside and platinum by suction in the normal manner. Dry the crucible and cathode to constant weight as previously described. The accuracy of this method is limited only by the care of manipulation and the accuracy of weighing (0.001118 g of silver is deposited by 1 coulomb of electricity).

Table 11　　Standard electrode potentials in aqueous solution at 298 K

	$E^{\ominus}/V$		$E^{\ominus}/V$
Li^+, Li	−3.024	Calomel (0.1 mol dm^{-3})	0.3335
K^+, K	−2.924	Cu^{++}, Cu	0.339
Na^+, Na	−2.714	$Fe(CN)_6^{---}$, $Fe(CN)_6^{----}$	0.356
Zn^{++}, Zn	−0.761	I_2, Pt, I^-	0.54
Fe^{++}, Fe	−0.441	Quinone, hydroquinone	0.70*
Cr^{+++}, Cr^{++}	−0.41	Fe^{+++}, Fe^{++}	0.783
Co^{++}, Co	−0.283	Ag^+, Ag	0.799
AgI, Ag, I^-	−0.15	Hg_2^{++}, Hg	0.799
Sn^{++}, Sn	−0.140	Hg^{++}, Hg_2^{++}	0.91
H_3O^+, Pt, H_2	0.000	Br_2, Pt, Br^-	1.09
AgBr, Ag, Br^-	0.07	Cl_2, Pt, Cl^-	1.358
Sn^{++++}, Sn^{++}	0.15	$Cr_2O_7^-$, Cr^{+++}	1.36
AgCl, Ag, Cl^-	0.2225	MnO_4^-, H_3O^+, Mn^{++}	1.52*
Calomel (satd.)	0.242	Ce^{++++}, Ce^{+++}	1.61
Calomel (1 mol dm^{-3})	0.281		

* $a_{H_3O^+} = 1$

Table 12　　Value of the factor 2.303 $RT/\mathscr{F}$ in volts at different temperatures

$t/°C$	$\dfrac{2.303\,RT}{\mathscr{F}}$	$t/°C$	$\dfrac{2.303\,RT}{\mathscr{F}}$
0	0.05420	22	0.05857
10	0.05619	24	0.05897
12	0.05658	25	0.05916
14	0.05698	26	0.05936
16	0.05738	28	0.05976
18	0.05777	30	0.06016
20	0.05817	40	0.06214

Standardization of pH scale and calibration of pH meters

On account of the importance of accurate pH measurements, and also to ensure that a given solution has the same pH value in any laboratory using any pH meter, the British Standards Institution (1961) has adopted the following definition: the difference in pH between two solutions A and B measured at constant temperature (T) is given by:

$$pH_A - pH_B = \frac{(E_A - E_B)\mathscr{F}}{2.3026RT}$$

where E_A and E_B are the emf. values of the cells:

$$\text{Pt, } H_2 \mid \text{Soln A} \mid 3.5 \text{ mol dm}^{-3} \text{ KCl} \mid \text{Reference electrode}$$

$$\text{Pt, } H_2 \mid \text{Soln B} \mid 3.5 \text{ mol dm}^{-3} \text{ KCl} \mid \text{Reference electrode}$$

respectively. This definition, while to some extent an arbitrary one, is nevertheless operational. Further, the primary standard selected is 0.05 mol dm^{-3} solution of pure potassium hydrogen phthalate. This has a pH of 4.000 at 15 °C, the pH at any temperature t (0—55 °C) is given by

$$pH_t = 4.000 + \frac{1}{2}\left\{\frac{t-15}{100}\right\}^2$$

and (55—95 °C) by:

$$pH = 4.000 + \frac{1}{2}\left\{\frac{t-15}{100}\right\}^2 - \frac{t-55}{500}$$

The pH values of a range of solutions, determined in accordance with the above specification, are listed Table 13. These are known with an accuracy of ±0.005 pH unit and should be used for accurate calibration of the glass electrode assembly; other solutions of known pH may be used when an accuracy of ±0.03 pH units is sufficient.

Table 13   pH values of aqueous solutions recommended for calibration of glass electrodes (BS 1647:1961)

		pH *values at*	
		25 °C	38 °C
0.1 mol dm^{-3}	Potassium tetraoxalate	1.48	1.50
0.01 mol dm^{-3}	Hydrochloric acid and 0.09 mol dm^{-3} potassium chloride	2.07	2.08
0.05 mol dm^{-3}	Potassium hydrogen phthalate (primary standard)	4.005	4.026
0.1 mol dm^{-3}	Acetic acid and 0.1 mol dm^{-3} sodium acetate*	4.64	4.65
0.01 mol dm^{-3}	Acetic acid and 0.01 mol dm^{-3} sodium acetate*	4.70	4.72
0.025 mol dm^{-3}	Disodium hydrogen phosphate and 0.025 mol dm^{-3} potassium dihydrogen phosphate	6.85	6.84
0.05 mol dm^{-3}	Sodium tetraborate	9.18	9.07
0.025 mol dm^{-3}	Sodium bicarbonate and 0.025 mol dm^{-3} sodium carbonate	10.00	

* Prepared from pure acetic acid, diluted and half neutralized with sodium hydroxide. It must *not* be prepared from sodium acetate.

Table 14   pH range: 0.65–5.20; 18 °C (Walpole).
         Sodium acetate–Hydrochloric acid

1.0 mol dm^{-3} sodium acetate (136.09 g CH_3COONa . $3H_2O$ in 1000 cm^3).
1.0 mol dm^{-3} hydrochloric acid.
50 cm^3 of sodium acetate + x cm^3 HCl made up to 250 cm^3.

x	pH	x	pH	x	pH
100	0.65	52.5	1.99	42.5	3.79
90	0.75	51.0	2.32	40.0	3.95
80	0.91	50.0	2.64	35.0	4.19
70	1.09	49.75	2.72	30.0	4.39
65	1.24	48.5	3.09	25.0	4.58
60	1.42	47.5	3.29	20.0	4.76
55	1.71	46.25	3.49	15.0	4.92
53.5	1.85	45.0	3.61	10.0	5.20

The ionic strength of these solutions is not constant.

Table 15   pH range: 2.0–12.0 (Davies).
         Universal buffer solution for use in ultraviolet spectrophotometry

(These solutions are very useful for measurements in the ultraviolet region, since, unlike barbitone solutions, they do not absorb strongly. The absorbance in the range 220–260 nm relative to air is almost independent of pH value.)

Solutions:

A.	Citric acid	21.01 g	
	Crystalline potassium dihydrogen phosphate	16.61 g	made
	Sodium tetraborate	19.07 g	up to
	Trishydroxymethylaminomethane	12.11 g	1000 cm^3
	Potassium chloride.	7.46 g	
B.	0.4 mol dm^{-3} hydrochloric acid.		
C.	0.4 mol dm^{-3} sodium hydroxide solution.		

Preparation: 50 cm^3 of solution A + specified amount of B or C diluted to 200 cm^3. Only selected volumes of B or C are listed, intermediate pH values can be prepared by titration.

Volume of B (HCl)/cm^3	Volume of C (NaOH)/cm^3	pH at 25 °C	I/mol dm^{-3}
34.8	0	2.0	0.1
19.6	0	3.0	0.09
10.0	0	4.0	0.09
0	0.4	5.0	0.11
0	11.4	6.0	0.14
0	22.4	7.0	0.21
0	33.2	8.0	0.25
0	46.2	9.0	0.27
0	59.0	10.0	0.29
0	65.6	11.0	0.30
0	77.2	12.0	0.33

Table 16   pH range: 2.20–8.0 (McIlvaine).
    Disodium hydrogen phosphate—citric acid

0.2 mol dm^{-3} disodium hydrogen phosphate (71.64 g Na$_2$HPO$_4$. 12H$_2$O in 1 dm^3 of solution).
0.1 mol dm^{-3} citric acid (19.21 g in 1/dm^3).

Vol of sodium phosphate/cm^3	*Vol of citric acid*/cm^3	*pH*	*Vol of sodium phosphate*/cm^3	*Vol of citric acid*/cm^3	*pH*
0.40	19.60	2.2	10.72	9.28	5.2
1.24	18.76	2.4	11.15	8.85	5.4
2.18	17.82	2.6	11.60	8.40	5.6
3.17	16.83	2.8	12.09	7.91	5.8
4.11	15.89	3.0	12.63	7.37	6.0
4.94	15.06	3.2	13.22	6.78	6.2
5.70	14.30	3.4	13.85	6.15	6.4
6.44	13.56	3.6	14.55	5.45	6.6
7.10	12.90	3.8	15.45	4.55	6.8
7.71	12.29	4.0	16.47	3.53	7.0
8.28	11.72	4.2	17.39	2.61	7.2
8.82	11.18	4.4	18.17	1.83	7.4
9.35	10.65	4.6	18.73	1.27	7.6
9.86	10.14	4.8	19.15	0.85	7.8
10.30	9.70	5.0	19.45	0.55	8.0

These solutions are not of constant ionic strength.

Table 17   pH range: 2.6–9.6 (Michaelis).
    Acetate — veronal — hydrochloric acid

Stock solutions: A. 9.714 g CH$_3$COONa . 3H$_2$O   made up to
            14.714 g sodium diethylbarbiturate   500 cm^3.

        B. 8.5 g sodium chloride in 100 cm^3.

        C. 0.1 mol dm^{-3} hydrochloric acid.

Preparation: To 5.0 cm^3 of A add 2.0 cm^3 of B, c cm^3 of 0.1 mol dm^{-3} HCl and $(18 - c)$ cm^3 water.

c	*pH*	c	*pH*	c	*pH*
0	9.64	4.0	7.66	11.0	4.33
0.25	9.16	5.0	7.42	12.0	4.13
0.50	8.90	6.0	6.99	13.0	3.88
0.75	8.68	7.0	6.75	14.0	3.62
1.0	8.55	8.0	6.12	15.0	3.20
2.0	8.18	9.0	5.32	16.0	2.62
3.0	7.90	10.0	4.66		

All these solutions have constant ionic strength (0.173 mol dm^{-3}) and are isotonic with blood.

Table 18   pH range: 3.42—5.89; 18 °C (Walpole).
           Acetic acid — sodium acetate

0.2 mol dm^{-3} sodium acetate (27.22 g CH$_3$COONa . 3H$_2$O in 1 dm^3 of solution).
0.2 mol dm^{-3} acetic acid.

Vol of acetic acid/cm³	Vol of sodium acetate/cm³	pH	Vol of acetic acid/cm³	Vol of sodium acetate/cm³	pH
9.5	0.5	3.42	4.0	6.0	4.80
9.0	1.0	3.72	3.0	7.0	4.99
8.0	2.0	4.05	2.0	8.0	5.23
7.0	3.0	4.27	1.5	8.5	5.37
6.0	4.0	4.45	1.0	9.0	5.57
5.0	5.0	4.63	0.5	9.5	5.89

These solutions are not of constant ionic strength.

Table 19   pH range: 5.29—8.04; 18 °C (Sørensen).
           Phosphate buffer solution

0.0667 mol dm^{-3} disodium hydrogen phosphate (23.88 g Na$_2$HPO$_4$. 12H$_2$O in 1 dm^3 of solution).
0.0667 mol dm^{-3} potassium dihydrogen phosphate (9.074 g KH$_2$PO$_4$ in 1 dm^3 of solution).

Vol of Na₂HPO₄ /cm³	Vol of KH₂PO₄ /cm³	pH	Vol of Na₂HPO₄ /cm³	Vol of KH₂PO₄ /cm³	pH
0.25	9.75	5.29	5.0	5.0	6.81
0.50	9.50	5.59	6.0	4.0	6.98
1.0	9.0	5.91	7.0	3.0	7.17
2.0	8.0	6.24	8.0	2.0	7.38
3.0	7.0	6.47	9.0	1.0	7.73
4.0	6.0	6.64	9.5	0.5	8.04

These solutions are not of constant ionic strength.

Table 20   pH range: 6.80–9.60; 25 °C (Michaelis).
         Sodium diethylbarbiturate — hydrochloric acid

0.1 mol dm^{-3} sodium diethylbarbiturate (20.618 g in 1 dm^3 of solution).
0.1 mol dm^{-3} hydrochloric acid.

Vol of sodium diethylbarbiturate /cm^3	Vol of HCl /cm^3	pH	Vol of sodium diethylbarbiturate /cm^3	Vol of HCl /cm^3	pH
5.22	4.78	6.80	8.23	1.77	8.40
5.36	4.64	7.00	8.71	1.29	8.60
5.54	4.46	7.20	9.08	0.92	8.80
5.81	4.19	7.40	9.36	0.64	9.00
6.15	3.85	7.60	9.52	0.48	9.20
6.62	3.38	7.80	9.74	0.26	9.40
7.16	2.84	8.00	9.85	0.15	9.60

These solutions are all of constant ionic strength 0.1 mol dm^{-3}

Table 21   pH range: 9.24–12.39 (Nalgeli and Tyabji).
         Borax-sodium hydroxide

0.05 mol dm^{-3} borax (19.072 g Na$_2$B$_4$O$_7$. 10H$_2$O in 1 dm^3 of solution).
0.1 mol dm^{-3} sodium hydroxide.

Preparation: To 10 cm^3 of borax solution add x cm^3 NaOH.

x/cm^3	pH	x/cm^3	pH
0	9.24	11.20	11.72
6.66	9.94	11.96	11.98
8.24	10.21	12.68	12.12
9.25	10.49	13.97	12.28
9.96	10.85	15.03	12.39
10.54	11.34		

These solutions are not of constant ionic strength.

Table 22   pH range: 10.97—12.06 (Ringer).
Disodium hydrogen phosphate—sodium hydroxide

0.15 mol dm^{-3} disodium hydrogen phosphate (53.733 g Na$_2$HPO$_4$. 12H$_2$O in 1 dm^3 solution).
0.1 mol dm^{-3} sodium hydroxide.

Preparation: To 50 cm^3 of the phosphate solution add x cm^3 sodium hydroxide.

x/cm^3	pH
15.0	10.97
25.0	11.29
50.0	11.77
75.0	12.06

These solutions are not of constant ionic strength.

Table 23   Colour changes and pH range of some acid-base indicators

	Colour		Approx. pH range	pK_a
	Acid	*Alkaline*		
Methyl violet	yellow	violet	0.1—2.0	—
Thymol blue	red	yellow	1.2—2.8	1.7
Bromophenol blue	yellow	blue	2.9—4.6	4.0
Methyl orange	red	yellow	3.1—4.4	3.7
Methyl red	red	yellow	4.2—6.3	5.1
Bromocresol purple	yellow	violet	5.2—6.8	6.3
Bromothymol blue	yellow	blue	6.0—7.6	6.3
p-nitrophenol	colourless	yellow	5.6—7.6	7.1
Phenol red	yellow	red	6.8—8.4	7.9
Thymol blue	yellow	blue	8.0—9.6	8.9
Phenolphthalein	colourless	red	8.3—10.0	9.6
Thymolphthalein	colourless	blue	8.3—10.5	9.2
Alizarin yellow R	yellow	red	10.0—12.0	—
Tropaeolin O	yellow	orange	11.1—12.7	—

Table 24   Some oxidation-reduction indicators

Indicator	Colour change		$E^{\ominus}$/V (pH = 0)
	Oxidized form	*Reduced form*	
Phenosafranine	red	colourless	0.28
Methylene blue	blue	colourless	0.52
Diphenylamine	violet	colourless	0.76
Diphenylamine-sulphonic acid	red-violet	colourless	0.85
Lissamine green	orange	green	0.99
N-Phenylanthranilic acid	purple-red	colourless	1.08
o-Phenanthroline	blue	red	1.08

Table 25 Stability constants of some metal chelates

	Mg^{++}	Ca^{++}	Sr^{++}	Fe^{++}	Fe^{+++}	Ni^{++}	Cu^{++}	Hg^{++}
Trien	–	–	–	7.8	–	14.0	20.1	25.0
EDTA	8.7	10.7	18.7	14.4	25.1	18.6	18.8	22.1
DTPA	9.0	10.7	9.7	16.7	27.5	20.2	21.0	27.0

The values are log K, where K = stability constant. Trien: triethylenetetramine; DTPA; diethylenetriamine-pentaacetic acid.

Table 26 Typical metallochromic indicators, conditions of use and applications in EDTA titrations

Metal ion	Indicator	Conditions	Colour change at end-point
Bi	Xylenol Orange	$0.1-0.2$ mol dm^{-3} HNO$_3$	Red to yellow
Ca	Calcon	NaOH at pH 12.5	Red to blue
Pb	Eriochrome Black T	Tartrate, NH$_3$ buffer pH 10.0	Violet to blue
	Xylenol Orange	Acetate buffer, pH 5.0	Red to yellow
Mg	Eriochrome Black T	Ammonia buffer, pH 10.0	Red to blue
Zn	Eriochrome Black T	Ammonia buffer, pH 10.0	Red to blue
	Zincon	Ammonia buffer pH $9.0-10.0$	Blue to yellow

Activity coefficients

Many useful forms of the Debye–Hückel equation for activity coefficients exist. The simple forms used throughout this book are:

(*a*) Single electrolyte:

$$\log \gamma = -A z_1 z_2 \sqrt{I}$$

$$\log \gamma = \frac{-A z_1 z_2 \sqrt{I}}{1 + B \mathring{a} \sqrt{I}}$$

(*b*) Individual ion:

$$\log \gamma_i = -A z^2 \sqrt{I}$$

$$\log \gamma_i = \frac{-A z^2 \sqrt{I}}{1 + B \mathring{a} \sqrt{I}}$$

where $I/\text{mol dm}^{-3}$ is the ionic strength, z, z_1 or z_2 the charge of the ion(s),

$$A = \frac{1.825 \times 10^6}{(\epsilon_r T)^{3/2}} = 0.509 \ \text{dm}^{3/2} \, \text{mol}^{-1/2}$$

at 25 °C in water, and

$$B = \frac{50.29 \times 10^8}{(\epsilon_r T)^{1/2}} = 0.33 \ \text{dm}^{1/2} \, \text{mol}^{-1/2}$$

at 25 °C, in water and $\mathring{a}$, which may be regarded as the effective diameter of the hydrated ion, is expressed in Angström units, ϵ_r is the relative permittivity. The mean ionic activity coefficient (γ) is related to the activity coefficients of the individual ions (γ_+, γ_-) by the relationship: $\gamma = (\gamma_+^{\nu_+} \gamma_-^{\nu_-})^{1/\nu}$ for an electrolyte with ν ions, ν_+ cations and ν_- anions.

Table 27 lists the values of activity coefficients of selected individual ions at different ionic strengths (a total of 130 ions are listed by Kielland, *J. Am. Chem. Soc.* (1937), **59**, 1675). The parameter $\mathring{a}$, is the approximate value of the effective diameter obtained by various methods.

Table 27   Activity coefficients of individual ions in water at 25 °C

Parameter $\overset{\circ}{a}$	Ionic strength $I/mol\ dm^{-3}$							
	0.0005	0.001	0.0025	0.005	0.01	0.025	0.05	0.1
				Ion Charge 1				
9	0.975	0.967	0.950	0.933	0.914	0.88	0.86	0.83
6	0.975	0.965	0.948	0.929	0.907	0.87	0.835	0.80
5	0.975	0.964	0.947	0.928	0.904	0.865	0.83	0.79
4.5	0.975	0.964	0.947	0.928	0.902	0.86	0.82	0.775
4	0.975	0.964	0.947	0.927	0.901	0.855	0.815	0.77
3.5	0.975	0.964	0.946	0.926	0.900	0.855	0.81	0.760
3	0.975	0.964	0.945	0.925	0.899	0.85	0.805	0.755
2.5	0.975	0.964	0.945	0.924	0.898	0.85	0.80	0.75
				Ion Charge 2				
8	0.906	0.872	0.813	0.755	0.69	0.595	0.52	0.45
6	0.905	0.870	0.809	0.749	0.675	0.57	0.485	0.405
5	0.903	0.868	0.805	0.744	0.67	0.555	0.465	0.38
4.5	0.903	0.868	0.805	0.742	0.665	0.55	0.455	0.37
4	0.903	0.867	0.803	0.740	0.66	0.545	0.445	0.355
				Ion Charge 3				
9	0.802	0.738	0.632	0.54	0.445	0.325	0.245	0.18
5	0.796	0.728	0.616	0.51	0.405	0.27	0.18	0.115
4	0.796	0.725	0.612	0.505	0.395	0.25	0.16	0.095
				Ion Charge 4				
11	0.678	0.588	0.455	0.35	0.255	0.155	0.10	0.065
5	0.668	0.57	0.425	0.31	0.20	0.10	0.048	0.021

Values of the parameter, $\overset{\circ}{a}$, for selected ions

$\overset{\circ}{a}$	Charge 1
9	H^+
6	Li^+
5	$CHCl_2COO^-$, CCl_3COO^-
4–4.5	Na^+, IO_3^-, HCO_3^-, $H_2PO_4^-$, CH_3COO^-, $NH_2CH_2COO^-$, $C_2H_5NH_3^+$
3.5	OH^-, F^-, CNS^-, CNO^-, BrO_3^-, MnO_4^-, $HCOO^-$, $H_2\ citrate^-$, $CH_3NH_3^+$
3	K^+, Cl^-, Br^-, I^-, CN^-, NO_2^-, NO_3^-
2.5	NH_4^+, Ag^+

	Charge 2
8	Mg^{++}
6	Ca^{++}, Cu^{++}, Zn^{++}, Sn^{++}, Mn^{++}, Fe^{++}, Ni^{++}, Co^{++}
5	Sr^{++}, Ba^{++}, Cd^{++}, Hg^{++}
4.5	Pb^{++}, CO_3^{--}, SO_3^{--}, $(COO^-)_2$, $H\ citrate^{--}$
4	Hg_2^{++}, SO_4^{--}, $S_2O_3^{--}$, CrO_4^{--}, HPO_4^{--}

	Charge 3
9	Al^{+++}, Fe^{+++}, Cr^{+++}, Ce^{+++}
5	$Citrate^{---}$
4	PO_4^{---}, $Fe(CN)_6^{---}$, $Cr(NH_3)_6^{+++}$

	Charge 4
11	Ce^{++++}, Sn^{++++}
5	$Fe(CN)_6^{----}$

Table 28   Values of the absorbance of unfiltered solutions of thickness 10 mm at 25 °C, used as spectral transmission standards

Solution A: Copper sulphate ($CuSO_4$. $5H_2O$)—20.000 g, sulphuric acid (specific gravity 1.835)—10.0 cm^3, distilled water to 1000 cm^3.

Solution B: * Cobalt ammonium sulphate ($CoSO_4(NH_4)_2SO_4$. $6H_2O$—14.481 g, sulphuric acid (specific gravity 1.835)—10.0 cm^3, distilled water to 1000 cm^3.
(* 10.3 g cobalt sulphate ($CoSO_4$. $7H_2O$) may be used in place, the data quoted then only apply to the range 400—750 nm).

Solution C: 0.0400 g potassium chromate (K_2CrO_4) in 0.05 mol dm^{-3} potassium hydroxide solution (3.3 g KOH sticks to 1000 cm^3).

Wavelength /nm	Absorbance			Wavelength /nm	Absorbance		
	Soln. A	Soln. B	Soln. C†		Soln. A	Soln. B	Soln. C†
215	—	—	1.4318	510	0.0038	0.1742	—
220	—	—	0.4559	520	0.0055	0.1689	—
230	—	—	0.1675	530	0.0079	0.1452	—
240	—	—	0.2933	540	0.0111	0.1113	—
250	—	—	0.4962	550	0.0155	0.0775	—
260	—	—	0.6345	560	0.0216	0.0496	—
270	—	—	0.7447	570	0.0292	0.0308	—
280	—	—	0.7235	580	0.0390	0.0207	—
290	—	—	0.4295	590	0.0518	0.0158	—
300	—	—	0.1518	600	0.0680	0.0137	—
310	—	—	0.0458	610	0.0885	0.0124	—
320	—	—	0.0620	620	0.1125	0.0115	—
330	—	—	0.1457	630	0.143	0.0112	—
340	—	—	0.3143	640	0.180	0.0110	—
350	0.0090	0.0038	0.5528	650	0.224	0.0105	—
360	0.0063	0.0040	‚0.8297	660	0.274	0.0097	—
370	0.0046	0.0050	0.9914	670	0.332	0.0087	—
380	0.0035	0.0065	0.9281	680	0.392	0.0076	—
390	0.0028	0.0088	0.6840	690	0.459	0.0066	—
400	0.0023	0.0125	0.3872	700	0.527	0.0054	—
410	0.0019	0.0168	0.1972	710	0.592	0.0046	—
420	0.0016	0.0224	0.1261	720	0.656	0.0038	—
430	0.0014	0.0340	0.0841	730	0.715	0.0032	—
440	0.0012	0.0522	0.0535	740	0.768	0.0030	—
450	0.0011	0.0773	0.0325	750	0.817	0.0028	—
460	0.0011	0.1031	0.0173				
470	0.0012	0.1213	0.0083	Hg 404.7	0.0021	0.0144	0.2840
480	0.0014	0.1349	0.0035	Hg 435.8	0.0013	0.0437	0.0650
490	0.0018	0.1472	0.0009	Hg 491.6	0.0019	0.1497	—
500	0.0026	0.1635	0.0000	He 501.6	0.0028	0.1661	—
				Hg 546.1	0.0135	0.0901	—
				Hg 578.0	0.0368	0.0219	—
				He 587.6	0.0487	0.0167	—
				He 667.8	0.319	0.0089	—

† Tentative values, derived from values of transmittance.

Table 29 Flame spectra

The prominent lines and bands, in the visible region, given when volatile compounds are introduced into a gas flame are listed. Weaker lines are often observed. Wavelengths (nm) for bands (b) refer to the position of the band head. The most persistent lines or bands are designated by P.

Barium	513.7	(b)		*Potassium*	404.416	
	534.7	(b)			404.722	
	553.5	(b) P			766.494	P
					769.901	P
Boron	519.3	(b)				
	544.0	(b)		*Rubidium*	420.18	P
	548.1	(b) P			421.56	P
					780.03	
Calcium	554.4	(b)			794.763	
	618.2	(b) P				
	620.3	(b) P		*Sodium*	588.9965	P
					589.5932	P
Caesium	455.55	P		*Strontium*	606.0	(b) P
	459.318	P			662.8	(b)
	621.3				674.7	(b)
	672.33					
	697.33			*Thallium*	535.047	
Lithium	610.359					
	670.786	P				

Table 30 Complementary colours and Ilford spectrum filters

Solution colour	Filter colour	Transmitting in range (nm)	Corresponding Ilford spectrum filter		
			No.	Colour name	Approx. peak of transmission /nm
Yellowish green	Violet	400–435	601*	Spectrum violet	430
Yellow	Blue	435–480	602	Spectrum blue	470
Orange	Greenish blue	480–490	603	Spectrum blue green	490
Red	Bluish green	490–500			
			604	Spectrum green	515
Purple	Green	500–560		Spectrum yellow green	545
			605		
Violet	Yellowish green	560–580	606	Spectrum yellow	570
Blue	Yellow	580–595			
Greenish blue	Orange	595–610	607	Spectrum orange	600
Bluish green	Red	610–750	608	Spectrum red	680

* In increasing order of percentage transmission, Ilford series are 621, 622 . . .; 611, 612 . . .; 601, 602 . . .; every filter with the same final digit has the same relative spectral distribution.

Table 31   Spectroscopic solvents and materials — ultraviolet and visible spectroscopy

SOLVENT	Cut-off‡	Minimum percentage transmittance in a 1 cm cell relative to distilled water ('ANALAR') at the following wavelengths (nm)														
		195	200	210	220	230	240	250	260	270	280	290	300	320	350	400
Acetonitrile†	<210	–	–	–	80	90	93	–	96	–	98	–	99	–	–	–
Benzene	278	–	–	–	–	–	–	–	–	–	25	80	90	95	99	–
Carbon tetrachloride	265	–	–	–	–	–	–	–	–	45	78	91	97	–	–	–
Chloroform	245	–	–	–	–	–	–	20	60	85	92	95	97	–	–	–
Cyclohexane†	200	–	–	15	50	75	90	97	–	–	–	–	–	–	–	–
Dimethylformamide	268	–	–	–	–	–	–	–	–	25	72	85	90	95	99	–
Dimethyl sulphoxide*	261	–	–	–	–	–	–	–	10 at 261	50 (58)	76 (83)	84 (92)	89 (95)	–	–	–
1,4-Dioxan†	215	–	–	–	20	35	47	59	70	79	87	95	–	–	–	–
Ethanol 95%†	205	–	–	30	52	73	85	90	–	–	–	–	–	–	–	–
Ethanol absolute†	205	–	–	30	52	73	85	90	–	–	–	–	–	–	–	–
Ethyl acetate	250	–	–	–	–	–	–	–	40	67	81	88	91	–	–	–
Hexane fraction from petroleum†	210	–	–	10	70	86	92	95	–	–	–	–	–	–	–	–
1,1,1,3,3,3-Hexafluoropropan-2-ol‡	<190	90	93	95	96	98	99	99	–	–	–	–	–	–	–	–
Methanol†	205	–	–	15	40	65	83	90	–	–	–	–	–	–	–	–
Propan-2-ol†	205	–	–	30	62	83	94	99	–	–	–	–	–	–	–	–
Tetrachloroethylene	290	–	–	–	–	–	–	–	–	–	–	10	75	85	–	–
Trifluoroacetic acid	260	–	–	–	–	–	–	–	10	80	90	93	95	95	–	95
2,2,2-Trifluoroethanol†	<190	70	90	92	94	95	96	97	98	–	–	–	–	–	–	–
2,2,4-Trimethylpentane†	200	–	10	30	60	85	95	97	–	–	–	–	–	–	–	–

* The figures in parentheses apply after purging with nitrogen for fifteen minutes.
† For use at short wavelengths the solvent should be purged with nitrogen immediately before use.
‡ The approximate wavelength at which the relative transmittance in a 1 cm cell falls to 10%. Users are reminded that the appropriate precautions should be taken when making measurements below 215 nm if errors are to be avoided.

Table 32   Characteristics of window material for infrared sample cells

Material	Range/μm	General properties
Fused silica	0.22−3.5	Unaffected by most solvents
NaCl	0.2−16	Hygroscopic
KCl	0.21−30	Hygroscopic
KBr	1.0−25	Hygroscopic
BaF$_2$	0.2−10	Insol. in H$_2$O, sol. in acids, resistant to fluorine and fluorides
CaF$_2$	0.21−8	Insol. in H$_2$O, resistant to acids and bases
CsBr	0.5−35	Hygroscopic
CsI	0.24−55	Hygroscopic, sol. in alcohol
LiF	0.12−6	Slightly sol. in H$_2$O
Irtran−2	0.6−14	Insol. in H$_2$O, acids, bases and most organic solvents. High refractive index.
AgCl	2.5−22.5	Insol. in H$_2$O, sol. in acids and NH$_4$Cl
AgBr	0.45−35	Insol. in H$_2$O, acetone and alcohols
KRS−5	0.5−40	Not hygroscopic, poisonous, high refractive index.
Polyethylene	16−1000	Inexpensive far infrared window material. Insol. in H$_2$O, swells in organic solvents.

Table 33   Substances used as wave-length standards

Substance	Wavenumber range/cm^{-1}
Hg	310−4290
H$_2$O,CO$_2$	5435−5200
CH$_4$	4550−4120
H$_2$O,CO$_2$	3900−3560
NH$_3$	3510−3170
CH$_4$	3170−2880
HCl	3060−2725
HBr	2675−2410
CO$_2$	2380−2290
CO	2220−2040
H$_2$O,CO$_2$	1990−1360
NH$_3$	1613−1580
CH$_4$	1360−1250
NH$_3$	1230− 720
H$_2$O,CO$_2$	720− 280
Indene	3929− 693
Polystyrene	3082− 699

Table 34   Spectroscopic solvents and materials — infrared spectroscopy

The white windows in the chart below show the regions in which the materials listed have at least 25% transmittance at the path length indicated. The diagrams also indicate the regions over which these materials have been examined.

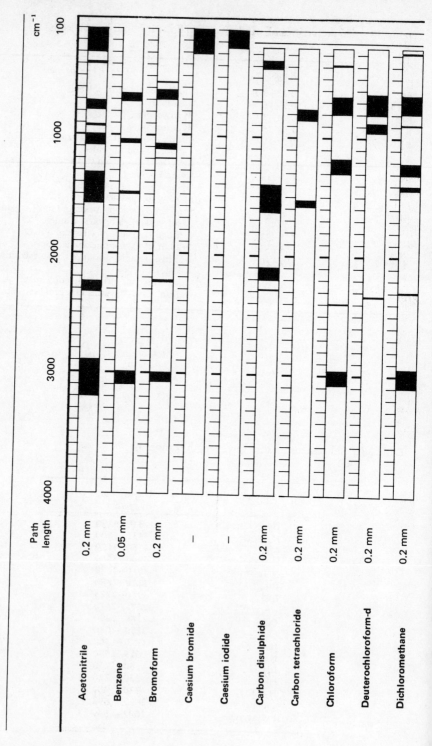

Table 34 (*continued*)

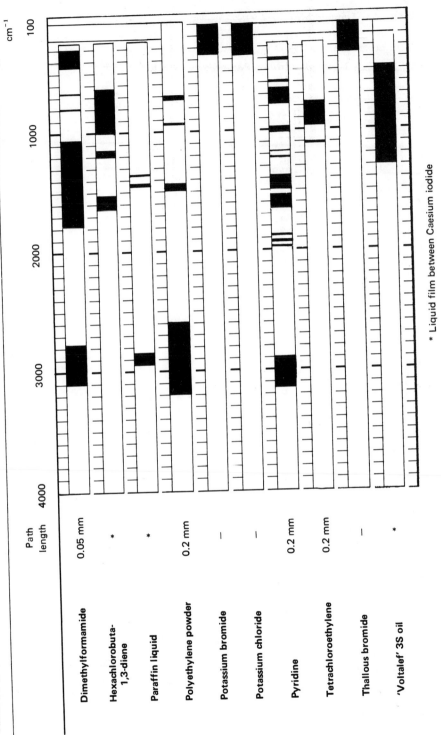

* Liquid film between Caesium iodide

Table 35   Temperature—emf. tables for thermocouples

These tables represent the temperature—emf. functions for various thermocouples, when one junction is at 0 °C. For accurate work the individual couples must be calibrated.

Temperature /°C	Copper-constantan	Chromel-P—alumel	Platinum-13% rhodium platinum
0	0 mV	0 mV	0 μV
10	0.39	0.40	54
20	0.79	0.80	111
30	1.19	1.20	170
40	1.61	1.61	231
50	2.03	2.02	295
60	2.47	2.43	361
70	2.91	2.85	429
80	3.36	3.26	499
90	3.81	3.68	571
100	4.28	4.10	644
200	9.29	8.13	1.463 mV
300	14.86	12.21	2.392
400	20.87	16.39	3.397
500	—	20.64	4.460
600	—	24.90	5.571
700	—	29.14	6.735
800	—	33.31	7.952
900	—	37.36	9.209
1000	—	41.31	10.510
1100	—	45.14	11.850
1200	—	48.85	13.222
1300	—	52.41	14.617
1400	—	55.81	16.039

Table 36 Preparation of volumetric solutions

Reagent	Relative molecular mass	Concn. /mol dm^{-3}	Weight of reagent (g) to make 1 dm^3 of solution	Remarks
Ammonium Cerium (IV)-Sulphate (NH$_4$)$_4$ Ce(SO$_4$)$_2$. 2H$_2$O	632.6	0.1	63.26	Use 1 mol dm^{-3} H$_2$SO$_4$
* Ammonium Iron(II)-Sulphate (NH$_4$)$_2$ Fe(SO$_4$)$_2$. 6H$_2$O	392.14	0.1	39.214	Use 1 mol dm^{-3} H$_2$SO$_4$
* Arsenic trioxide As$_2$O$_3$	197.84	0.025	4.946	Dissolve in 200 cm^3 20% NaOH and make up to 1 dm^3 with water
Barium Hydroxide Ba(OH)$_2$. 8H$_2$O	315.48	0.05	15.77	
Cerium(IV)Sulphate Ce(SO$_4$)$_2$	332.24	0.1	33.224	Use 1 mol dm^{-3} H$_2$SO$_4$
* Copper Sulphate CuSO$_4$. 5H$_2$O	249.68	0.1	24.968	
Iodine I$_2$	253.81	0.05	12.7	Use 2% KI solution
* Oxalic Acid H$_2$C$_2$O$_4$. 2H$_2$O	126.7	0.05	6.303	
Potassium Dichromate K$_2$Cr$_2$O$_7$	294.19	0.02	5.884	
* Potassium Hydrogen Phthalate KHC$_8$H$_4$O$_4$	204.23	0.1	20.423	
* Potassium iodate KIO$_3$	214.00	0.02	4.28	
Potassium Permanganate KMnO$_4$	158.04	0.02	3.16	
Potassium Thiocyanate KCNS	97.18	0.1	9.718	
* Silver Nitrate AgNO$_3$	169.87	0.1	16.987	
* Sodium Carbonate Na$_2$CO$_3$	105.99	0.05	5.2995	
* Sodium Chloride NaCl	58.44	0.1	5.844	
Sodium salt of Ethylene-Diamine Tetra-Acetic Acid [CH$_2$N(CH$_2$COONa) CH$_2$COOH]$_2$. 2H$_2$O	372.1	0.01	3.721	Use deionized water
Sodium Hydroxide NaOH	40.00	0.1	4.000	
* Sodium Oxalate Na$_2$C$_2$O$_4$	134.00	0.05	6.700	
Sodium Thiosulphate Na$_2$S$_2$O$_3$. 5H$_2$O	248.18	0.1	24.82	
* Zinc Sulphate ZnSO$_4$. 6H$_2$O	269.43	0.01	2.6943	Use deionized water

* These solutions can be prepared as standard solutions.
(All the remainder must be standardized.)

Table 37   Preparation of aqueous solutions of the common acids and ammonia

Reagent	*Relative molecular mass*	*Concentration of concentrated solution* /mol dm^{-3}	*Vol/cm^3 required to make 1 dm^3 of approx. 0.1 mol dm^{-3} solution*
Hydrochloric acid	36.46	11.6	8.6
Nitric acid	63.01	15.4	6.5
Sulphuric acid	98.08	17.8	5.6
Perchloric acid	100.45	11.6	8.6
Phosphoric acid	97.99	14.6	6.8
Acetic acid	60.05	17.4	5.8
Ammonia	17.03	14.8	6.8

Four figure logarithm and antilogarithm tables

FOUR-FIGURE LOGARITHMS

	0	1	2	3	4	5	6	7	8	9	1	2	3	4	5	6	7	8	9
10	·0000	0043	0086	0128	0170	0212	0253	0294	0334	0374	4	8	12	17	21	25	29	33	37
11	·0414	0453	0492	0531	0569	0607	0645	0682	0719	0755	4	8	11	15	19	23	26	30	34
12	·0792	0828	0864	0899	0934	0969	1004	1038	1072	1106	3	7	10	14	17	21	24	28	31
13	·1139	1173	1206	1239	1271	1303	1335	1367	1399	1430	3	6	10	13	16	19	23	26	29
14	·1461	1492	1523	1553	1584	1614	1644	1673	1703	1732	3	6	9	12	15	18	21	24	27
15	·1761	1790	1818	1847	1875	1903	1931	1959	1987	2014	3	6	8	11	14	17	20	22	25
16	·2041	2068	2095	2122	2148	2175	2201	2227	2253	2279	3	5	8	11	13	16	18	21	24
17	·2304	2330	2355	2380	2405	2430	2455	2480	2504	2529	2	5	7	10	12	15	17	20	22
18	·2553	2577	2601	2625	2648	2672	2695	2718	2742	2765	2	5	7	9	12	14	16	19	21
19	·2788	2810	2833	2856	2878	2900	2923	2945	2967	2989	2	4	7	9	11	13	16	18	20
20	·3010	3032	3054	3075	3096	3118	3139	3160	3181	3201	2	4	6	8	11	13	15	17	19
21	·3222	3243	3263	3284	3304	3324	3345	3365	3385	3404	2	4	6	8	10	12	14	16	18
22	·3424	3444	3464	3483	3502	3522	3541	3560	3579	3598	2	4	6	8	10	12	14	15	17
23	·3617	3636	3655	3674	3692	3711	3729	3747	3766	3784	2	4	6	7	9	11	13	15	17
24	·3802	3820	3838	3856	3874	3892	3909	3927	3945	3962	2	4	5	7	9	11	12	14	16
25	·3979	3997	4014	4031	4048	4065	4082	4099	4116	4133	2	3	5	7	9	10	12	14	15
26	·4150	4166	4183	4200	4216	4232	4249	4265	4281	4298	2	3	5	7	8	10	11	13	15
27	·4314	4330	4346	4362	4378	4393	4409	4425	4440	4456	2	3	5	6	8	9	11	13	14
28	·4472	4487	4502	4518	4533	4548	4564	4579	4594	4609	2	3	5	6	8	9	11	12	14
29	·4624	4639	4654	4669	4683	4698	4713	4728	4742	4757	1	3	4	6	7	9	10	12	13
30	·4771	4786	4800	4814	4829	4843	4857	4871	4886	4900	1	3	4	6	7	9	10	11	13
31	·4914	4928	4942	4955	4969	4983	4997	5011	5024	5038	1	3	4	6	7	8	10	11	12
32	·5051	5065	5079	5092	5105	5119	5132	5145	5159	5172	1	3	4	5	7	8	9	11	12
33	·5185	5198	5211	5224	5237	5250	5263	5276	5289	5302	1	3	4	5	7	8	9	10	12
34	·5315	5328	5340	5353	5366	5378	5391	5403	5416	5428	1	3	4	5	6	8	9	10	11
35	·5441	5453	5465	5478	5490	5502	5514	5527	5539	5551	1	2	4	5	6	7	9	10	11
36	·5563	5575	5587	5599	5611	5623	5635	5647	5658	5670	1	2	4	5	6	7	8	10	11
37	·5682	5694	5705	5717	5729	5740	5752	5763	5775	5786	1	2	3	5	6	7	8	9	10
38	·5798	5809	5821	5832	5843	5855	5866	5877	5888	5899	1	2	3	5	6	7	8	9	10
39	·5911	5922	5933	5944	5955	5966	5977	5988	5999	6010	1	2	3	4	6	7	8	9	10
40	·6021	6031	6042	6053	6064	6075	6085	6096	6107	6117	1	2	3	4	5	6	7	9	10
41	·6128	6138	6149	6160	6170	6180	6191	6201	6212	6222	1	2	3	4	5	6	7	8	9
42	·6232	6243	6253	6263	6274	6284	6294	6304	6314	6325	1	2	3	4	5	6	7	8	9
43	·6335	6345	6355	6365	6375	6385	6395	6405	6415	6425	1	2	3	4	5	6	7	8	9
44	·6435	6444	6454	6464	6474	6484	6493	6503	6513	6522	1	2	3	4	5	6	7	8	9
45	·6532	6542	6551	6561	6571	6580	6590	6599	6609	6618	1	2	3	4	5	6	7	8	9
46	·6628	6637	6646	6656	6665	6675	6684	6693	6702	6712	1	2	3	4	5	6	7	7	8
47	·6721	6730	6739	6749	6758	6767	6776	6785	6794	6803	1	2	3	4	5	5	6	7	8
48	·6812	6821	6830	6839	6848	6857	6866	6875	6884	6893	1	2	3	4	5	5	6	7	8
49	·6902	6911	6920	6928	6937	6946	6955	6964	6972	6981	1	2	3	4	4	5	6	7	8
50	·6990	6998	7007	7016	7024	7033	7042	7050	7059	7067	1	2	3	3	4	5	6	7	8
51	·7076	7084	7093	7101	7110	7118	7126	7135	7143	7152	1	2	3	3	4	5	6	7	8
52	·7160	7168	7177	7185	7193	7202	7210	7218	7226	7235	1	2	2	3	4	5	6	7	7
53	·7243	7251	7259	7267	7275	7284	7292	7300	7308	7316	1	2	2	3	4	5	6	6	7
54	·7324	7332	7340	7348	7356	7364	7372	7380	7388	7396	1	2	2	3	4	5	6	6	7
	0	1	2	3	4	5	6	7	8	9	1	2	3	4	5	6	7	8	9

FOUR-FIGURE LOGARITHMS

	0	1	2	3	4	5	6	7	8	9	1	2	3	4	5	6	7	8	9
55	·7404	7412	7419	7427	7435	7443	7451	7459	7466	7474	1	2	2	3	4	5	5	6	7
56	·7482	7490	7497	7505	7513	7520	7528	7536	7543	7551	1	2	2	3	4	5	5	6	7
57	·7559	7566	7574	7582	7589	7597	7604	7612	7619	7627	1	2	2	3	4	5	5	6	7
58	·7634	7642	7649	7657	7664	7672	7679	7686	7694	7701	1	2	2	3	4	5	5	6	7
59	·7709	7716	7723	7731	7738	7745	7752	7760	7767	7774	1	1	2	3	4	4	5	6	7
60	·7782	7789	7796	7803	7810	7818	7825	7832	7839	7846	1	1	2	3	4	4	5	6	6
61	·7853	7860	7868	7875	7882	7889	7896	7903	7910	7917	1	1	2	3	4	4	5	6	6
62	·7924	7931	7938	7945	7952	7959	7966	7973	7980	7987	1	1	2	3	3	4	5	6	6
63	·7993	8000	8007	8014	8021	8028	8035	8041	8048	8055	1	1	2	3	3	4	5	6	6
64	·8062	8069	8075	8082	8089	8096	8102	8109	8116	8122	1	1	2	3	3	4	5	5	6
65	·8129	8136	8142	8149	8156	8162	8169	8176	8182	8189	1	1	2	3	3	4	5	5	6
66	·8195	8202	8209	8215	8222	8228	8235	8241	8248	8254	1	1	2	3	3	4	5	5	6
67	·8261	8267	8274	8280	8287	8293	8299	8306	8312	8319	1	1	2	3	3	4	4	5	6
68	·8325	8331	8338	8344	8351	8357	8363	8370	8376	8382	1	1	2	3	3	4	4	5	6
69	·8388	8395	8401	8407	8414	8420	8426	8432	8439	8445	1	1	2	3	3	4	4	5	6
70	·8451	8457	8463	8470	8476	8482	8488	8494	8500	8506	1	1	2	2	3	4	4	5	6
71	·8513	8519	8525	8531	8537	8543	8549	8555	8561	8567	1	1	2	2	3	4	4	5	5
72	·8573	8579	8585	8591	8597	8603	8609	8615	8621	8627	1	1	2	2	3	4	4	5	5
73	·8633	8639	8645	8651	8657	8663	8669	8675	8681	8686	1	1	2	2	3	4	4	5	5
74	·8692	8698	8704	8710	8716	8722	8727	8733	8739	8745	1	1	2	2	3	4	4	5	5
75	·8751	8756	8762	8768	8774	8779	8785	8791	8797	8802	1	1	2	2	3	3	4	5	5
76	·8808	8814	8820	8825	8831	8837	8842	8848	8854	8859	1	1	2	2	3	3	4	5	5
77	·8865	8871	8876	8882	8887	8893	8899	8904	8910	8915	1	1	2	2	3	3	4	4	5
78	·8921	8927	8932	8938	8943	8949	8954	8960	8965	8971	1	1	2	2	3	3	4	4	5
79	·8976	8982	8987	8993	8998	9004	9009	9015	9020	9025	1	1	2	2	3	3	4	4	5
80	·9031	9036	9042	9047	9053	9058	9063	9069	9074	9079	1	1	2	2	3	3	4	4	5
81	·9085	9090	9096	9101	9106	9112	9117	9122	9128	9133	1	1	2	2	3	3	4	4	5
82	·9138	9143	9149	9154	9159	9165	9170	9175	9180	9186	1	1	2	2	3	3	4	4	5
83	·9191	9196	9201	9206	9212	9217	9222	9227	9232	9238	1	1	2	2	3	3	4	4	5
84	·9243	9248	9253	9258	9263	9269	9274	9279	9284	9289	1	1	2	2	3	3	4	4	5
85	·9294	9299	9304	9309	9315	9320	9325	9330	9335	9340	1	1	2	2	3	3	4	4	5
86	·9345	9350	9355	9360	9365	9370	9375	9380	9385	9390	1	1	2	2	3	3	4	4	5
87	·9395	9400	9405	9410	9415	9420	9425	9430	9435	9440	1	1	2	2	3	3	4	4	5
88	·9445	9450	9455	9460	9465	9469	9474	9479	9484	9489	0	1	1	2	2	3	3	4	4
89	·9494	9499	9504	9509	9513	9518	9523	9528	9533	9538	0	1	1	2	2	3	3	4	4
90	·9542	9547	9552	9557	9562	9566	9571	9576	9581	9586	0	1	1	2	2	3	3	4	4
91	·9590	9595	9600	9605	9609	9614	9619	9624	9628	9633	0	1	1	2	2	3	3	4	4
92	·9638	9643	9647	9652	9657	9661	9666	9671	9675	9680	0	1	1	2	2	3	3	4	4
93	·9685	9689	9694	9699	9703	9708	9713	9717	9722	9727	0	1	1	2	2	3	3	4	4
94	·9731	9736	9741	9745	9750	9754	9759	9763	9768	9773	0	1	1	2	2	3	3	4	4
95	·9777	9782	9786	9791	9795	9800	9805	9809	9814	9818	0	1	1	2	2	3	3	4	4
96	·9823	9827	9832	9836	9841	9845	9850	9854	9859	9863	0	1	1	2	2	3	3	4	4
97	·9868	9872	9877	9881	9886	9890	9894	9899	9903	9908	0	1	1	2	2	3	3	4	4
98	·9912	9917	9921	9926	9930	9934	9939	9943	9948	9952	0	1	1	2	2	3	3	4	4
99	·9956	9961	9965	9969	9974	9978	9983	9987	9991	9996	0	1	1	2	2	3	3	4	4
	0	1	2	3	4	5	6	7	8	9	1	2	3	4	5	6	7	8	9

ANTILOGARITHMS

	0	1	2	3	4	5	6	7	8	9		1	2	3	4	5	6	7	8	9
·00	1000	1002	1005	1007	1009	1012	1014	1016	1019	1021		0	0	1	1	1	1	2	2	2
·01	1023	1026	1028	1030	1033	1035	1038	1040	1042	1045		0	0	1	1	1	1	2	2	2
·02	1047	1050	1052	1054	1057	1059	1062	1064	1067	1069		0	0	1	1	1	1	2	2	2
·03	1072	1074	1076	1079	1081	1084	1086	1089	1091	1094		0	0	1	1	1	1	2	2	2
·04	1096	1099	1102	1104	1107	1109	1112	1114	1117	1119		0	1	1	1	1	2	2	2	2
·05	1122	1125	1127	1130	1132	1135	1138	1140	1143	1146		0	1	1	1	1	2	2	2	2
·06	1148	1151	1153	1156	1159	1161	1164	1167	1169	1172		0	1	1	1	1	2	2	2	2
·07	1175	1178	1180	1183	1186	1189	1191	1194	1197	1199		0	1	1	1	1	2	2	2	2
·08	1202	1205	1208	1211	1213	1216	1219	1222	1225	1227		0	1	1	1	1	2	2	2	3
·09	1230	1233	1236	1239	1242	1245	1247	1250	1253	1256		0	1	1	1	1	2	2	2	3
·10	1259	1262	1265	1268	1271	1274	1276	1279	1282	1285		0	1	1	1	1	2	2	2	3
·11	1288	1291	1294	1297	1300	1303	1306	1309	1312	1315		0	1	1	1	2	2	2	2	3
·12	1318	1321	1324	1327	1330	1334	1337	1340	1343	1346		0	1	1	1	2	2	2	2	3
·13	1349	1352	1355	1358	1361	1365	1368	1371	1374	1377		0	1	1	1	2	2	2	2	3
·14	1380	1384	1387	1390	1393	1396	1400	1403	1406	1409		0	1	1	1	2	2	2	3	3
·15	1413	1416	1419	1422	1426	1429	1432	1435	1439	1442		0	1	1	1	2	2	2	3	3
·16	1445	1449	1452	1455	1459	1462	1466	1469	1472	1476		0	1	1	1	2	2	2	3	3
·17	1479	1483	1486	1489	1493	1496	1500	1503	1507	1510		0	1	1	1	2	2	2	3	3
·18	1514	1517	1521	1524	1528	1531	1535	1538	1542	1545		0	1	1	1	2	2	2	3	3
·19	1549	1552	1556	1560	1563	1567	1570	1574	1578	1581		0	1	1	1	2	2	3	3	3
·20	1585	1589	1592	1596	1600	1603	1607	1611	1614	1618		0	1	1	1	2	2	3	3	3
·21	1622	1626	1629	1633	1637	1641	1644	1648	1652	1656		0	1	1	2	2	2	3	3	3
·22	1660	1663	1667	1671	1675	1679	1683	1687	1690	1694		0	1	1	2	2	2	3	3	3
·23	1698	1702	1706	1710	1714	1718	1722	1726	1730	1734		0	1	1	2	2	2	3	3	4
·24	1738	1742	1746	1750	1754	1758	1762	1766	1770	1774		0	1	1	2	2	2	3	3	4
·25	1778	1782	1786	1791	1795	1799	1803	1807	1811	1816		0	1	1	2	2	3	3	3	4
·26	1820	1824	1828	1832	1837	1841	1845	1849	1854	1858		0	1	1	2	2	3	3	3	4
·27	1862	1866	1871	1875	1879	1884	1888	1892	1897	1901		0	1	1	2	2	3	3	4	4
·28	1905	1910	1914	1919	1923	1928	1932	1936	1941	1945		0	1	1	2	2	3	3	4	4
·29	1950	1954	1959	1963	1968	1972	1977	1982	1986	1991		0	1	1	2	2	3	3	4	4
·30	1995	2000	2004	2009	2014	2018	2023	2028	2032	2037		0	1	1	2	2	3	3	4	4
·31	2042	2046	2051	2056	2061	2065	2070	2075	2080	2084		0	1	1	2	2	3	3	4	4
·32	2089	2094	2099	2104	2109	2113	2118	2123	2128	2133		0	1	1	2	2	3	3	4	4
·33	2138	2143	2148	2153	2158	2163	2168	2173	2178	2183		1	1	2	2	3	3	4	4	5
·34	2188	2193	2198	2203	2208	2213	2218	2223	2228	2234		1	1	2	2	3	3	4	4	5
·35	2239	2244	2249	2254	2259	2265	2270	2275	2280	2286		1	1	2	2	3	3	4	4	5
·36	2291	2296	2301	2307	2312	2317	2323	2328	2333	2339		1	1	2	2	3	3	4	4	5
·37	2344	2350	2355	2360	2366	2371	2377	2382	2388	2393		1	1	2	2	3	3	4	4	5
·38	2399	2404	2410	2415	2421	2427	2432	2438	2443	2449		1	1	2	2	3	3	4	4	5
·39	2455	2460	2466	2472	2477	2483	2489	2495	2500	2506		1	1	2	2	3	3	4	5	5
·40	2512	2518	2523	2529	2535	2541	2547	2553	2559	2564		1	1	2	2	3	4	4	5	5
·41	2570	2576	2582	2588	2594	2600	2606	2612	2618	2624		1	1	2	2	3	4	4	5	5
·42	2630	2636	2642	2649	2655	2661	2667	2673	2679	2685		1	1	2	2	3	4	4	5	6
·43	2692	2698	2704	2710	2716	2723	2729	2735	2742	2748		1	1	2	2	3	4	4	5	6
·44	2754	2761	2767	2773	2780	2786	2793	2799	2805	2812		1	1	2	3	3	4	4	5	6
·45	2818	2825	2831	2838	2844	2851	2858	2864	2871	2877		1	1	2	3	3	4	5	5	6
·46	2884	2891	2897	2904	2911	2917	2924	2931	2938	2944		1	1	2	3	3	4	5	5	6
·47	2951	2958	2965	2972	2979	2985	2992	2999	3006	3013		1	1	2	3	3	4	5	6	6
·48	3020	3027	3034	3041	3048	3055	3062	3069	3076	3083		1	1	2	3	4	4	5	6	6
·49	3090	3097	3105	3112	3119	3126	3133	3141	3148	3155		1	1	2	3	4	4	5	6	6
	0	1	2	3	4	5	6	7	8	9		1	2	3	4	5	6	7	8	9

ANTILOGARITHMS

	0	1	2	3	4	5	6	7	8	9	1	2	3	4	5	6	7	8	9
·50	3162	3170	3177	3184	3192	3199	3206	3214	3221	3228	1	1	2	3	4	4	5	6	7
·51	3236	3243	3251	3258	3266	3273	3281	3289	3296	3304	1	2	2	3	4	5	5	6	7
·52	3311	3319	3327	3334	3342	3350	3357	3365	3373	3381	1	2	2	3	4	5	5	6	7
·53	3388	3396	3404	3412	3420	3428	3436	3443	3451	3459	1	2	2	3	4	5	6	6	7
·54	3467	3475	3483	3491	3499	3508	3516	3524	3532	3540	1	2	2	3	4	5	6	6	7
·55	3548	3556	3565	3573	3581	3589	3597	3606	3614	3622	1	2	2	3	4	5	6	7	7
·56	3631	3639	3648	3656	3664	3673	3681	3690	3698	3707	1	2	3	3	4	5	6	7	8
·57	3715	3724	3733	3741	3750	3758	3767	3776	3784	3793	1	2	3	3	4	5	6	7	8
·58	3802	3811	3819	3828	3837	3846	3855	3864	3873	3882	1	2	3	4	4	5	6	7	8
·59	3890	3899	3908	3917	3926	3936	3945	3954	3963	3972	1	2	3	4	5	5	6	7	8
·60	3981	3990	3999	4009	4018	4027	4036	4046	4055	4064	1	2	3	4	5	6	7	7	8
·61	4074	4083	4093	4102	4111	4121	4130	4140	4150	4159	1	2	3	4	5	6	7	8	9
·62	4169	4178	4188	4198	4207	4217	4227	4236	4246	4256	1	2	3	4	5	6	7	8	9
·63	4266	4276	4285	4295	4305	4315	4325	4335	4345	4355	1	2	3	4	5	6	7	8	9
·64	4365	4375	4385	4395	4406	4416	4426	4436	4446	4457	1	2	3	4	5	6	7	8	9
·65	4467	4477	4487	4498	4508	4519	4529	4539	4550	4560	1	2	3	4	5	6	7	8	9
·66	4571	4581	4592	4603	4613	4624	4634	4645	4656	4667	1	2	3	4	5	6	7	8	10
·67	4677	4688	4699	4710	4721	4732	4742	4753	4764	4775	1	2	3	4	5	7	8	9	10
·68	4786	4797	4808	4819	4831	4842	4853	4864	4875	4887	1	2	3	4	6	7	8	9	10
·69	4898	4909	4920	4932	4943	4955	4966	4977	4989	5000	1	2	3	5	6	7	8	9	10
·70	5012	5023	5035	5047	5058	5070	5082	5093	5105	5117	1	2	4	5	6	7	8	9	11
·71	5129	5140	5152	5164	5176	5188	5200	5212	5224	5236	1	2	4	5	6	7	8	10	11
·72	5248	5260	5272	5284	5297	5309	5321	5333	5346	5358	1	2	4	5	6	7	9	10	11
·73	5370	5383	5395	5408	5420	5433	5445	5458	5470	5483	1	3	4	5	6	8	9	10	11
·74	5495	5508	5521	5534	5546	5559	5572	5585	5598	5610	1	3	4	5	6	8	9	10	12
·75	5623	5636	5649	5662	5675	5689	5702	5715	5728	5741	1	3	4	5	7	8	9	10	12
·76	5754	5768	5781	5794	5808	5821	5834	5848	5861	5875	1	3	4	5	7	8	9	11	12
·77	5888	5902	5916	5929	5943	5957	5970	5984	5998	6012	1	3	4	6	7	8	10	11	12
·78	6026	6039	6053	6067	6081	6095	6109	6124	6138	6152	1	3	4	6	7	8	10	11	13
·79	6166	6180	6194	6209	6223	6237	6252	6266	6281	6295	1	3	4	6	7	9	10	12	13
·80	6310	6324	6339	6353	6368	6383	6397	6412	6427	6442	1	3	4	6	7	9	10	12	13
·81	6457	6471	6486	6501	6516	6531	6546	6561	6577	6592	2	3	5	6	8	9	11	12	14
·82	6607	6622	6637	6653	6668	6683	6699	6714	6730	6745	2	3	5	6	8	9	11	12	14
·83	6761	6776	6792	6808	6823	6839	6855	6871	6887	6902	2	3	5	6	8	9	11	13	14
·84	6918	6934	6950	6966	6982	6998	7015	7031	7047	7063	2	3	5	6	8	10	11	13	14
·85	7079	7096	7112	7129	7145	7161	7178	7194	7211	7228	2	3	5	7	8	10	12	13	15
·86	7244	7261	7278	7295	7311	7328	7345	7362	7379	7396	2	3	5	7	8	10	12	14	15
·87	7413	7430	7447	7464	7482	7499	7516	7534	7551	7568	2	3	5	7	9	10	12	14	16
·88	7586	7603	7621	7638	7656	7674	7691	7709	7727	7745	2	4	5	7	9	11	12	14	16
·89	7762	7780	7798	7816	7834	7852	7870	7889	7907	7925	2	4	5	7	9	11	13	14	16
·90	7943	7962	7980	7998	8017	8035	8054	8072	8091	8110	2	4	6	7	9	11	13	15	17
·91	8128	8147	8166	8185	8204	8222	8241	8260	8279	8299	2	4	6	8	10	11	13	15	17
·92	8318	8337	8356	8375	8395	8414	8433	8453	8472	8492	2	4	6	8	10	12	14	15	17
·93	8511	8531	8551	8570	8590	8610	8630	8650	8670	8690	2	4	6	8	10	12	14	16	18
·94	8710	8730	8750	8770	8790	8810	8831	8851	8872	8892	2	4	6	8	10	12	14	16	18
·95	8913	8933	8954	8974	8995	9016	9036	9057	9078	9099	2	4	6	8	10	12	14	17	19
·96	9120	9141	9162	9183	9204	9226	9247	9268	9290	9311	2	4	6	9	11	13	15	17	19
·97	9333	9354	9376	9397	9419	9441	9462	9484	9506	9528	2	4	7	9	11	13	15	17	20
·98	9550	9572	9594	9616	9638	9661	9683	9705	9727	9750	2	4	7	9	11	13	16	18	20
·99	9772	9795	9817	9840	9863	9886	9908	9931	9954	9977	2	5	7	9	11	14	16	18	21
	0	1	2	3	4	5	6	7	8	9	1	2	3	4	5	6	7	8	9

Table 38   Relative atomic masses of the elements (*Recommended by IUPAC in 1967*)

	Symbol	Proton No.	Relative atomic mass		Symbol	Proton No.	Relative atomic mass
Actinium	Ac	89	(227)	Mercury	Hg	80	200.59
Aluminium	Al	13	26.9815	Molybdenum	Mo	42	95.94
Americium	Am	95	(243)	Neodymium	Nd	60	144.24
Antimony	Sb	51	121.75	Neon	Ne	10	20.179
Argon	Ar	18	39.948	Neptunium	Np	93	(237)
Arsenic	As	33	74.9216	Nickel	Ni	28	58.71
Astatine	At	85	(210)	Niobium	Nb	41	92.906
Barium	Ba	56	137.34	Nitrogen	N	7	14.0067
Berkelium	Bk	97	(249)	Nobelium	No	102	(253)
Beryllium	Be	4	9.0122	Osmium	Os	76	190.2
Bismuth	Bi	83	208.980	Oxygen	O	8	15.9994
Boron	B	5	10.811	Palladium	Pd	46	106.4
Bromine	Br	35	79.904	Phosphorus	P	15	30.9738
Cadmium	Cd	48	112.40	Platinum	Pt	78	195.09
Calcium	Ca	20	40.08	Plutonium	Pu	94	(242)
Californium	Cf	98	(251)	Polonium	Po	84	(210)
Carbon	C	6	12.01115	Potassium	K	19	39.102
Cerium	Ce	58	140.12	Praseodymium	Pr	59	140.907
Cesium	Cs	55	132.905	Promethium	Pm	61	(147)
Chlorine	Cl	17	35.453	Protactinium	Pa	91	(231)
Chromium	Cr	24	51.996	Radium	Ra	88	(226)
Cobalt	Co	27	58.9332	Radon	Rn	86	(222)
Copper	Cu	29	63.546	Rhenium	Re	75	186.2
Curium	Cm	96	(247)	Rhodium	Rh	45	102.905
Dysprosium	Dy	66	162.50	Rubidium	Rb	37	85.47
Einsteinium	Es	99	(254)	Ruthenium	Ru	44	101.07
Erbium	Er	68	167.26	Samarium	Sm	62	150.35
Europium	Eu	63	151.96	Scandium	Sc	21	44.956
Fermium	Fm	100	(253)	Selenium	Se	34	78.96
Fluorine	F	9	18.9984	Silicon	Si	14	28.086
Francium	Fr	87	(223)	Silver	Ag	47	107.868
Gadolinium	Gd	64	157.25	Sodium	Na	11	22.9898
Gallium	Ga	31	69.72	Strontium	Sr	38	87.62
Germanium	Ge	32	72.59	Sulphur	S	16	32.064
Gold	Au	79	196.967	Tantalum	Ta	73	180.948
Hafnium	Hf	72	178.49	Technetium	Tc	43	(99)
Helium	He	2	4.0026	Tellurium	Te	52	127.60
Holmium	Ho	67	164.930	Terbium	Tb	65	158.924
Hydrogen	H	1	1.00797	Thallium	Tl	81	204.37
Indium	In	49	114.82	Thorium	Th	90	232.038
Iodine	I	53	126.9044	Thulium	Tm	69	168.934
Iridium	Ir	77	192.2	Tin	Sn	50	118.69
Iron	Fe	26	55.85	Titanium	Ti	22	47.90
Krypton	Kr	36	83.80	Tungsten	W	74	183.85
Lanthanum	La	57	138.91	Uranium	U	92	238.03
Lawrencium	Lw	103	(257)	Vanadium	V	23	50.942
Lead	Pb	82	207.19	Xenon	Xe	54	131.30
Lithium	Li	3	6.939	Ytterbium	Yb	70	173.04
Lutetium	Lu	71	174.97	Yttrium	Y	39	88.905
Magnesium	Mg	12	24.305	Zinc	Zn	30	65.37
Manganese	Mn	25	54.938	Zirconium	Zr	40	91.22
Mendelevium	Md	101	(256)				

Index